Secular Humanism

" Organizing The Best of Humanity "

Contents

1 Secular humanism   **1**
   1.1 Terminology 1
   1.2 History 2
     1.2.1 Secularism 2
     1.2.2 Positivism & the Church of Humanity 2
     1.2.3 Ethical movement 2
     1.2.4 Secular humanism 3
   1.3 Manifestos and declarations 4
     1.3.1 International Humanist and Ethical Union 4
     1.3.2 Council for Secular Humanism 4
     1.3.3 American Humanist Association 5
   1.4 Ethics and relationship to religious belief 5
   1.5 Modern context 6
   1.6 Humanist celebrations 6
   1.7 Legal mentions in the United States 7
     1.7.1 Hatch amendment 7
     1.7.2 Case law 7
     1.7.3 Controversy 8
   1.8 Notable humanists 8
   1.9 Manifestos 8
   1.10 Related organizations 8
   1.11 See also 9
     1.11.1 Related philosophies 9
     1.11.2 Wikibooks 10
   1.12 Notes and references 10
   1.13 Further reading 12
     1.13.1 Primary sources 12

2 Humanism and Its Aspirations   **14**
   2.1 Signatories 14

		2.1.1	Notable signatories	14
		2.1.2	Nobel laureates	16
		2.1.3	Past AHA presidents	16
		2.1.4	AHA board	16
		2.1.5	Drafting committee	16
	2.2		See also	17
	2.3		References	17
	2.4		External links	17
3	**Humanist Manifesto**			**18**
	3.1		Humanist Manifesto I	18
	3.2		Humanist Manifesto II	18
	3.3		Humanist Manifesto III	19
	3.4		Other Manifestos for Humanism	19
		3.4.1	A Secular Humanist Declaration	19
		3.4.2	Humanist Manifesto 2000	20
		3.4.3	Amsterdam Declaration	20
	3.5		References	20
	3.6		External links	20
		3.6.1	Manifestos	20
		3.6.2	Miscellaneous	20
4	**Reason**			**21**
	4.1		Etymology and related words	21
	4.2		Philosophical history	22
		4.2.1	Classical philosophy	22
		4.2.2	Subject-centred reason in early modern philosophy	22
		4.2.3	Substantive and formal reason	23
		4.2.4	The critique of reason	24
	4.3		Reason compared to related concepts	25
		4.3.1	Compared to logic	25
		4.3.2	Reason compared to cause-and-effect thinking, and symbolic thinking	25
		4.3.3	Reason, imagination, mimesis, and memory	25
		4.3.4	Logical reasoning methods and argumentation	26
	4.4		Traditional problems raised concerning reason	28
		4.4.1	Reason versus truth, and "first principles"	28
		4.4.2	Reason versus emotion or passion	28
		4.4.3	Reason versus faith or tradition	29
	4.5		Reason in particular fields of study	30

		4.5.1	Reason in political philosophy and ethics	30
		4.5.2	Psychology	31
		4.5.3	Computer science	31
		4.5.4	Evolution of reason	32
	4.6	See also		32
	4.7	References		32
	4.8	Further reading		35

5 Ethics — 37

- 5.1 Defining ethics — 37
- 5.2 Meta-ethics — 37
- 5.3 Normative ethics — 38
    - 5.3.1 Virtue ethics — 38
    - 5.3.2 Hedonism — 40
    - 5.3.3 State consequentialism — 40
    - 5.3.4 Consequentialism/Teleology — 41
    - 5.3.5 Deontology — 42
    - 5.3.6 Pragmatic ethics — 43
    - 5.3.7 Ethics of care — 43
    - 5.3.8 Role ethics — 43
    - 5.3.9 Anarchist ethics — 43
    - 5.3.10 Postmodern ethics — 44
- 5.4 Applied ethics — 44
    - 5.4.1 Specific questions — 45
    - 5.4.2 Particular fields of application — 45
- 5.5 Moral psychology — 47
    - 5.5.1 Evolutionary ethics — 47
- 5.6 Descriptive ethics — 47
- 5.7 See also — 48
- 5.8 Notes — 48
- 5.9 References — 50
- 5.10 Further reading — 50
- 5.11 External links — 51

6 Social justice — 52

- 6.1 History — 52
- 6.2 Contemporary theory — 54
    - 6.2.1 Philosophical perspectives — 54
- 6.3 Religious perspectives — 56

		6.3.1	Hinduism	56
		6.3.2	Islam	56
		6.3.3	Judaism	57
		6.3.4	Christianity	57
	6.4		Criticism	58
	6.5		Social justice movements	59
		6.5.1	Liberation theology	59
		6.5.2	Health care	59
		6.5.3	Human rights education	59
	6.6		See also	60
	6.7		Notes	60
	6.8		References	62
		6.8.1	Articles	62
		6.8.2	Books	62
	6.9		External links	63
7	**Naturalism (philosophy)**			**64**
	7.1		Origins and history	64
		7.1.1	Etymology	65
	7.2		Metaphysical naturalism	65
	7.3		Methodological naturalism	65
	7.4		Views	66
		7.4.1	Alvin Plantinga	66
		7.4.2	Robert T. Pennock	66
		7.4.3	W. V. O. Quine	67
		7.4.4	Karl Popper	67
	7.5		See also	67
	7.6		Notes	67
	7.7		References	69
	7.8		Further reading	69
	7.9		External links	69
		7.9.1	Supportive	69
		7.9.2	Neutral	69
		7.9.3	Critical	69
8	**Utilitarianism**			**70**
	8.1		Etymology	70
	8.2		Historical background	70
		8.2.1	Chinese philosophy	70

	8.2.2	Western philosophy	70
8.3	Classical utilitarianism		72
	8.3.1	Jeremy Bentham	72
	8.3.2	John Stuart Mill	73
8.4	Twentieth-century developments		75
	8.4.1	Ideal utilitarianism	75
	8.4.2	Act and rule utilitarianism	76
	8.4.3	Two-level utilitarianism	77
	8.4.4	Preference utilitarianism	77
8.5	More varieties of utilitarianism		78
	8.5.1	Negative utilitarianism	78
	8.5.2	Motive utilitarianism	78
8.6	Criticisms		79
	8.6.1	Ignores justice	79
	8.6.2	Predicting consequences	79
	8.6.3	Demandingness objection	80
	8.6.4	Aggregating utility	80
	8.6.5	Calculating utility is self-defeating	81
	8.6.6	Karl Marx's criticism	81
	8.6.7	John Paul II's personalist criticism	82
8.7	Additional considerations		82
	8.7.1	Average v. total happiness	82
	8.7.2	Motives, intentions, and actions	82
8.8	Application to specific issues		83
	8.8.1	Nonhuman animals	83
	8.8.2	World poverty	84
8.9	See also		84
8.10	Notes		85
8.11	References		89
8.12	Further reading		91
8.13	External links		92

9 Ethical movement — **93**

9.1	History		93
	9.1.1	Background	93
	9.1.2	Ethical movement	94
	9.1.3	In Britain	96
9.2	Ethical perspective		96
9.3	Religious aspect		97

	9.4	Key ideas 97
	9.5	Locations 98
	9.6	Structure and events ... 98
	9.7	Legal challenges 98
	9.8	Advocates 98
	9.9	See also 98
	9.10	References 98
	9.11	Further reading 99
	9.12	External links 99

10 Ethical naturalism — 100

- 10.1 Ethical theories that can be naturalistic ... 100
- 10.2 Criticisms ... 101
  - 10.2.1 Moral relativism ... 101
  - 10.2.2 Moral nihilism ... 101
- 10.3 Morality as a science ... 101
- 10.4 References ... 102
- 10.5 Other sources ... 102
- 10.6 External links ... 102

11 Evolutionary ethics — 103

- 11.1 History ... 103
- 11.2 Descriptive evolutionary ethics ... 103
- 11.3 Normative evolutionary ethics ... 104
- 11.4 Evolutionary Metaethics ... 104
- 11.5 See also ... 105
- 11.6 Notes ... 105
- 11.7 References ... 105
- 11.8 Further reading ... 105
- 11.9 External links ... 106

12 Secular ethics — 107

- 12.1 Tenets of secular ethics ... 107
  - 12.1.1 Humanist ethics ... 107
  - 12.1.2 Secular ethics and religion ... 108
- 12.2 Examples of secular ethical codes ... 108
  - 12.2.1 Humanist Manifestos ... 108
  - 12.2.2 Alternatives to the Ten Commandments ... 108
  - 12.2.3 Girl Scout law ... 108

12.2.4　United States Naval Academy honor concept 108
　　　12.2.5　Minnesota Principles 109
　　　12.2.6　Rotary Four-Way Test 109
　　　12.2.7　Military codes 109
　12.3　Nature and ethics 109
　12.4　Key philosophers and philosophical texts 109
　　　12.4.1　Valluvar 110
　　　12.4.2　Holyoake 110
　　　12.4.3　Nietzsche 110
　　　12.4.4　Kant 111
　　　12.4.5　Utilitarianism 111
　12.5　See also 112
　12.6　References 112
　12.7　Bibliography 113

13 Law of three stages　　114
　13.1　The progression of the three stages of Sociology 114
　13.2　Critiques of the law 115
　13.3　See also 115
　13.4　References 115
　13.5　External links 115

14 Science of morality　　116
　14.1　Overview 116
　14.2　History 116
　　　14.2.1　In philosophy 116
　　　14.2.2　Popular literature 116
　14.3　Views in Scientific Morality 117
　　　14.3.1　Training to promote good behaviour 117
　　　14.3.2　Research 118
　　　14.3.3　Other implications 118
　14.4　Criticisms 118
　14.5　See also 119
　14.6　Notes 119
　14.7　References 119
　14.8　Further reading 120

15 Secular morality　　122
　15.1　Secular moral frameworks 122

	15.1.1 Consequentialism	122
	15.1.2 Freethinking	122
	15.1.3 Secular humanism	122
15.2	Positions on religion and morality	123
	15.2.1 Morality requires religious tenets	123
	15.2.2 Morality does not rely on religion	123
	15.2.3 Religion is a poor moral guide	125
	15.2.4 Evidential findings	125
	15.2.5 Other views	125
15.3	See also	126
15.4	Notes	126
15.5	References	126
15.6	External links	128

16 Metaphysical naturalism — 129

16.1	Definition	129
	16.1.1 Methodological Naturalism	130
	16.1.2 Lack of necessity for worship	130
16.2	Science and naturalism	130
16.3	Various associated beliefs	131
	16.3.1 Undesigned universe	131
	16.3.2 Abiogenesis and evolution	131
	16.3.3 Ethics and meta-ethics	131
	16.3.4 The mind is a natural phenomenon	131
	16.3.5 Utility of reason	132
	16.3.6 Value of society	132
16.4	History	132
	16.4.1 Ancient period	132
	16.4.2 Middle ages to modernity	132
	16.4.3 Marxism, Objectivism, and secular humanism	133
16.5	Arguments for metaphysical naturalism	133
	16.5.1 Argument from physical minds	134
	16.5.2 Cosmological argument for naturalism	134
16.6	Arguments against metaphysical naturalism	134
	16.6.1 Evolutionary argument against naturalism	134
	16.6.2 Pre-Modern philosophy	135
16.7	See also	135
16.8	Further reading	135
	16.8.1 Historical overview	135

16.8.2 Pro	135
16.8.3 Con	135
16.9 Notes	136
16.10 References	137
16.11 External links	138

17 Religion of Humanity — 139

- 17.1 Origins . . . 139
- 17.2 Tenets . . . 139
- 17.3 Liturgy and priesthood . . . 140
- 17.4 Influence . . . 140
    - 17.4.1 Religion of Humanity in Brazil . . . 140
- 17.5 Other examples . . . 140
- 17.6 See also . . . 141
- 17.7 References . . . 141
- 17.8 External links . . . 141

18 International Humanist and Ethical Union — 142

- 18.1 Humanism as a life stance . . . 142
    - 18.1.1 Other major resolutions . . . 142
- 18.2 Organisation . . . 143
    - 18.2.1 Founding in 1952 . . . 143
    - 18.2.2 Current structure . . . 143
- 18.3 Strategy and activities . . . 143
    - 18.3.1 The Freedom of Thought Report . . . 143
    - 18.3.2 Focus of advocacy and campaigns . . . 145
- 18.4 Historical dates and figures . . . 146
    - 18.4.1 Chairs and presidents . . . 146
    - 18.4.2 Awards . . . 146
- 18.5 See also . . . 148
- 18.6 Footnotes . . . 148
- 18.7 External links . . . 150

19 Renaissance humanism — 151

- 19.1 Origin . . . 151
- 19.2 Paganism and Christianity in the Renaissance . . . 152
- 19.3 Humanists . . . 153
- 19.4 See also . . . 153
- 19.5 Notes . . . 153

19.6 Further reading 154

19.7 External links 155

20 Renaissance humanism in Northern Europe **156**

20.1 Italian roots of the humanism in Germany 157

20.2 Universities 158

20.3 Education 158

    20.3.1 See also 159

20.4 Leaders of humanism 159

20.5 See also 160

20.6 References 160

20.7 Sources 160

20.8 Literature 160

20.9 External links 160

21 Humanism in France **161**

21.1 See also 162

21.2 Further reading 162

21.3 External links and references 162

22 Center for Inquiry **163**

22.1 History 163

    22.1.1 Expansion 163

    22.1.2 Departure of founder 164

22.2 Paranormal and fringe science claims 164

    22.2.1 The Independent Investigations Group 165

22.3 Religion, ethics, and society 165

22.4 Publications 165

22.5 Projects and programs 166

    22.5.1 Center for Inquiry On Campus 166

    22.5.2 Skeptic's Toolbox 166

    22.5.3 Center for Inquiry Libraries 166

    22.5.4 Secular Rescue 167

    22.5.5 Office of Public Policy 167

    22.5.6 "Science and the Public" Master of Education program 167

    22.5.7 Skeptics and Humanist Aid and Relief Effort 167

22.6 Past projects and programs 167

    22.6.1 Camp Inquiry 168

    22.6.2 CFI Institute 168

- 22.6.3 Medicine and health ... 168
- 22.6.4 Naturalism Research Project ... 168
- 22.7 CFI organization and locations ... 168
  - 22.7.1 International activities ... 168
  - 22.7.2 University exchange programs ... 168
  - 22.7.3 Centre for Inquiry Canada ... 169
- 22.8 Affiliate organizations ... 169
- 22.9 In the media ... 169
  - 22.9.1 Wyndgate Country Club and Richard Dawkins, 2011 ... 169
  - 22.9.2 CSH actions against faith-based initiatives ... 169
  - 22.9.3 Heckled at the UN ... 170
  - 22.9.4 Blasphemy Day ... 170
- 22.10 See also ... 170
- 22.11 References ... 170
- 22.12 External links ... 173

23 A Secular Humanist Declaration — 174
- 23.1 Table of Contents ... 174
- 23.2 Signatories ... 174
  - 23.2.1 United States ... 174
  - 23.2.2 Canada ... 175
  - 23.2.3 France ... 175
  - 23.2.4 Great Britain (i.e. Scotland, Wales and England) ... 175
  - 23.2.5 India ... 176
  - 23.2.6 Israel ... 176
  - 23.2.7 Norway ... 176
  - 23.2.8 Yugoslavia ... 176
- 23.3 See also ... 176
- 23.4 External links ... 176

24 Amsterdam Declaration — 177
- 24.1 Humanist principles ... 177
- 24.2 History ... 178
- 24.3 References ... 178
- 24.4 External links ... 178

25 American Humanist Association — 179
- 25.1 Background ... 179
- 25.2 History ... 179

- 25.3 Adjuncts and affiliates 180
  - 25.3.1 Black Humanist Alliance 180
  - 25.3.2 Feminist Humanist Alliance 180
  - 25.3.3 LGBTQ Humanist Alliance 180
  - 25.3.4 Humanist Charities 180
  - 25.3.5 Appignani Humanist Legal Center 181
- 25.4 Advertising campaigns 182
- 25.5 National Day of Reason 182
- 25.6 Reason Rally 183
- 25.7 Famous awardees 183
- 25.8 AHA's Humanists of the Year 183
- 25.9 See also 184
- 25.10 References 184
- 25.11 External links 186

26 Humanists UK — 187
- 26.1 Aims 187
- 26.2 History 187
- 26.3 Campaigns 188
  - 26.3.1 Schools 188
  - 26.3.2 Constitutional reform 189
  - 26.3.3 Ethical issues 189
  - 26.3.4 Public awareness 189
- 26.4 Organisation 190
  - 26.4.1 Presidents 190
  - 26.4.2 Staff 191
  - 26.4.3 Humanist celebrants 191
  - 26.4.4 Pastoral carers 191
  - 26.4.5 Young Humanists 191
  - 26.4.6 Patrons 191
  - 26.4.7 Affiliations 192
- 26.5 Annual award 192
- 26.6 Criticism 192
- 26.7 See also 192
- 26.8 References 193
- 26.9 External links 195

27 Humanistischer Verband Deutschlands — 196
- 27.1 Aims 196

	27.2 Activities	196
	27.3 References	196
	27.4 External links	196

28 Humanist Society Scotland — 197
- 28.1 History and aims — 197
- 28.2 Campaigns — 197
  - 28.2.1 Weddings — 197
  - 28.2.2 Other issues — 198
- 28.3 References — 198
- 28.4 External links — 198

29 Norwegian Humanist Association — 199
- 29.1 See also — 199
- 29.2 References — 199
- 29.3 External links — 200

30 Alternatives to the Ten Commandments — 201
- 30.1 Examples — 201
  - 30.1.1 Christopher Hitchens — 201
  - 30.1.2 Richard Dawkins — 201
  - 30.1.3 Bertrand Russell — 202
  - 30.1.4 Bayer and Figdor's Ten Non-Commandments — 202
  - 30.1.5 The Atheists' New Ten Commandments — 202
  - 30.1.6 Ten Indian Commandments — 203
  - 30.1.7 Selig Starr — 203
  - 30.1.8 Ted Kaczynski — 203
  - 30.1.9 Summum — 204
  - 30.1.10 Hu Jintao — 204
- 30.2 See also — 204
- 30.3 References — 204
- 30.4 External links — 205

31 Maslow's hierarchy of needs — 206
- 31.1 Hierarchy — 206
  - 31.1.1 Physiological needs — 206
  - 31.1.2 Safety needs — 207
  - 31.1.3 Social belonging — 207
  - 31.1.4 Esteem — 207
  - 31.1.5 Self-actualization — 208

31.1.6 Self-transcendence 208

31.2 Application to Nursing 208

    31.2.1 The Human Dimensions and Basic Human Needs 208

31.3 Research 208

31.4 Criticism 208

    31.4.1 Ranking 208

    31.4.2 Definition of terms 209

31.5 See also 209

31.6 References 209

31.7 Further reading 211

31.8 External links 211

32 Positive psychology   212

32.1 Definition and basic assumptions 212

    32.1.1 Definition 212

    32.1.2 Basic concepts 212

    32.1.3 Research topics 212

    32.1.4 Basic assumptions 213

32.2 Origins and development 213

    32.2.1 Origin 213

    32.2.2 Development 213

    32.2.3 Historical antecedents 214

32.3 Theory and methods 215

    32.3.1 Flow 215

    32.3.2 PERMA 215

    32.3.3 *Character Strengths and Virtues* 216

32.4 Applications and research findings 217

32.5 Criticism 217

32.6 See also 217

32.7 Notes 218

32.8 References 218

32.9 Sources 220

32.10 Further reading 221

32.11 External links 221

33 Posthumanism   222

33.1 Philosophical posthumanism 222

33.2 Emergence of philosophical posthumanism 223

33.3 Contemporary posthuman discourse 223

33.4	Relationship with transhumanism	224
33.5	Criticism	224
33.6	See also	224
33.7	References	224
	33.7.1 Works cited	225
33.8	Text and image sources, contributors, and licenses	226
	33.8.1 Text	226
	33.8.2 Images	236
	33.8.3 Content license	241

Chapter 1

Secular humanism

The philosophy or life stance of **secular humanism** embraces human reason, ethics, social justice, and philosophical naturalism while specifically rejecting religious dogma, supernaturalism, pseudoscience, and superstition as the bases of morality and decision making.[1][2][3][4]

Secular humanism posits that human beings are capable of being ethical and moral without religion or a god. It does not, however, assume that humans are either inherently evil or innately good, nor does it present humans as being superior to nature. Rather, the humanist life stance emphasizes the unique responsibility facing humanity and the ethical consequences of human decisions. Fundamental to the concept of secular humanism is the strongly held viewpoint that ideology—be it religious or political—must be thoroughly examined by each individual and not simply accepted or rejected on faith. Along with this, an essential part of secular humanism is a continually adapting search for truth, primarily through science and philosophy. Many secular humanists derive their moral codes from a philosophy of utilitarianism, ethical naturalism, or evolutionary ethics, and some advocate a science of morality.

The International Humanist and Ethical Union (IHEU) is the world union of more than one hundred Humanist, rationalist, irreligious, atheistic, Bright, secular, Ethical Culture, and freethought organizations in more than 40 countries. The "Happy Human" is the official symbol of the IHEU as well as being regarded as a universally recognised symbol for those who call themselves Humanists. Secular humanist organizations are found in all parts of the world. Those who call themselves humanists are estimated to number between four and five million people worldwide.

1.1 Terminology

The meaning of the phrase *secular humanism* has evolved over time. The phrase has been used since at least the 1930s by Anglican priests,[5] and in 1943, the then Archbishop of Canterbury, William Temple, was reported as warning that the "Christian tradition... was in danger of being undermined by a 'Secular Humanism' which hoped to retain Christian values without Christian faith."[6] During the 1960s and 1970s the term was embraced by some humanists who considered themselves anti-religious,[7] as well as those who, although not critical of religion in its various guises, preferred a non-religious approach.[8] The release in 1980 of *A Secular Humanist Declaration* by the newly formed Council for Democratic and Secular Humanism (CODESH, now the Council for Secular Humanism) gave secular humanism an organisational identity within the United States.

However, many adherents of the approach reject the use of the word *secular* as obfuscating and confusing, and consider that the term *secular humanism* has been "demonized by the religious right... All too often secular humanism is reduced to a sterile outlook consisting of little more than secularism slightly broadened by academic ethics. This kind of 'hyphenated humanism' easily becomes more about the adjective than its referent".[9] Adherents of this view, including the International Humanist and Ethical Union and the American Humanist Association, consider that the unmodified but capitalised word Humanism should be used. The endorsement by the IHEU of the capitalization of the word *Humanism*, and the dropping of any adjective such as *secular*, is quite recent. The American Humanist Association began to adopt this view in 1973, and the IHEU formally endorsed the position in 1989. In 2002 the IHEU General Assembly unanimously adopted the Amsterdam Declaration, which represents the official defining statement of World Humanism for Humanists. This declaration makes exclusive use of capitalized *Humanist* and *Humanism*, which is consistent with IHEU's general practice and recommendations for promoting a unified Humanist identity. To further promote Humanist identity, these words are also free of any adjectives, as recommended by prominent members of IHEU. Such usage is not universal among IHEU member organizations, though most of them do observe these conventions.

1.2 History

Historical use of the term humanism (reflected in some current academic usage), is related to the writings of pre-Socratic philosophers. These writings were lost to European societies until Renaissance scholars rediscovered them through Muslim sources and translated them from Arabic into European languages.[10] Thus the term humanist can mean a humanities scholar, as well as refer to The Enlightenment/ Renaissance intellectuals, and those who have agreement with the pre-Socratics, as distinct from secular humanists.

1.2.1 Secularism

George Holyoake coined the term "secularism" and led the secular movement in Britain from the mid-19th century.

In 1851 George Holyoake coined the term "secularism"[11] to describe "a form of opinion which concerns itself only with questions, the issues of which can be tested by the experience of this life".[12]

The modern secular movement coalesced around Holyoake, Charles Bradlaugh and their intellectual circle. The first secular society, the Leicester Secular Society, dates from 1851. Similar regional societies came together to form the National Secular Society in 1866.

1.2.2 Positivism & the Church of Humanity

Holyoake's secularism was strongly influenced by Auguste Comte, the founder of positivism and of modern sociology. Comte believed human history would progress in a "law of three stages" from a theological phase, to the "metaphysical", toward a fully rational "positivist" society. In later life, Comte had attempted to introduce a "religion of humanity" in light of growing anti-religious sentiment and social malaise in revolutionary France. This religion would necessarily fulfil the functional, cohesive role that supernatural religion once served.

Although Comte's religious movement was unsuccessful in France, the positivist philosophy of science itself played a major role in the proliferation of secular organizations in the 19th century in England. Richard Congreve visited Paris shortly after the French Revolution of 1848 where he met Auguste Comte and was heavily influenced by his positivist system. He founded the London Positivist Society in 1867, which attracted Frederic Harrison, Edward Spencer Beesly, Vernon Lushington, and James Cotter Morison amongst others.

In 1878, the Society established the Church of Humanity under Congreve's direction. There they introduced sacraments of the Religion of Humanity and published a cooperative translation of Comte's Positive Polity. When Congreve repudiated their Paris co-religionists in 1878, Beesly, Harrison, Bridges, and others formed their own positivist society, with Beesly as president, and opened a rival centre, Newton Hall, in a courtyard off Fleet Street.

The New York City version of the church was established by English immigrant Henry Edger. The American version of the "Church of Humanity". was largely modeled on the English church. Like the English version it wasn't atheistic and had sermons and sacramental rites.[13] At times the services included readings from conventional religious works like the Book of Isaiah.[14] It was not as significant as the church in England, but did include several educated people.[15]

1.2.3 Ethical movement

Another important precursor was the ethical movement of the 19th century. The South Place Ethical Society was founded in 1793 as the South Place Chapel on Finsbury Square, on the edge of the City of London,[16] and in the early nineteenth century was known as "a radical gathering-place".[17] At that point it was a Unitarian chapel, and that movement, like Quakers, supported female equality.[18] Under the leadership of Reverend William Johnson Fox,[19] it lent its pulpit to activists such as Anna Wheeler, one of the first women to campaign for feminism at public meet-

Felix Adler, founder of the ethical movement

ings in England, who spoke in 1829 on "rights of women". In later decades, the chapel changed its name to the South Place Ethical Society, now the Conway Hall Ethical Society. Today Conway Hall explicitly identifies itself as a humanist organisation, albeit one primarily focused on concerts, events, and the maintenance of its humanist library and archives. It bills itself as "The landmark of London's independent intellectual, political and cultural life."

In America, the ethical movement was propounded by Felix Adler, who established the New York Society for Ethical Culture in 1877.[20] By 1886, similar societies had sprouted up in Philadelphia, Chicago and St. Louis.[21]

These societies all adopted the same statement of principles:

- The belief that morality is independent of theology;

- The affirmation that new moral problems have arisen in modern industrial society which have not been adequately dealt with by the world's religions;

- The duty to engage in philanthropy in the advancement of morality;

- The belief that self-reform should go in lock step with social reform;

- The establishment of republican rather than monarchical governance of Ethical societies

- The agreement that educating the young is the most important aim.

In effect, the movement responded to the religious crisis of the time by replacing theology with unadulterated morality. It aimed to "disentangle moral ideas from religious doctrines, metaphysical systems, and ethical theories, and to make them an independent force in personal life and social relations."[21] Adler was also particularly critical of the religious emphasis on creed, believing it to be the source of sectarian bigotry. He therefore attempted to provide a universal fellowship devoid of ritual and ceremony, for those who would otherwise be divided by creeds. For the same reasons the movement at that time adopted a neutral position on religious beliefs, advocating neither atheism nor theism, agnosticism nor deism.[21]

The first ethical society along these lines in Britain was founded in 1886. By 1896 the four London societies formed the Union of Ethical Societies, and between 1905 and 1910 there were over fifty societies in Great Britain, seventeen of which were affiliated with the Union. The Union of Ethical Societies would later rename itself the British Humanist Association.

1.2.4 Secular humanism

In the 1930s, "humanism" was generally used in a religious sense by the Ethical movement in the United States, and not much favoured among the non-religious in Britain. Yet "it was from the Ethical movement that the non-religious philosophical sense of *Humanism* gradually emerged in Britain, and it was from the convergence of the Ethical and Rationalist movements that this sense of *Humanism* eventually prevailed throughout the Freethought movement".[22]

As an organized movement, Humanism itself is quite recent – born at the University of Chicago in the 1920s, and made public in 1933 with the publication of the first Humanist Manifesto.[23] The American Humanist Association was incorporated as an Illinois non-profit organization in 1943. The International Humanist and Ethical Union was founded in 1952, when a gathering of world Humanists met under the leadership of Sir Julian Huxley. The British Humanist Association took that name in 1967, but had developed from the Union of Ethical Societies which had been founded by Stanton Coit in 1896.[24]

1.3 Manifestos and declarations

Humanists have put together various Humanist Manifestos, in attempts to unify the Humanist identity.

The original signers of the first Humanist Manifesto of 1933, declared themselves to be religious humanists. Because, in their view, traditional religions were failing to meet the needs of their day, the signers of 1933 declared it a necessity to establish a religion that was a dynamic force to meet the needs of the day. However, this "religion" did not profess a belief in any god. Since then two additional Manifestos were written to replace the first. In the Preface of Humanist Manifesto II, in 1973, the authors Paul Kurtz and Edwin H. Wilson assert that faith and knowledge are required for a hopeful vision for the future. Manifesto II references a section on Religion and states traditional religion renders a disservice to humanity. Manifesto II recognizes the following groups to be part of their naturalistic philosophy: "scientific", "ethical", "democratic", "religious", and "Marxist" humanism.

1.3.1 International Humanist and Ethical Union

In 2002, the IHEU General Assembly unanimously adopted the Amsterdam Declaration 2002 which represents the official defining statement of World Humanism.[25]

All member organisations of the International Humanist and Ethical Union are required by bylaw 5.1[26] to accept the *Minimum Statement on Humanism*:

> Humanism is a democratic and ethical life stance, which affirms that human beings have the right and responsibility to give meaning and shape to their own lives. It stands for the building of a more humane society through an ethic based on human and other natural values in the spirit of reason and free inquiry through human capabilities. It is not theistic, and it does not accept supernatural views of reality.

To promote and unify "Humanist" identity, prominent members of the IHEU have endorsed the following statements on Humanist identity:

- All Humanists, nationally and internationally, should always use the one word Humanism as the name of Humanism: no added adjective, and the initial letter capital (by life stance orthography);
- All Humanists, nationally and internationally, should use a clear, recognizable and uniform symbol on their publications and elsewhere: our Humanist symbol the "Happy Human";
- All Humanists, nationally and internationally, should seek to establish recognition of the fact that Humanism is a life stance.

1.3.2 Council for Secular Humanism

According to the Council for Secular Humanism, within the United States, the term "secular humanism" describes a world view with the following elements and principles:[8]

- **Need to test beliefs** – A conviction that dogmas, ideologies and traditions, whether religious, political or social, must be weighed and tested by each individual and not simply accepted by faith.
- **Reason, evidence, scientific method** – A commitment to the use of critical reason, factual evidence and scientific method of inquiry in seeking solutions to human problems and answers to important human questions.
- **Fulfillment, growth, creativity** – A primary concern with fulfillment, growth and creativity for both the individual and humankind in general.
- **Search for truth** – A constant search for objective truth, with the understanding that new knowledge and experience constantly alter our imperfect perception of it.
- **This life** – A concern for this life (as opposed to an afterlife) and a commitment to making it meaningful through better understanding of ourselves, our history, our intellectual and artistic achievements, and the outlooks of those who differ from us.
- **Ethics** – A search for viable individual, social and political principles of ethical conduct, judging them on their ability to enhance human well-being and individual responsibility.
- **Justice and fairness** – an interest in securing justice and fairness in society and in eliminating discrimination and intolerance.[27]
- **Building a better world** – A conviction that with reason, an open exchange of ideas, good will, and tolerance, progress can be made in building a better world for ourselves and our children.

A Secular Humanist Declaration was issued in 1980 by the Council for Secular Humanism's predecessor, CODESH. It lays out ten ideals: Free inquiry as opposed to censorship and imposition of belief; separation of church and state; the ideal of freedom from religious control and from jingoistic government control; ethics based on critical intelligence rather than that deduced from religious belief; moral education; religious skepticism; reason; a belief in science and technology as the best way of understanding the world; evolution; and education as the essential method of building humane, free, and democratic societies.[28]

1.3.3 American Humanist Association

A general outline of Humanism is also set out in the *Humanist Manifesto* prepared by the American Humanist Association.[29]

1.4 Ethics and relationship to religious belief

See also: Secular ethics

In the 20th and 21st centuries, members of Humanist organizations have disagreed as to whether Humanism is a religion. They categorize themselves in one of three ways. Religious Humanism, in the tradition of the earliest Humanist organizations in the UK and US, attempts to fulfill the traditional social role of religion.[30] Secular humanism considers all forms of religion, including religious Humanism, to be superseded.[31] In order to sidestep disagreements between these two factions, recent Humanist proclamations define Humanism as a "life stance"; proponents of this view making up the third faction. All three types of Humanism (and all three of the American Humanist Association's manifestos) reject deference to supernatural beliefs; promoting the practical, methodological naturalism of science, but also going further and supporting the philosophical stance of metaphysical naturalism.[32] The result is an approach to issues in a secular way. Humanism addresses ethics without reference to the supernatural as well, attesting that ethics is a human enterprise (see naturalistic ethics).[2][3][4]

Secular humanism does not prescribe a specific theory of morality or code of ethics. As stated by the Council for Secular Humanism,

> It should be noted that Secular Humanism is not so much a specific morality as it is a method for the explanation and discovery of rational moral principles.[33]

Secular humanism affirms that with the present state of scientific knowledge, dogmatic belief in an absolutist moral/ethical system (e.g. Kantian, Islamic, Christian) is unreasonable. However, it affirms that individuals engaging in rational moral/ethical deliberations can discover some universal "objective standards".

> We are opposed to absolutist morality, yet we maintain that objective standards emerge, and ethical values and principles may be discovered, in the course of ethical deliberation.[33]

Many Humanists adopt principles of the Golden Rule. Some believe that universal moral standards are required for the proper functioning of society. However, they believe such necessary universality can and should be achieved by developing a richer notion of morality through reason, experience and scientific inquiry rather than through faith in a supernatural realm or source.[34]

> Fundamentalists correctly perceive that universal moral standards are required for the proper functioning of society. But they erroneously believe that God is the only possible source of such standards. Philosophers as diverse as Plato, Immanuel Kant, John Stuart Mill, George Edward Moore, and John Rawls have demonstrated that it is possible to have a universal morality without God. Contrary to what the fundamentalists would have us believe, then, what our society really needs is not more religion but a richer notion of the nature of morality.[35]

Humanism is compatible with atheism[36] and agnosticism,[37] but being atheist or agnostic does not automatically make one a humanist. Nevertheless, humanism is diametrically opposed to state atheism.[38][39] According to Paul Kurtz, considered by some to be the founder of the American secular humanist movement,[40] one of the differences between Marxist–Leninist atheists and humanists is the latter's commitment to "human freedom and democracy" while stating that the militant atheism of the Soviet Union consistently violated basic human rights.[41] Kurtz also stated that the "defense of religious liberty is as precious to the humanist as are the rights of the believers".[41] Greg M. Epstein states that, "modern, organized Humanism began, in the minds of its founders, as nothing more nor less than a religion without a God".[42]

Many Humanists address ethics from the point of view of ethical naturalism, and some support an actual science of morality.[43]

1.5 Modern context

Secular humanist organizations are found in all parts of the world. Those who call themselves humanists are estimated to number between four[44] and five[45] million people worldwide in 31 countries, but there is uncertainty because of the lack of universal definition throughout censuses. Humanism is a non-theistic belief system and, as such, it could be a sub-category of "Religion" only if that term is defined to mean "Religion and (any) belief system". This is the case in the International Covenant on Civil and Political Rights on freedom of religion *and* beliefs. Many national censuses contentiously define Humanism as a further sub-category of the sub-category "No Religion", which typically includes atheist, rationalist and agnostic thought. In England, Wales[46] 25% of people specify that they have 'No religion' up from 15% in 2001 and in Australia,[47][48] around 15% of the population specifies "No Religion" in the national census. However, in its 2006 and 2011 census Australia used Humanism as an example of "other religions". In the USA, the decennial census does not inquire about religious affiliation or its lack; surveys report the figure at roughly 13%.[49] In the 2001 Canadian census, 16.5% of the populace reported having no religious affiliation.[50] In the 2011 Scottish census, 37% stated they had no religion up from 28% in 2001.[51] One of the largest Humanist organizations in the world (relative to population) is Norway's *Human-Etisk Forbund*,[52] which had over 86,000 members out of a population of around 4.6 million in 2013 – approximately 2% of the population.[53]

The International Humanist and Ethical Union (IHEU) is the worldwide umbrella organization for those adhering to the Humanist life stance. It represents the views of over three million Humanists organized in over 100 national organizations in 30 countries.[54] Originally based in the Netherlands, the IHEU now operates from London. Some regional groups that adhere to variants of the Humanist life stance, such as the humanist subgroup of the Unitarian Universalist Association, do not belong to the IHEU. Although the European Humanist Federation is also separate from the IHEU, the two organisations work together and share an agreed protocol.[55]

Starting in the mid-20th century, religious fundamentalists and the religious right began using the term "secular humanism" in hostile fashion. Francis A. Schaeffer, an American theologian based in Switzerland, seizing upon the exclusion of the divine from most humanist writings, argued that rampant secular humanism would lead to moral relativism and ethical bankruptcy in his book *How Should We Then Live: The Rise and Decline of Western Thought and Culture* (1976). Schaeffer portrayed secular humanism as pernicious and diabolical, and warned it would undermine the moral and spiritual tablet of America. His themes have been very widely repeated in Fundamentalist preaching in North America.[56] Toumey (1993) found that secular humanism is typically portrayed as a vast evil conspiracy, deceitful and immoral, responsible for feminism, pornography, abortion, homosexuality, and New Age spirituality.[57] In certain areas of the world, Humanism finds itself in conflict with religious fundamentalism, especially over the issue of the separation of church and state. Many Humanists see religions as superstitious, repressive and closed-minded, while religious fundamentalists may see Humanists as a threat to the values set out in their sacred texts.[58]

In recent years, humanists such as Dwight Gilbert Jones and R. Joseph Hoffmann have decried the over-association of Humanism with affirmations of non-belief and atheism. Jones cites a lack of new ideas being presented or debated outside of secularism,[59] while Hoffmann is unequivocal: "I regard the use of the term 'humanism' to mean secular humanism or atheism to be one of the greatest tragedies of twentieth century movementology, perpetrated by second-class minds and perpetuated by third-class polemicists and village atheists. The attempt to sever humanism from the religious and the spiritual was a flatfooted, largely American way of taking on the religious right. It lacked finesse, subtlety, and the European sense of history."[60]

1.6 Humanist celebrations

Some Humanists celebrate official religion-based public holidays, such as Christmas or Easter, but as secular holidays rather than religious ones.[61] Many Humanists also celebrate the winter and summer solstice, the former of which (in the northern hemisphere) coincides closely with the religiously-oriented celebration of Christmas, and the equinoxes, of which the vernal equinox is associated with Christianity's Easter and indeed with all other springtime festivals of renewal, and the autumnal equinox which is related to such celebrations such as Halloween and All Souls' Day. The Society for Humanistic Judaism celebrates most Jewish holidays in a secular manner.

The IHEU endorses World Humanist Day (21 June), Darwin Day (12 February), Human Rights Day (10 December) and HumanLight (23 December) as official days of Humanist celebration, though none are yet a public holiday.

In many countries, Humanist officiants (or celebrants) perform celebrancy services for weddings, funerals, child namings, coming of age ceremonies, and other rituals.

1.7 Legal mentions in the United States

The issue of whether and in what sense secular humanism might be considered a religion, and what the implications of this would be has become the subject of legal maneuvering and political debate in the United States. The first reference to "secular humanism" in a US legal context was in 1961, although church-state separation lawyer Leo Pfeffer had referred to it in his 1958 book, *Creeds in Competition*.

1.7.1 Hatch amendment

The Education for Economic Security Act of 1984 included a section, Section 20 U.S.C.A. 4059, which initially read: "Grants under this subchapter ['Magnet School Assistance'] may not be used for consultants, for transportation or for any activity which does not augment academic improvement." With no public notice, Senator Orrin Hatch tacked onto the proposed exclusionary subsection the words "or for any course of instruction the substance of which is Secular Humanism". Implementation of this provision ran into practical problems because neither the Senator's staff, nor the Senate's Committee on Labor and Human Resources, nor the Department of Justice could propose a definition of what would constitute a "course of instruction the substance of which is Secular Humanism". So, this determination was left up to local school boards. The provision provoked a storm of controversy which within a year led Senator Hatch to propose, and Congress to pass, an amendment to delete from the statute all reference to secular humanism. While this episode did not dissuade fundamentalists from continuing to object to what they regarded as the "teaching of Secular Humanism", it did point out the vagueness of the claim.

1.7.2 Case law

Torcaso v. Watkins

The phrase "secular humanism" became prominent after it was used in the United States Supreme Court case *Torcaso v. Watkins*. In the 1961 decision, Justice Hugo Black commented in a footnote, "Among religions in this country which do not teach what would generally be considered a belief in the existence of God are Buddhism, Taoism, Ethical Culture, Secular Humanism, and others."

Fellowship of Humanity v. County of Alameda

The footnote in *Torcaso v. Watkins* referenced *Fellowship of Humanity v. County of Alameda*,[62] a 1957 case in which an organization of humanists[63] sought a tax exemption on the ground that they used their property "solely and exclusively for religious worship." Despite the group's non-theistic beliefs, the court determined that the activities of the *Fellowship of Humanity*, which included weekly Sunday meetings, were analogous to the activities of theistic churches and thus entitled to an exemption. The *Fellowship of Humanity* case itself referred to *Humanism* but did not mention the term *secular humanism*. Nonetheless, this case was cited by Justice Black to justify the inclusion of secular humanism in the list of religions in his note. Presumably Justice Black added the word *secular* to emphasize the non-theistic nature of the *Fellowship of Humanity* and distinguish their brand of humanism from that associated with, for example, Christian humanism.

Washington Ethical Society v. District of Columbia

Another case alluded to in the *Torcaso v. Watkins* footnote, and said by some to have established secular humanism as a religion under the law, is the 1957 tax case of *Washington Ethical Society v. District of Columbia*, 249 F.2d 127 (D.C. Cir. 1957). The *Washington Ethical Society* functions much like a church, but regards itself as a non-theistic religious institution, honoring the importance of ethical living without mandating a belief in a supernatural origin for ethics. The case involved denial of the Society's application for tax exemption as a religious organization. The U.S. Court of Appeals reversed the Tax Court's ruling, defined the Society as a religious organization, and granted its tax exemption. The Society terms its practice Ethical Culture. Though Ethical Culture is based on a humanist philosophy, it is regarded by some as a type of religious humanism. Hence, it would seem most accurate to say that this case affirmed that a religion need not be theistic to qualify as a religion under the law, rather than asserting that it established generic secular humanism as a religion.

In the cases of both the *Fellowship of Humanity* and the *Washington Ethical Society,* the court decisions turned not so much on the particular beliefs of practitioners as on the function and form of the practice being similar to the function and form of the practices in other religious institutions.

Peloza v. Capistrano School District

The implication in Justice Black's footnote that secular humanism is a religion has been seized upon by religious opponents of the teaching of evolution, who have made the argument that teaching evolution amounts to teaching a religious idea. The claim that secular humanism could be considered a religion for legal purposes was examined by the United States Court of Appeals for the Ninth Circuit in *Peloza v. Capistrano School District*, 37 F.3d 517 (9th Cir. 1994), *cert. denied*, 515 U.S. 1173 (1995). In this case, a science teacher argued that, by requiring him to teach evolution, his school district was forcing him to teach the "religion" of secular humanism. The Court responded, "We reject this claim because neither the Supreme Court, nor this circuit, has ever held that evolutionism or Secular Humanism are 'religions' for Establishment Clause purposes." The Supreme Court refused to review the case.

The decision in a subsequent case, *Kalka v. Hawk et al.*, offered this commentary:[63]

> The Court's statement in *Torcaso* does not stand for the proposition that humanism, no matter in what form and no matter how practiced, amounts to a religion under the First Amendment. The Court offered no test for determining what system of beliefs qualified as a "religion" under the First Amendment. The most one may read into the *Torcaso* footnote is the idea that a particular non-theistic group calling itself the "Fellowship of Humanity" qualified as a religious organization under California law.

1.7.3 Controversy

Decisions about tax status have been based on whether an organization functions like a church. On the other hand, Establishment Clause cases turn on whether the ideas or symbols involved are inherently religious. An organization can function like a church while advocating beliefs that are not necessarily inherently religious. Author Marci Hamilton has pointed out: "Moreover, the debate is not between secularists and the religious. The debate is believers and non-believers on the one side debating believers and non-believers on the other side. You've got citizens who are [...] of faith who believe in the separation of church and state and you have a set of believers who do not believe in the separation of church and state."[64]

In the 1987 case of *Smith v. Board of School Commissioners of Mobile County* a group of plaintiffs brought a case alleging that the school system was teaching the tenets of an anti-religious religion called "secular humanism" in violation of the Establishment Clause. The complainants asked that 44 different elementary through high school level textbooks (including books on home economics, social science and literature) be removed from the curriculum. Federal judge William Brevard Hand ruled for the plaintiffs agreeing that the books promoted secular humanism, which he ruled to be a religion. The Eleventh Circuit Court unanimously reversed him, with Judge Frank stating that Hand held a "misconception of the relationship between church and state mandated by the establishment clause," commenting also that the textbooks did not show "an attitude antagonistic to theistic belief. The message conveyed by these textbooks is one of neutrality: the textbooks neither endorse theistic religion as a system of belief, nor discredit it".[65]

1.8 Notable humanists

Further information: List of secular humanists

1.9 Manifestos

There are numerous Humanist Manifestos and Declarations, including the following:

- Humanist Manifesto I (1933)
- Humanist Manifesto II (1973)
- A Secular Humanist Declaration (1980)
- A Declaration of Interdependence (1988)
- IHEU Minimum Statement on Humanism (1996)
- HUMANISM: Why, What, and What For, In 882 Words (1996)
- Humanist Manifesto 2000: A Call For A New Planetary Humanism (2000)
- The Affirmations of Humanism: A Statement of Principles
- Amsterdam Declaration (2002)
- Humanism and Its Aspirations
- Humanist Manifesto III (Humanism And Its Aspirations) (2003)
- Alternatives to the Ten Commandments

1.10 Related organizations

See also: List of secularist organizations

- American Atheists
- American Humanist Association
- Brights
- British Humanist Association
- Camp Quest
- Campus Freethought Alliance
- Center for Inquiry
- City Congregation for Humanistic Judaism
- Committee for the Scientific Investigation of Claims of the Paranormal
- Council for Secular Humanism (formerly CODESH)
- Council of Australian Humanist Societies
- Ethical Culture
- European Humanist Federation
- Federation of Indian Rationalist Associations
- Freedom From Religion Foundation
- Godless Americans Political Action Committee
- Humani (the Humanist Association of Northern Ireland)
- Humanist Association of Canada
- Humanist Association of Ireland
- Humanist Society Scotland
- Institute for Humanist Studies
- International Humanist and Ethical Union
- Internet Infidels
- Military Association of Atheists and Freethinkers
- National Center for Science Education
- National Secular Society (UK)
- New Zealand Association of Rationalists and Humanists
- Philippine Atheists and Agnostics Society
- Quackwatch
- Scouting for All
- Secular Student Alliance
- Secular Web
- Sidmennt (Iceland)
- The Skeptics Society
- Society for Humanistic Judaism
- Swedish Humanist Association
- Washington Area Secular Humanists
- World Transhumanist Association

1.11 See also

- List of official religions
- Comparative religion
- Renaissance humanism, the Renaissance liberal arts movement
- Maslow's Hierarchy of Needs
- Positive Psychology
- John Henry Silva

1.11.1 Related philosophies

- Effective altruism
- Empiricism
- Epicureanism
- Eupraxsophy
- Extropianism
- Marxist humanism
- Morality without religion
- Naturalistic pantheism
- Nontheism
- Objectivism
- Rationalism
- Secular religion
- Skepticism
- Stoicism
- Transhumanism

1.11.2 Wikibooks

- *Thinking And Moral Problems*
- *Religions And Their Source*
- *Purpose*
- *Developing A Universal Religion*, four parts of a Wikibook

1.12 Notes and references

[1] Council for Secular Humanism. "10 Myths About Secular Humanism". Retrieved 12 June 2015.

[2] Edwords, Fred (1989). "What Is Humanism?". American Humanist Association. Retrieved 19 August 2009. Secular Humanism is an outgrowth of eighteenth century enlightenment rationalism and nineteenth century freethought... A decidedly anti-theistic version of secular humanism, however, is developed by Adolf Grünbaum, 'In Defense of Secular Humanism' (1995), in his *Collected Works* (edited by Thomas Kupka), vol. I, New York: Oxford University Press 2013, ch. 6 (pp. 115–48)

[3] *Compact Oxford English dictionary*. Oxford University Press. 2007. humanism *n.* 1 a rationalistic system of thought attaching prime importance to human rather than divine or supernatural matters.

[4] "Definitions of humanism (subsection)". Institute for Humanist Studies. Retrieved 16 January 2007.

[5] See "Unemployed at service: church and the world", *The Guardian*, 25 May 1935, p. 18: citing the comments of Rev. W.G. Peck, rector of St. John the Baptist, Hulme Manchester, concerning "The modern age of secular humanism". Guardian and Observer Digital Archive

[6] "Free Church ministers in Anglican pulpits. Dr Temple's call: the South India Scheme." *The Guardian*, 26 May 1943, p. 6 Guardian and Observer Digital Archive

[7] See Mouat, Kit (1972) *An Introduction to Secular Humanism*. Haywards Heath: Charles Clarke Ltd. Also, *The Freethinker* began to use the phrase "secular humanist monthly" on its front page masthead.

[8] "What Is Secular Humanism?". Council for Secular Humanism.

[9] Humanism Unmodified By Edd Doerr. Published in the *Humanist* (November/December 2002)

[10] "Islamic political philosophy: Al-Farabi, Avicenna, Averroes". Fordham.edu. Retrieved 13 November 2011.

[11] Holyoake, G. J. (1896). *The Origin and Nature of Secularism*. London: Watts & Co., p. 50.

[12] "Secularism 101: Defining Secularism: Origins with George Jacob Holyoake". Atheism.about.com. 2 September 2011. Retrieved 13 November 2011.

[13] *Positivist Republic*. Retrieved 12 June 2015.

[14] New York Times: 16 January 1881

[15] "The Church of Humanity": New York's Worshipping Positivists American Society of Church History.

[16] , City of London page on Finsbury Circus Conservation Area Character Summary.

[17] *The Sexual Contract*, by Carole Patema. p. 160

[18] "Women's Politics in Britain 1780–1870: Claiming Citizenship" by Jane Rendall, esp. "72. The religious backgrounds of feminist activists"

[19] "Ethical Society history page". Ethicalsoc.org.uk. Retrieved 2013-09-29.

[20] Howard B. Radest. 1969. *Toward Common Ground: The Story of the Ethical Societies in the United States*. New York: Fredrick Unger Publishing Co.

[21] Colin Campbell. 1971. *Towards a Sociology of Irreligion*. London: McMillan Press.

[22] Walter, Nicolas (1997). *Humanism: what's in the word?* London: RPA/BHA/Secular Society Ltd, p. 43.

[23] "Text of Humanist Manifesto I". Americanhumanist.org. Retrieved 13 November 2011.

[24] "British Humanist Association: History". Humanism.org.uk. Retrieved 13 November 2011.

[25] "Amsterdam Declaration 2002". International Humanist and Ethical Union. Retrieved 5 July 2008.

[26] "IHEU's Bylaws". International Humanist and Ethical Union. Retrieved 5 July 2008.

[27] "The Affirmations of Humanism: A Statement of Principles". *secularhumanism.org*. The Council for Secular Humanism. Retrieved 28 May 2012.

[28] the Council for Secular Humanism (1980). "A Secular Humanist Declaration". the Council for Secular Humanism. Retrieved 27 November 2008.

[29] "Humanism and Its Aspirations – Humanist Manifesto III, a successor to the Humanist Manifesto of 1933". Americanhumanist.org. Retrieved 13 November 2011.

[30] Wilson, Edwin H. (1995). *The Genesis of a Humanist Manifesto*. Amherst, NY: Humanist Press. This book quotes the constitution of the Humanistic Religious Association of London, founded in 1853, as saying, "In forming ourselves into a progressive religious body, we have adopted the name 'Humanistic Religious Association' to convey the idea that Religion is a principle inherent in man and is a means of

developing his being towards greater perfection. We have emancipated ourselves from the ancient compulsory dogmas, myths and ceremonies borrowed of old from Asia and still pervading the ruling churches of our age".

[31] Kurtz, Paul (1995). *Living Without Religion: Eupraxophy*. Amherst, NY: Prometheus Books. p. 8.

[32] Eugenie C. Scott, National Centre for Science and Education, "Science and Religion, Methodology and Humanism". Example quote: "The same principle applies to philosophical materialism, the view at the foundation of our Humanism; we may derive this view from science, but an ideology drawn from science is not the same as science itself... I have argued that a clear distinction must be drawn between science as a way of knowing about the natural world and science as a foundation for philosophical views. One should be taught to our children in school, and the other can optionally be taught to our children at home."

[33] "A Secular Humanist Declaration". Secularhumanism.org. 29 July 2005. Retrieved 13 November 2011.

[34] Norman, Richard (2004). *On Humanism*. New York: Routledge. ISBN 9780415305228.

[35] Theodore Schick, Jr (29 July 2005). "Morality Requires God ... or Does It?". Secularhumanism.org. Retrieved 13 November 2011.

[36] Baggini, Julian (2003). *Atheism: A Very Short Introduction*. Oxford: Oxford University Press. pp. 3–4. ISBN 0-19-280424-3. The atheist's rejection of belief in God is usually accompanied by a broader rejection of any supernaturalor transcendental reality. For example, an atheist does not usually believe in the existence of immortal souls, life after death, ghosts, or supernatural powers. Although strictly speaking an atheist could believe in any of these things and still remain an atheist... the arguments and ideas that sustain atheism tend naturally to rule out other beliefs in the supernatural or transcendental.

[37] Winston, Robert (Ed.) (2004). *Human*. New York: DK Publishing, Inc. p. 299. ISBN 0-7566-1901-7. Neither atheism nor agnosticism is a full belief system, because they have no fundamental philosophy or lifestyle requirements. These forms of thought are simply the absence of belief in, or denial of, the existence of deities.

[38] Paul Kurtz; Vern L. Bullough; Tim Madigan (19 October 2009). *Toward a New Enlightenment: the Philosophy of Paul Kurtz*. Transaction Books. ISBN 978-1-56000-118-8. In the past, the Communist Party of the Soviet Union waged unremitting warfare against religion. It persecuted religious believers, confiscated church properties, executed or exiled tens of thousands of clerics, and prohibited believers to engage in religious instruction or publish religious materials. It has also carried on militant pro-atheist propaganda campaigns as part of the official ideology of the state, in an effort to establish a "new Soviet man" committed to the ideals of Communist society. Mikhail Gorbachev is dismantling such policies by permitting greater freedom of religious conscience. If his reforms proceed unabated, they could have dramatic implications for the entire Communist world, for the Russians may be moving from militant atheism to tolerant humanism.

[39] Paul Kurtz; Vern L. Bullough; Tim Madigan (19 October 2009). *Toward a New Enlightenment: the Philosophy of Paul Kurtz*. Transaction Books. ISBN 978-1-56000-118-8. Ranged against the true believer are the militant atheists, who adamantly reject the faith as false stupid, and reactionary. They consider all religious believers to be gullible fools and claim that they are given to accepting gross exaggerations and untenable premises. Historic religious claims, they think, are totally implausible, unbelievable, disreputable, and controvertible, for they go beyond the bounds of reason. Militant atheists can find no value at all to any religious beliefs or institutions. They resist any effort to engage in inquiry or debate. Madalyn Murray O'Hair is as arrogant in her rejection of religion as is the true believer in his or her profession of faith. This form of atheism thus becomes mere dogma.

[40] *The New Atheism and Secular Humanism*. Center for Inquiry. 19 October 2009. Paul Kurtz, considered by many the father of the secular humanist movement, is Professor Emeritus of Philosophy at the State University of New York at Buffalo.

[41] Paul Kurtz; Vern L. Bullough; Tim Madigan (19 October 2009). *Toward a New Enlightenment: the Philosophy of Paul Kurtz*. Transaction Books. ISBN 978-1-56000-118-8. There have been fundamental and irreconcilable differences between humanists and atheists, particularly Marxist-Leninists. The defining characteristic of humanism is its commitment to human freedom and democracy; the kind of atheism practiced in the Soviet Union has consistently violated basic human rights. Humanists believe first and foremost in the freedom of conscience, the free mind, and the right of dissent. The defense of religious liberty is as precious to the humanist as are the rights of the believers.

[42] Esptein, Greg M. (2010). *Good Without God: What a Billion Nonreligious People Do Believe*. New York: HarperCollins. ISBN 978-0-06-167011-4.

[43] "The Science of Morality". Center for Inquiry. Retrieved 12 June 2015.

[44] "American humanist association – Publications – Chapter eight: The Development of Organization". Americanhumanist.org. Retrieved 13 November 2011.

[45] "India humanist". India.humanists.net. 25 June 1997. Retrieved 13 November 2011.

[46] "Census 2011 – Ethnicity and religion in England and Wales". Statistics.gov.uk. 27 March 2011. Retrieved 13 November 2011.

[47] Religious Affiliation Australian Bureau of Statistics

[48] RELP Religious Affiliation – 1st Release Australian Bureau of Statistics

[49] "Top Twenty Religions in the United States, 2001 (self-identification, ARIS)". Adherents.com. Retrieved 13 November 2011.

[50] "Statistics Canada – Population by religion, by province and territory (2001 Census)". 0.statcan.ca. 25 January 2005. Retrieved 13 November 2011.

[51] "Scotland's Census - Ethnicity, Identity, Language and Religion". scotlandscensus.gov.uk. 2017. Retrieved 26 May 2017.

[52] "Human-Etisk Forbund". Retrieved 12 June 2015.

[53] Norway – Members of philosophical2 communities outside the Church of Norway. 1990–2013.

[54] "International Humanist and Ethical Union - Our members". Retrieved 12 June 2015.

[55] International Humanist and Ethical Union. ""IHEU and EHF agree revised protocol", 24 February 2009". Iheu.org. Retrieved 13 November 2011.

[56] Randall Balmer, *Encyclopedia of Evangelicalism* 2002 p. 516

[57] Christopher P. Toumey, "Evolution and secular humanism," *Journal of the American Academy of Religion,* Summer 1993, Vol. 61 Issue 2, pp. 275–301

[58] "IslamWay Radio". English.islamway.com. Retrieved 13 November 2011.

[59] Jones, Dwight (2009). *Essays in the Philosophy of Humanism*. 17 (1).

[60] R. Joseph Hoffmann, *Humanism – What it isn't,* posted 7 July 2012 on "@Humanism" blog

[61] "A humanist discussion of… Religious Festivals and Ceremonies"

[62] *Fellowship of Humanity v. County of Alameda,* 153 Cal.App.2d 673, 315 P.2d 394 (1957).

[63] *Ben Kalka v Kathleen Hawk, et al.* (US D.C. Appeals No. 98-5485, 2000)

[64] Point of Inquiry podcast (17:44), 3 February 2006.

[65] Ivers, Greg (1992). *Redefining the First Freedom: The Supreme Court and the Consolidation of State Power, 1980–1990*. Transaction Books. pp. 47–48. ISBN 978-1560000549.

1.13 Further reading

- Bullock, Alan. *The Humanist Tradition in the West* (1985), by a leading historian.
- Coleman, T. J. III, (interviewer), Tom Flynn (interviewee) (2014, January), "Tom Flynn on 'Secular Humanism,'" The Religious Studies Project Podcast Series, http://www.religiousstudiesproject.com/podcast/tom-flynn-on-secular-humanism/
- Friess, Horace L. *Felix Adler and Ethical Culture* (1981).
- Pfeffer, Leo. "The 'Religion' of Secular Humanism," *Journal of Church and State,* Summer 1987, Vol. 29 Issue 3, pp. 495–507
- Radest, Howard B. *The Devil and Secular Humanism: The Children of the Enlightenment* (1990) online edition a favorable account
- Toumey, Christopher P. "Evolution and secular humanism," *Journal of the American Academy of Religion,* Summer 1993, Vol. 61 Issue 2, pp. 275–301, focused on fundamentalist attacks

1.13.1 Primary sources

- Adler, Felix. *An Ethical Philosophy of Life* (1918).
- Ericson, Edward L. *The Humanist Way: An introduction to ethical humanist religion* (1988).
- Frankel, Charles. *The Case for Modern Man* (1956).
- Hook, Sidney. *Out of Step: An Unquiet Life in the 20th century* (1987).
- Huxley, Julian. *Essay of a Humanist* (1964).
- Russell, Bertrand. *Why I Am Not a Christian* (1957).

1.13. FURTHER READING

David Niose, president of the American Humanist Association, speaks at a 2012 conference.

Levi Fragell, former Secretary General of the Norwegian Humanist Association and former president of the International Humanist and Ethical Union, at the World Humanist Congress 2011 in Oslo

Organizations like the International Humanist and Ethical Union use the "Happy Human" symbol, based on a 1965 design by Denis Barrington

Chapter 2

Humanism and Its Aspirations

Humanism and Its Aspirations subtitled ***Humanist Manifesto III, a successor to the Humanist Manifesto of 1933*** is the most recent of the *Humanist Manifestos* published in 2003 by the American Humanist Association (AHA).[1] The newest one is much shorter, listing six primary beliefs, which echo themes from its predecessors:

- Knowledge of the world is derived by observation, experimentation, and rational analysis. (See empiricism.)

- Humans are an integral part of nature, the result of unguided evolutionary change.

- Ethical values are derived from human need and interest as tested by experience. (See ethical naturalism.)

- Life's fulfillment emerges from individual participation in the service of humane ideals.

- Humans are social by nature and find meaning in relationships.

- Working to benefit society maximises individual happiness.

2.1 Signatories

The following academics and other prominent persons were signatories to the document, who signed the statement "We who sign *Humanism and Its Aspirations* declare ourselves in general agreement with its substance":

2.1.1 Notable signatories

- Philip Appleman (Poet and distinguished professor emeritus of English, Indiana University)

- Khoren Arisian (Senior Leader, New York Society for Ethical Culture)

- Bill Baird (Reproductive rights pioneer)

- Frank Berger (Pharmacologist, developer of anti-anxiety drugs)

- Howard Box (Minister emeritus, Oak Ridge Unitarian Universalist Church, Tennessee)

- Lester R. Brown (Founder and president, Earth Policy Institute)

- August E. Brunsman IV (Executive director, Secular Student Alliance)

- Rob Buitenweg (Vice president, International Humanist and Ethical Union)

- Vern Bullough (Sexologist and former co-president of the International Humanist and Ethical Union)

- David Bumbaugh (Professor, Meadville Lombard Theological School)

- Matthew Cherry (Executive director, Institute for Humanist Studies)

- Joseph Chuman (Visiting professor of religion, Columbia University, and leader, Ethical Culture Society of Bergen County, New Jersey)

- Curt Collier (leader, Riverdale-Yonkers Society for Ethical Culture, New York)

- Fred Cook (Retired executive committee member, International Humanist and Ethical Union)

- Carleton Coon (Former U.S. Ambassador to Nepal)

- Richard Dawkins (Charles Simonyi professor, University of Oxford)

- Charles Debrovner (President, NACH/The Humanist Institute)

- Arthur Dobrin (Professor of humanities, Hofstra University and leader emeritus Ethical Humanist Society of Long Island, New York)

- Margaret Downey (President, Freethought Society of Greater Philadelphia)
- Sonja Eggerickx (Vice president, Unie Vrijzinnige Verenigingen, Belgium, and vice president International Humanist and Ethical Union)
- Riane Eisler (President, Center for Partnership Studies)
- Albert Ellis (Creator of Rational Emotive Behavior Therapy and founder of the Albert Ellis Institute)
- Edward L. Ericson (Leader emeritus, Ethical Culture)
- Roy P. Fairfield (Co-founder, Union Graduate School)
- Antony Flew (Philosopher)
- Levi Fragell (President, International Humanist and Ethical Union)
- Arun Gandhi (Co-founder, M.K. Gandhi Institute for Nonviolence)
- Kendyl Gibbons (President, Unitarian Universalist Ministers Association)
- Babu R.R. Gogineni (Executive director, International Humanist and Ethical Union)
- Sol Gordon (Sexologist)
- Ethelbert Haskins (Retired treasurer of the Humanist Foundation)
- Jim Herrick (Editor, the New Humanist)
- Pervez Hoodbhoy (Professor of physics at Quaid-e-Azam University, Islamabad, Pakistan)
- Fran P. Hosken (Editor, Women's International Network News)
- Joan Johnson Lewis (President, National Leaders Council of the American Ethical Union)
- Stefan Jonasson (Immediate past president, HUUmanists)
- Larry Jones (President, Institute for Humanist Studies)
- Edwin Kagin (Founder and director, Camp Quest)
- Beth Lamont (AHA NGO representative to the United Nations)
- Gerald A. Larue (Professor emeritus of biblical history and archaeology, University of Southern California)
- Joseph Levee (Board member, Council for Secular Humanism)
- Ellen McBride (Immediate past president, American Ethical Union)
- Lester Mondale (Retired Unitarian Universalist minister and signer of Humanist Manifestos I and II)
- Henry Morgentaler (Abortion rights pioneer)
- Stephen Mumford (President, Center for Research on Population and Security)
- William Murry (President and dean, Meadville Lombard Theological School)
- Sarah Oelberg (President, HUUmanists)
- Indumati Parikh (President, Center for the Study of Social Change, India)
- Philip Paulson (Church-state activist)
- Katha Pollitt (Columnist, the Nation)
- Howard Radest (Dean emeritus, the Humanist Institute)
- James "Amazing" Randi (Magician, founder of the James Randi Educational Foundation)
- Larry Reyka (President, the Humanist Society)
- David Schafer (Retired research physiologist, U.S. Veterans Administration)
- Eugenie Scott (Executive director, National Center for Science Education)
- Michael Shermer (Editor of Skeptic magazine)
- James R. Simpson (Professor of international agricultural economics, Ryukoku University, Japan)
- Warren Allen Smith (Editor and author)
- Matthew les Spetter (Associate professor in social psychology at the Peace Studies Institute of Manhattan College, NY)
- Oliver Stone (Academy award-winning filmmaker)
- John Swomley (Professor emeritus of social ethics, St. Paul School of Theology)
- Robert Tapp (Dean, the Humanist Institute)
- Carl Thitchener (Co-minister, Unitarian Universalist Church of Amherst and of Canandaigua, New York)
- Maureen Thitchener (Co-minister, Unitarian Universalist Church of Amherst and of Canandaigua, New York)

- Rodrigue Tremblay (Emeritus professor of economics and of international finance, Universite de Montreal, Quebec, Canada)
- Kurt Vonnegut (Novelist)
- John Weston (Ministerial settlement director, Unitarian Universalist Association)
- Edward O. Wilson (Professor, Harvard University, and two-time Pulitzer Prize winner)
- Sherwin Wine (Founder and president, Society for Humanistic Judaism)

2.1.2 Nobel laureates

22 Nobel laureates signed the statement, these being:

- Philip W. Anderson (Physics, 1977)
- Paul D. Boyer (Chemistry, 1997)
- Owen Chamberlain (Physics, 1959)
- Francis Crick (Medicine, 1962)
- Paul J. Crutzen (Chemistry, 1995)
- Pierre-Gilles de Gennes (Physics, 1991)
- Johann Deisenhofer (Chemistry, 1988)
- Jerome I. Friedman (Physics, 1990)
- Sheldon Glashow (Physics, 1979)
- David J. Gross (Physics, 2004)
- Herbert A. Hauptman (Chemistry, 1985)
- Dudley Herschbach (Chemistry, 1986)
- Harold W. Kroto (Chemistry, 1996)
- Yuan T. Lee (Chemistry, 1986)
- Mario J. Molina (Chemistry, 1995)
- Erwin Neher (Medicine, 1991)
- Ilya Prigogine (Chemistry, 1977)
- Richard J. Roberts (Medicine, 1993)
- John E. Sulston (Medicine, 2002)
- Henry Taube (Chemistry, 1983)
- E. Donnall Thomas (Medicine, 1990)
- James Dewey Watson (Medicine, 1962)

2.1.3 Past AHA presidents

- Edd Doerr
- Michael W. Werner
- Kurt Vonnegut
- Suzanne I. Paul
- Lyle L. Simpson
- Bette Chambers
- Isaac Asimov
- Lloyd L. Morain
- Robert W. McCoy
- Vashti McCollum

2.1.4 AHA board

The then-current AHA board all signed, these being:

- Melvin Lipman (president)
- Lois Lyons (vice president)
- Ronald W. Fegley (secretary)
- John Nugent (treasurer)
- Wanda Alexander
- John R. Cole|John Cole
- Tom Ferrick
- Robert D. Finch
- John M. Higgins
- Herb Silverman
- Maddy Urken
- Mike Werner

2.1.5 Drafting committee

Finally, there was the drafting committee of:

- Fred Edwords (chair)
- Edd Doerr (also included above as a past president of the AHA)
- Tony Hileman
- Pat Duffy Hutcheon
- Maddy Urken

2.2 See also

- Amsterdam Declaration 2002, a similar document from the International Humanist and Ethical Union.

2.3 References

[1] "Humanism and its Aspirations". American Humanist Association, 2003. Retrieved 2 July 2017.

2.4 External links

- *Humanism and Its Aspirations*
- *Humanism and Its Aspirations* — Speech by Maddy Urken
- Notable Signers
- Critical commentary on the Humanist Manifesto III

Chapter 3

Humanist Manifesto

Humanist Manifesto is the title of three manifestos laying out a Humanist worldview. They are the original *Humanist Manifesto* (1933, often referred to as Humanist Manifesto I), the *Humanist Manifesto II* (1973), and *Humanism and Its Aspirations* (2003, a.k.a. *Humanist Manifesto III*). The Manifesto originally arose from religious Humanism, though secular Humanists also signed.

The central theme of all three *manifestos* is the elaboration of a philosophy and value system which does not necessarily include belief in any personal deity or "higher power", although the three differ considerably in their tone, form, and ambition. Each has been signed at its launch by various prominent members of academia and others who are in general agreement with its principles.

In addition, there is a similar document entitled *A Secular Humanist Declaration* published in 1980 by the Council for Secular Humanism.

3.1 Humanist Manifesto I

Main article: Humanist Manifesto I

The first manifesto, entitled simply *A Humanist Manifesto*, was written in 1933 primarily by Roy Wood Sellars and Raymond Bragg and was published with 34 signatories including philosopher John Dewey. Unlike the later ones, the first Manifesto talked of a new "religion", and referred to Humanism as a religious movement to transcend and replace previous religions based on allegations of supernatural revelation. The document outlines a fifteen-point belief system, which, in addition to a secular outlook, opposes "acquisitive and profit-motivated society" and outlines a worldwide egalitarian society based on voluntary mutual cooperation, language which was considerably softened by the Humanists' board, owners of the document, twenty years later.

The title "A Humanist Manifesto"—rather than "The Humanist Manifesto"—was intentional, predictive of later Manifestos to follow, as indeed has been the case. Unlike the creeds of major organized religions, the setting out of Humanist ideals in these Manifestos is an ongoing process. Indeed, in some communities of Humanists the compilation of personal Manifestos is actively encouraged, and throughout the Humanist movement it is accepted that the Humanist Manifestos are not permanent or authoritative dogmas but are to be subject to ongoing critique.

3.2 Humanist Manifesto II

Main article: Humanist Manifesto II

The second Manifesto was written in 1973 by Paul Kurtz and Edwin H. Wilson, and was intended to update and replace the previous one. It begins with a statement that the excesses of Nazism and World War II had made the first seem "far too optimistic", and indicated a more hard-headed and realistic approach in its seventeen-point statement, which was much longer and more elaborate than the previous version. Nevertheless, much of the unbridled optimism of the first remained, with hopes stated that war would become obsolete and poverty would be eliminated.

Many of the proposals in the document, such as opposition to racism and weapons of mass destruction and support of strong human rights, are fairly uncontroversial, and its prescriptions that divorce and birth control should be legal and that technology can improve life are widely accepted today in much of the Western world. Furthermore, its proposal of an international court has since been implemented. However, in addition to its rejection of supernaturalism, various controversial stances are strongly supported, notably the right to abortion.

Initially published with a small number of signatures, the document was circulated and gained thousands more, and indeed the AHA website encourages visitors to add their own name. A provision at the end noted that signators do "not necessarily endors[e] every detail" of the document.

Among the oft-quoted lines from this 1973 Manifesto are, "No deity will save us; we must save ourselves," and "We are responsible for what we are and for what we will be," both of which may present difficulties for members of certain Christian, Jewish, and Muslim sects, or other believers in doctrines of submission to the will of an all-powerful God.

Expanding upon the role the public education establishment should play to bring about the goals described in the Humanist Manifesto II, John Dunphy wrote: "I am convinced that the battle for humankind's future must be waged and won in the public school classroom by teachers that correctly perceive their role as proselytizers of a new faith: a religion of humanity that recognizes and respects the spark of what theologians call divinity in every human being...The classroom must and will become an arena of conflict between the old and new -- the rotting corpse of Christianity, together with all its adjacent evils and misery, and the new faith of humanism, resplendent with the promise of a world in which the never-realized Christian ideal of 'love thy neighbor' will finally be achieved." [1]

3.3 Humanist Manifesto III

Main article: Humanism and Its Aspirations

Humanism and Its Aspirations, subtitled *Humanist Manifesto III, a successor to the Humanist Manifesto of 1933*, was published in 2003 by the AHA, and was written by committee. Signatories included 21 Nobel laureates. The new document is the successor to the previous ones, and the name "Humanist Manifesto" is the property of the American Humanist Association.

The newest manifesto is deliberately much shorter, listing seven primary themes, which echo those from its predecessors :

- Knowledge of the world is derived by observation, experimentation, and rational analysis. (See empiricism.)
- Humans are an integral part of nature, the result of evolutionary change, an unguided process.
- Ethical values are derived from human need and interest as tested by experience. (See ethical naturalism.)
- Life's fulfillment emerges from individual participation in the service of humane ideals.
- Humans are social by nature and find meaning in relationships.
- Working to benefit society maximizes individual happiness.
- Respect for differing yet humane views in an open, secular, democratic, environmentally sustainable society

3.4 Other Manifestos for Humanism

Aside from the official Humanist Manifestos of the American Humanist Association, there have been other similar documents. "Humanist Manifesto" is a trademark of the AHA. Formulation of new statements in emulation of the three Humanist Manifestoes is encouraged, and examples follow.

3.4.1 A Secular Humanist Declaration

Main article: A Secular Humanist Declaration

In 1980, the Council for Secular Humanism, founded by Paul Kurtz, which is typically more detailed in its discussions regarding the function of Humanism than the AHA, published what is in effect its manifesto, entitled *A Secular Humanist Declaration*. It has as its main points:

1. Free Inquiry
2. Separation of Church and State
3. The Ideal of Freedom
4. Ethics Based on Critical Intelligence
5. Moral Education
6. Religious Skepticism
7. Reason
8. Science and Technology
9. Evolution
10. Education

A Secular Humanist Declaration was an argument for and statement of support for democratic secular humanism. The document was issued in 1980 by the Council for Democratic and Secular Humanism ("CODESH"), now the Council for Secular Humanism ("CSH"). Compiled by Paul Kurtz, it is largely a restatement of the content of the American Humanist Association's 1973 Humanist Manifesto II, of which he was co-author with Edwin H. Wilson. Both Wilson and Kurtz had served as editors of *The Humanist*, from which Kurtz departed in 1979 and thereafter set about establishing his own movement and his own periodical. His Secular Humanist Declaration was the starting point for these enterprises.

3.4.2 Humanist Manifesto 2000

Humanist Manifesto 2000: A Call for New Planetary Humanism is a book by Paul Kurtz published in 2000. It differs from the other three in that it is a full-length book rather than essay-length, and was published not by the American Humanist Association but by the Council for Secular Humanism. In it, Kurtz argues for many of the points already formulated in Humanist Manifesto 2, of which he had been co-author in 1973.

3.4.3 Amsterdam Declaration

Main article: Amsterdam Declaration

The **Amsterdam Declaration 2002** is a statement of the fundamental principles of modern Humanism passed unanimously by the General Assembly of the International Humanist and Ethical Union (IHEU) at the 50th anniversary World Humanist Congress in 2002. According to the IHEU, the declaration "is the official statement of World Humanism."

It is officially supported by all member organisations of the IHEU including:

- American Humanist Association
- British Humanist Association
- Humanist Canada
- Council of Australian Humanist Societies
- Council for Secular Humanism
- Gay and Lesbian Humanist Association
- Human-Etisk Forbund, the Norwegian Humanist Association
- Humanist Association of Ireland
- Indian Humanist Union
- Philippine Atheists and Agnostics Society (PATAS)

A complete list of signatories can be found on the IHEU page (see references).

This declaration makes exclusive use of capitalized *Humanist* and *Humanism*, which is consistent with IHEU's general practice and recommendations for promoting a unified Humanist identity. To further promote Humanist identity, these words are also free of any adjectives, as recommended by prominent members of IHEU. Such usage is not universal among IHEU member organizations, though most of them do observe these conventions.

3.5 References

[1] Dunphy, J., A Religion for a New Age, The Humanist, January–February 1983

3.6 External links

3.6.1 Manifestos

- *Humanist Manifesto I* (1933)
- *Amsterdam Declaration 1952*
- *Humanist Manifesto II* (1973)
- *A Secular Humanist Declaration* (1980)
- *A Declaration of Interdependence*(1988)
- *IEHU Minimum Statement on Humanism* (1996)
- *Humanism: Why, What, and What For, In 882 Words* (1996)
- *Humanist Manifesto 2000: A Call For A New Planetary Humanism*
- *The Promise of Manifesto 2000*
- *Amsterdam Declaration 2002*
- *Humanism and Its Aspirations: Humanist Manifesto III* (2003)
    - PDF Printer Friendly Version
- *Manifeste pour un humanisme contemporain* (2012)

3.6.2 Miscellaneous

- The Genesis of a Humanist Manifesto by Edwin H. Wilson
- HUUmanists, an association of Unitarian Universalist Humanists
- Notable signers

Chapter 4

Reason

This article is about the human faculty of reason or rationality. For other uses, see Reason (disambiguation).

Reason is the capacity for consciously making sense of things, applying logic, establishing and verifying facts, and changing or justifying practices, institutions, and beliefs based on new or existing information.[1] It is closely associated with such characteristically human activities as philosophy, science, language, mathematics, and art and is normally considered to be a definitive characteristic of human nature.[2] Reason, or an aspect of it, is sometimes referred to as rationality.

Reasoning is associated with thinking, cognition, and intellect. Reasoning may be subdivided into forms of logical reasoning (forms associated with the strict sense): deductive reasoning, inductive reasoning, abductive reasoning; and other modes of reasoning considered more informal, such as intuitive reasoning and verbal reasoning. Along these lines, a distinction is often drawn between **discursive reason**, reason proper, and intuitive reason,[3] in which the reasoning process—however valid—tends toward the personal and the opaque. Although in many social and political settings logical and intuitive modes of reason may clash, in other contexts intuition and formal reason are seen as complementary, rather than adversarial as, for example, in mathematics, where intuition is often a necessary building block in the creative process of achieving the hardest form of reason, a formal proof.

Reason, like habit or intuition, is one of the ways by which thinking comes from one idea to a related idea. For example, it is the means by which rational beings understand themselves to think about cause and effect, truth and falsehood, and what is good or bad. It is also closely identified with the ability to self-consciously change beliefs, attitudes, traditions, and institutions, and therefore with the capacity for freedom and self-determination.[4]

In contrast to reason as an abstract noun, a reason is a consideration which explains or justifies some event, phenomenon, or behavior.[5] The field of logic studies ways in which human beings reason formally through argument.[6]

Psychologists and cognitive scientists have attempted to study and explain how people reason, e.g. which cognitive and neural processes are engaged, and how cultural factors affect the inferences that people draw. The field of automated reasoning studies how reasoning may or may not be modeled computationally. Animal psychology considers the question of whether animals other than humans can reason.

4.1 Etymology and related words

In the English language and other modern European languages, "reason", and related words, represent words which have always been used to translate Latin and classical Greek terms in the sense of their philosophical usage.

- The original Greek term was "λόγος" *logos*, the root of the modern English word "logic" but also a word which could mean for example "speech" or "explanation" or an "account" (of money handled).[7]

- As a philosophical term *logos* was translated in its non-linguistic senses in Latin as *ratio*. This was originally not just a translation used for philosophy, but was also commonly a translation for *logos* in the sense of an account of money.[8]

- French *raison* is derived directly from Latin, and this is the direct source of the English word "reason".[5]

The earliest major philosophers to publish in English, such as Francis Bacon, Thomas Hobbes, and John Locke also routinely wrote in Latin and French, and compared their terms to Greek, treating the words "*logos*", "*ratio*", "*raison*" and "reason" as inter-changeable. The meaning of the word "reason" in senses such as "human reason" also overlaps to a large extent with "rationality" and the adjective of "reason" in philosophical contexts is normally "rational", rather than "reasoned" or "reasonable".[9] Some philosophers, Thomas

Hobbes for example, also used the word *ratiocination* as a synonym for "reasoning".

4.2 Philosophical history

Francisco de Goya, The Sleep of Reason Produces Monsters *(El sueño de la razón produce monstruos), c. 1797*

The proposal that reason gives humanity a special position in nature has been argued to be a defining characteristic of western philosophy and later western modern science, starting with classical Greece. Philosophy can be described as a way of life based upon reason, and in the other direction reason has been one of the major subjects of philosophical discussion since ancient times. Reason is often said to be reflexive, or "self-correcting," and the critique of reason has been a persistent theme in philosophy.[10] It has been defined in different ways, at different times, by different thinkers about human nature.

4.2.1 Classical philosophy

For many classical philosophers, nature was understood teleologically, meaning that every type of thing had a definitive purpose which fit within a natural order that was itself understood to have aims. Perhaps starting with Pythagoras or Heraclitus, the cosmos is even said to have reason.[11] Reason, by this account, is not just one characteristic that humans happen to have, and that influences happiness amongst other characteristics. Reason was considered of higher stature than other characteristics of human nature, such as sociability, because it is something humans share with nature itself, linking an apparently immortal part of the human mind with the divine order of the cosmos itself. Within the human mind or soul (*psyche*), reason was described by Plato as being the natural monarch which should rule over the other parts, such as spiritedness (*thumos*) and the passions. Aristotle, Plato's student, defined human beings as rational animals, emphasizing reason as a characteristic of human nature. He *defined* the highest human happiness or well being (*eudaimonia*) as a life which is lived consistently, excellently and completely in accordance with reason.[12]

The conclusions to be drawn from the discussions of Aristotle and Plato on this matter are amongst the most debated in the history of philosophy.[13] But teleological accounts such as Aristotle's were highly influential for those who attempt to explain reason in a way which is consistent with monotheism and the immortality and divinity of the human soul. For example, in the neo-platonist account of Plotinus, the cosmos has one soul, which is the seat of all reason, and the souls of all individual humans are part of this soul. Reason is for Plotinus both the provider of form to material things, and the light which brings individuals souls back into line with their source.[14] Such neo-Platonist accounts of the rational part of the human soul were standard amongst medieval Islamic philosophers, and under this influence, mainly via Averroes, came to be debated seriously in Europe until well into the renaissance, and they remain important in Iranian philosophy.[13]

4.2.2 Subject-centred reason in early modern philosophy

The early modern era was marked by a number of significant changes in the understanding of reason, starting in Europe. One of the most important of these changes involved a change in the metaphysical understanding of human beings. Scientists and philosophers began to question the teleological understanding of the world.[15] Nature was no longer assumed to be human-like, with its own aims or reason, and human nature was no longer assumed to work according to anything other than the same "laws of nature" which affect inanimate things. This new understanding eventually displaced the previous world view that derived from a spiritual understanding of the universe.

René Descartes

Accordingly, in the 17th century, René Descartes explicitly rejected the traditional notion of humans as "rational animals," suggesting instead that they are nothing more than "thinking things" along the lines of other "things" in nature. Any grounds of knowledge outside that understanding was, therefore, subject to doubt.

In his search for a foundation of all possible knowledge, Descartes deliberately decided to throw into doubt *all* knowledge – *except* that of the mind itself in the process of thinking:

> At this time I admit nothing that is not necessarily true. I am therefore precisely nothing but a thinking thing; that is a mind, or intellect, or understanding, or reason – words of whose meanings I was previously ignorant.[16]

This eventually became known as epistemological or "subject-centred" reason, because it is based on the *knowing subject*, who perceives the rest of the world and itself as a set of objects to be studied, and successfully mastered by applying the knowledge accumulated through such study. Breaking with tradition and many thinkers after him, Descartes explicitly did not divide the incorporeal soul into parts, such as reason and intellect, describing them as one indivisible incorporeal entity.

A contemporary of Descartes, Thomas Hobbes described reason as a broader version of "addition and subtraction" which is not limited to numbers.[17] This understanding of reason is sometimes termed "calculative" reason. Similar to Descartes, Hobbes asserted that "No discourse whatsoever, can end in absolute knowledge of fact, past, or to come" but that "sense and memory" is absolute knowledge.[18]

In the late 17th century, through the 18th century, John Locke and David Hume developed Descartes' line of thought still further. Hume took it in an especially skeptical direction, proposing that there could be no possibility of deducing relationships of cause and effect, and therefore no knowledge is based on reasoning alone, even if it seems otherwise.[19][20]

Hume famously remarked that, "We speak not strictly and philosophically when we talk of the combat of passion and of reason. Reason is, and ought only to be the slave of the passions, and can never pretend to any other office than to serve and obey them."[21] Hume also took his definition of reason to unorthodox extremes by arguing, unlike his predecessors, that human reason is not qualitatively different from either simply conceiving individual ideas, or from judgments associating two ideas,[22] and that "reason is nothing but a wonderful and unintelligible instinct in our souls, which carries us along a certain train of ideas, and endows them with particular qualities, according to their particular situations and relations."[23] It followed from this that animals have reason, only much less complex than human reason.

In the 18th century, Immanuel Kant attempted to show that Hume was wrong by demonstrating that a "transcendental" self, or "I", was a necessary condition of all experience. Therefore, suggested Kant, on the basis of such a self, it is in fact possible to reason both about the conditions and limits of human knowledge. And so long as these limits are respected, reason can be the vehicle of morality, justice and understanding.

4.2.3 Substantive and formal reason

In the formulation of Kant, who wrote some of the most influential modern treatises on the subject, the great achievement of reason (German: *Vernunft*) is that it is able to exercise a kind of universal law-making. Kant was able therefore to re-formulate the basis of moral-practical, theoretical and aesthetic reasoning, on "universal" laws.

Here practical reasoning is the self-legislating or self-governing formulation of universal norms, and theoretical reasoning the way humans posit universal laws of nature.[24]

Under practical reason, the moral autonomy or freedom of human beings depends on their ability to behave according to laws that are given to them by the proper exercise of that reason. This contrasted with earlier forms of morality, which depended on religious understanding and interpretation, or nature for their substance.[25]

According to Kant, in a free society each individual must be able to pursue their goals however they see fit, so long as their actions conform to principles given by reason. He formulated such a principle, called the "categorical imperative", which would justify an action only if it could be universalized:

> Act only according to that maxim whereby you can, at the same time, will that it should become a universal law.[26]

In contrast to Hume then, Kant insists that reason itself (German *Vernunft*) has natural ends itself, the solution to the metaphysical problems, especially the discovery of the foundations of morality. Kant claimed that this problem could be solved with his "transcendental logic" which unlike normal logic is not just an instrument, which can be used indifferently, as it was for Aristotle, but a theoretical science in its own right and the basis of all the others.[27]

According to Jürgen Habermas, the "substantive unity" of reason has dissolved in modern times, such that it can no longer answer the question "How should I live?" Instead, the unity of reason has to be strictly formal, or "procedural." He thus described reason as a group of three autonomous spheres (on the model of Kant's three critiques):

1. **Cognitive-instrumental reason** is the kind of reason employed by the sciences. It is used to observe events, to predict and control outcomes, and to intervene in the world on the basis of its hypotheses;

2. **Moral-practical reason** is what we use to deliberate and discuss issues in the moral and political realm, according to universalizable procedures (similar to Kant's categorical imperative); and

3. **Aesthetic reason** is typically found in works of art and literature, and encompasses the novel ways of seeing the world and interpreting things that those practices embody.

For Habermas, these three spheres are the domain of experts, and therefore need to be mediated with the "lifeworld" by philosophers. In drawing such a picture of reason, Habermas hoped to demonstrate that the substantive unity of reason, which in pre-modern societies had been able to answer questions about the good life, could be made up for by the unity of reason's formalizable procedures.[28]

4.2.4 The critique of reason

Hamann, Herder, Kant, Hegel, Kierkegaard, Nietzsche, Heidegger, Foucault, Rorty, and many other philosophers have contributed to a debate about what reason means, or ought to mean. Some, like Kierkegaard, Nietzsche, and Rorty, are skeptical about subject-centred, universal, or instrumental reason, and even skeptical toward reason as a whole. Others, including Hegel, believe that it has obscured the importance of intersubjectivity, or "spirit" in human life, and attempt to reconstruct a model of what reason should be.

Some thinkers, e.g. Foucault, believe there are other *forms* of reason, neglected but essential to modern life, and to our understanding of what it means to live a life according to reason.[10]

In the last several decades, a number of proposals have been made to "re-orient" this critique of reason, or to recognize the "other voices" or "new departments" of reason:

For example, in opposition to subject-centred reason, Habermas has proposed a model of communicative reason that sees it as an essentially cooperative activity, based on the fact of linguistic intersubjectivity.[29]

Nikolas Kompridis has proposed a widely encompassing view of reason as "that ensemble of practices that contributes to the opening and preserving of openness" in human affairs, and a focus on reason's possibilities for social change.[30]

The philosopher Charles Taylor, influenced by the 20th century German philosopher Martin Heidegger, has proposed that reason ought to include the faculty of disclosure, which is tied to the way we make sense of things in everyday life, as a new "department" of reason.[31]

In the essay "What is Enlightenment?", Michel Foucault proposed a concept of critique based on Kant's distinction between "private" and "public" uses of reason. This distinction, as suggested, has two dimensions:

- **Private reason** is the reason that is used when an individual is "a cog in a machine" or when one "has a role to play in society and jobs to do: to be a soldier, to have taxes to pay, to be in charge of a parish, to be a civil servant."

- **Public reason** is the reason used "when one is reasoning as a reasonable being (and not as a cog in a machine), when one is reasoning as a member of reasonable humanity." In these circumstances, "the use of reason must be free and public."[32]

4.3 Reason compared to related concepts

4.3.1 Compared to logic

Main article: Logic

The terms "logic" or "logical" are sometimes used as if they were identical with the term "reason" or with the concept of being "rational", or sometimes logic is seen as the most pure or the defining form of reason. For example in modern economics, rational choice is assumed to equate to logically consistent choice.

Reason and logic can however be thought of as distinct, although logic is one important aspect of reason. Author Douglas Hofstadter, in *Gödel, Escher, Bach*, characterizes the distinction in this way. Logic is done inside a system while reason is done outside the system by such methods as skipping steps, working backward, drawing diagrams, looking at examples, or seeing what happens if you change the rules of the system.[33]

Reason is a type of thought, and the word "logic" involves the attempt to describe rules or norms by which reasoning operates, so that orderly reasoning can be taught. The oldest surviving writing to explicitly consider the rules by which reason operates are the works of the Greek philosopher Aristotle, especially *Prior Analysis* and *Posterior Analysis*.[34] Although the Ancient Greeks had no separate word for logic as distinct from language and reason, Aristotle's newly coined word "syllogism" (*syllogismos*) identified logic clearly for the first time as a distinct field of study. When Aristotle referred to "the logical" (*hē logikē*), he was referring more broadly to rational thought.[35]

4.3.2 Reason compared to cause-and-effect thinking, and symbolic thinking

Main articles: Causality and Symbols

As pointed out by philosophers such as Hobbes, Locke and Hume, some animals are also clearly capable of a type of "associative thinking", even to the extent of associating causes and effects. A dog once kicked, can learn how to recognize the warning signs and avoid being kicked in the future, but this does not mean the dog has reason in any strict sense of the word. It also does not mean that humans acting on the basis of experience or habit are using their reason.[36]

Human reason requires more than being able to associate two ideas, even if those two ideas might be described by a reasoning human as a cause and an effect, perceptions of smoke, for example, and memories of fire. For reason to be involved, the association of smoke and the fire would have to be thought through in a way which can be explained, for example as cause and effect. In the explanation of Locke, for example, reason requires the mental use of a third idea in order to make this comparison by use of syllogism.[37]

More generally, reason in the strict sense requires the ability to create and manipulate a system of symbols, as well as indices and icons, according to Charles Sanders Peirce, the symbols having only a nominal, though habitual, connection to either smoke or fire.[38] One example of such a system of artificial symbols and signs is language.

The connection of reason to symbolic thinking has been expressed in different ways by philosophers. Thomas Hobbes described the creation of "Markes, or Notes of remembrance" (*Leviathan* Ch.4) as *speech*. He used the word *speech* as an English version of the Greek word *logos* so that speech did not need to be communicated.[39] When communicated, such speech becomes language, and the marks or notes or remembrance are called "Signes" by Hobbes. Going further back, although Aristotle is a source of the idea that only humans have reason (*logos*), he does mention that animals with imagination, for whom sense perceptions can persist, come closest to having something like reasoning and *nous*, and even uses the word "logos" in one place to describe the distinctions which animals can perceive in such cases.[40]

4.3.3 Reason, imagination, mimesis, and memory

Main articles: Imagination, Mimesis, Memory, and Recollection

Reason and imagination rely on similar mental processes.[41] Imagination is not only found in humans. Aristotle, for example, stated that *phantasia* (imagination: that which can hold images or *phantasmata*) and *phronein* (a type of thinking that can judge and understand in some sense) also exist in some animals.[42] According to him, both are related to the primary perceptive ability of animals, which gathers the perceptions of different senses and defines the order of the things that are perceived without distinguishing universals, and without deliberation or *logos*. But this is not yet reason, because human imagination is different.

The recent modern writings of Terrence Deacon and Merlin Donald, writing about the origin of language, also connect reason connected to not only language, but also mimesis.[43] More specifically they describe the ability to create language as part of an internal modeling of reality specific to humankind. Other results are consciousness, and imagination or fantasy. In contrast, modern proponents of a genetic pre-disposition to language itself include Noam Chomsky and Steven Pinker, to whom Donald and Deacon can be contrasted.

As reason is symbolic thinking, and peculiarly human, then this implies that humans have a special ability to maintain a clear consciousness of the distinctness of "icons" or images and the real things they represent. Starting with a modern author, Merlin Donald writes[44]

> A dog might perceive the "meaning" of a fight that was realistically play-acted by humans, but it could not reconstruct the message or distinguish the representation from its referent (a real fight). [...] Trained apes are able to make this distinction; young children make this distinction early – hence, their effortless distinction between play-acting an event and the event itself

In classical descriptions, an equivalent description of this mental faculty is *eikasia*, in the philosophy of Plato.[45] This is the ability to perceive whether a perception is an image of something else, related somehow but not the same, and therefore allows humans to perceive that a dream or memory or a reflection in a mirror is not reality as such. What Klein refers to as *dianoetic eikasia* is the *eikasia* concerned specifically with thinking and mental images, such as those mental symbols, icons, *signes*, and marks discussed above as definitive of reason. Explaining reason from this direction: human thinking is special in the way that we often understand visible things as if they were themselves images of our intelligible "objects of thought" as "foundations" (*hypothēses* in Ancient Greek). This thinking (*dianoia*) is "...an activity which consists in making the vast and diffuse jungle of the visible world depend on a plurality of more 'precise' *noēta*."[46]

Both Merlin Donald and the Socratic authors such Plato and Aristotle emphasize the importance of *mimesis*, often translated as *imitation* or *representation*. Donald writes[47]

> Imitation is found especially in monkeys and apes [... but ...] Mimesis is fundamentally different from imitation and mimicry in that it involves the invention of intentional representations. [...] Mimesis is not absolutely tied to external communication.

Mimēsis is a concept, now popular again in academic discussion, that was particularly prevalent in Plato's works, and within Aristotle, it is discussed mainly in the *Poetics*.

In Michael Davis's account of the theory of man in this work.[48]

> It is the distinctive feature of human action, that whenever we choose what we do, we imagine an action for ourselves as though we were inspecting it from the outside. Intentions are nothing more than imagined actions, internalizings of the external. All action is therefore imitation of action; it is poetic...[49]

Donald like Plato (and Aristotle, especially in *On Memory and Recollection*), emphasizes the peculiarity in humans of voluntary initiation of a search through one's mental world. The ancient Greek *anamnēsis*, normally translated as "recollection" was opposed to *mneme* or *memory*. Memory, shared with some animals,[50] requires a consciousness not only of what happened in the past, but also *that* something happened in the past, which is in other words a kind of *eikasia*[51] "...but nothing except man is able to recollect."[52] Recollection is a deliberate effort to search for and recapture something once known. Klein writes that, "To become aware of our having forgotten something means to begin recollecting."[53] Donald calls the same thing *autocueing*, which he explains as follows:[54] "Mimetic acts are reproducible on the basis of internal, self-generated cues. This permits voluntary recall of mimetic representations, without the aid of external cues – probably the earliest form of representational *thinking*."

In a celebrated paper in modern times, the fantasy author and philologist J.R.R. Tolkien wrote in his essay "On Fairy Stories" that the terms "fantasy" and "enchantment" are connected to not only "....the satisfaction of certain primordial human desires...." but also "...the origin of language and of the mind."

4.3.4 Logical reasoning methods and argumentation

Looking at logical categorizations of different types of reasoning the traditional main division made in philosophy is between deductive reasoning and inductive reasoning. Formal logic has been described as *the science of deduction*.[55] The study of inductive reasoning is generally carried out within the field known as informal logic or critical thinking.

Deductive reasoning

Main article: Deductive reasoning

A subdivision of Philosophy is Logic. Logic is the study of reasoning. Deduction is a form of reasoning in which a conclusion follows necessarily from the stated premises. A deduction is also the conclusion reached by a deductive reasoning process. One classic example of deductive reasoning is that found in syllogisms like the following:

- Premise 1: All humans are mortal.
- Premise 2: Socrates is a human.
- Conclusion: Socrates is mortal.

The reasoning in this argument is valid, because there is no way in which the premises, 1 and 2, could be true and the conclusion, 3, be false.

Inductive reasoning

Main article: Inductive reasoning

Induction is a form of inference producing propositions about unobserved objects or types, either specifically or generally, based on previous observation. It is used to ascribe properties or relations to objects or types based on previous observations or experiences, or to formulate general statements or laws based on limited observations of recurring phenomenal patterns.

Inductive reasoning contrasts strongly with deductive reasoning in that, even in the best, or strongest, cases of inductive reasoning, the truth of the premises does not guarantee the truth of the conclusion. Instead, the conclusion of an inductive argument follows with some degree of probability. Relatedly, the conclusion of an inductive argument contains more information than is already contained in the premises. Thus, this method of reasoning is ampliative.

A classic example of inductive reasoning comes from the empiricist David Hume:

- Premise: The sun has risen in the east every morning up until now.
- Conclusion: The sun will also rise in the east tomorrow.

Abductive reasoning

Main article: Abductive reasoning

Abductive reasoning, or argument to the best explanation, is a form of inductive reasoning, since the conclusion in an abductive argument does not follow with certainty from its premises and concerns something unobserved. What distinguishes abduction from the other forms of reasoning is an attempt to favour one conclusion above others, by attempting to falsify alternative explanations or by demonstrating the likelihood of the favoured conclusion, given a set of more or less disputable assumptions. For example, when a patient displays certain symptoms, there might be various possible causes, but one of these is preferred above others as being more probable.

Analogical reasoning

Main article: Analogical reasoning

Analogical reasoning is reasoning from the particular to the particular. It is often used in case-based reasoning, especially legal reasoning.[56] An example follows:

- Premise 1: Socrates is human and mortal.
- Premise 2: Plato is human.
- Conclusion: Plato is mortal.

Analogical reasoning can be viewed as a form of inductive reasoning from a single example, but if it is intended as inductive reasoning it is a bad example, because inductive reasoning typically uses a large number of examples to reason from the particular to the general.[57] Analogical reasoning often leads to wrong conclusions. For example:

- Premise 1: Socrates is human and male.
- Premise 2: Ada Lovelace is human.
- Conclusion: Therefore Ada Lovelace is male.

Fallacious reasoning

Main articles: Fallacy, Formal fallacy, and Informal fallacy

Flawed reasoning in arguments is known as fallacious reasoning. Bad reasoning within arguments can be because it commits either a formal fallacy or an informal fallacy.

Formal fallacies occur when there is a problem with the form, or structure, of the argument. The word "formal" refers to this link to the *form* of the argument. An argument that contains a formal fallacy will always be invalid.

An informal fallacy is an error in reasoning that occurs due to a problem with the *content*, rather than mere *structure*, of the argument.

4.4 Traditional problems raised concerning reason

Philosophy is sometimes described as a life of reason, with normal human reason pursued in a more consistent and dedicated way than usual. Two categories of problem concerning reason have long been discussed by philosophers concerning reason, essentially being reasonings about reasoning itself as a human aim, or philosophizing about philosophizing. The first question is concerning whether we can be confident that reason can achieve knowledge of truth better than other ways of trying to achieve such knowledge. The other question is whether a life of reason, a life that aims to be guided by reason, can be expected to achieve a happy life more so than other ways of life (whether such a life of reason results in knowledge or not).

4.4.1 Reason versus truth, and "first principles"

See also: Truth, First principle, and Nous

Since classical times a question has remained constant in philosophical debate (which is sometimes seen as a conflict between movements called Platonism and Aristotelianism) concerning the role of reason in confirming truth. People use logic, deduction, and induction, to reach conclusions they think are true. Conclusions reached in this way are considered more certain than sense perceptions on their own.[58] On the other hand, if such reasoned conclusions are only built originally upon a foundation of sense perceptions, then, our most logical conclusions can never be said to be certain because they are built upon the very same fallible perceptions they seek to better.[59]

This leads to the question of what types of first principles, or starting points of reasoning, are available for someone seeking to come to true conclusions. In Greek, "first principles" are *archai*, "starting points",[60] and the faculty used to perceive them is sometimes referred to in Aristotle[61] and Plato[62] as *nous* which was close in meaning to *awareness* or *consciousness*.[63]

Empiricism (sometimes associated with Aristotle[64] but more correctly associated with British philosophers such as John Locke and David Hume, as well as their ancient equivalents such as Democritus) asserts that sensory impressions are the only available starting points for reasoning and attempting to attain truth. This approach always leads to the controversial conclusion that absolute knowledge is not attainable. Idealism, (associated with Plato and his school), claims that there is a "higher" reality, from which certain people can directly arrive at truth without needing to rely only upon the senses, and that this higher reality is therefore the primary source of truth.

Philosophers such as Plato, Aristotle, Al-Farabi, Avicenna, Averroes, Maimonides, Aquinas and Hegel are sometimes said to have argued that reason must be fixed and discoverable—perhaps by dialectic, analysis, or study. In the vision of these thinkers, reason is divine or at least has divine attributes. Such an approach allowed religious philosophers such as Thomas Aquinas and Étienne Gilson to try to show that reason and revelation are compatible. According to Hegel, "...the only thought which Philosophy brings with it to the contemplation of History, is the simple conception of reason; that reason is the Sovereign of the World; that the history of the world, therefore, presents us with a rational process."[65]

Since the 17th century rationalists, reason has often been taken to be a subjective faculty, or rather the unaided ability (pure reason) to form concepts. For Descartes, Spinoza and Leibniz, this was associated with mathematics. Kant attempted to show that pure reason could form concepts (time and space) that are the conditions of experience. Kant made his argument in opposition to Hume, who denied that reason had any role to play in experience.

4.4.2 Reason versus emotion or passion

See also: Emotion and Passion (emotion)

After Plato and Aristotle, western literature often treated reason as being the faculty that trained the passions and appetites. Stoic philosophy by contrast considered all passions bad. After the critiques of reason in the early Enlightenment the appetites were rarely discussed or conflated with the passions. Some Enlightenment camps took after the Stoics to say Reason should oppose Passion rather than order it, while others like the Romantics considered Passion the ruler over Reason or to the exclusion of Reason, thus the Modern colloquy of "follow your heart".

Reason has been seen as a slave, or judge, of the passions, notably in the work of David Hume, and more recently of Freud. Reasoning which claims that the object of a desire is demanded by logic alone is called *rationalization*.

Rousseau first proposed, in his second *Discourse*, that reason and political life is not natural and possibly harmful to mankind.[66] He asked what really can be said about what is natural to mankind. What, other than reason and civil society, "best suits his constitution"? Rousseau saw "two principles prior to reason" in human nature. First we hold an intense interest in our own well-being. Secondly we object to the suffering or death of any sentient being, especially one like ourselves.[67] These two passions lead us to desire more than we could achieve. We become dependent upon each other, and on relationships of authority and obedience. This effectively puts the human race into slavery. Rousseau says that he almost dares to assert that nature does not destine men to be healthy. According to Velkley, "Rousseau outlines certain programs of rational self-correction, most notably the political legislation of the *Contrat Social* and the moral education in *Émile*. All the same, Rousseau understands such corrections to be only ameliorations of an essentially unsatisfactory condition, that of socially and intellectually corrupted humanity."

This quandary presented by Rousseau led to Kant's new way of justifying reason as freedom to create good and evil. These therefore are not to be blamed on nature or God. In various ways, German Idealism after Kant, and major later figures such Nietzsche, Bergson, Husserl, Scheler, and Heidegger, remain pre-occupied with problems coming from the metaphysical demands or *urges* of *reason*.[68] The influence of Rousseau and these later writers is also large upon art and politics. Many writers (such as Nikos Kazantzakis) extol passion and disparage reason. In politics modern nationalism comes from Rousseau's argument that rationalist cosmopolitanism brings man ever further from his natural state.[69]

Another view on reason and emotion was proposed in the 1994 book titled *Descartes' Error* by Antonio Damasio. In it, Damasio presents the "Somatic Marker Hypothesis" which states that emotions guide behavior and decision-making. Damasio argues that these somatic markers (known collectively as "gut feelings") are "intuitive signals" that direct our decision making processes in a certain way that cannot be solved with rationality alone. Damasio further argues that rationality requires emotional input in order to function.

4.4.3 Reason versus faith or tradition

Main articles: Faith, Religion, and Tradition

There are many religious traditions, some of which are explicitly fideist and others of which claim varying degrees of rationalism. Secular critics sometimes accuse all religious adherents of irrationality, since they claim such adherents are guilty of ignoring, suppressing, or forbidding some kinds of reasoning concerning some subjects (such as religious dogmas, moral taboos, etc.).[70] Though the theologies and religions such as classical monotheism typically do not claim to be irrational, there is often a perceived conflict or tension between faith and tradition on the one hand, and reason on the other, as potentially competing sources of wisdom, law and truth.[71][72]

Religious adherents sometimes respond by arguing that faith and reason can be reconciled, or have different non-overlapping domains, or that critics engage in a similar kind of irrationalism:

- **Reconciliation:** Philosopher Alvin Plantinga argues that there is no real conflict between reason and classical theism because classical theism explains (among other things) why the universe is intelligible and why reason can successfully grasp it.[73][74]

- **Non-overlapping magisteria:** Evolutionary biologist Stephen Jay Gould argues that there need not be conflict between reason and religious belief because they are each authoritative in their own domain (or "magisterium").[75][76] For example, perhaps reason alone is not enough to explain such big questions as the origins of the universe, the origin of life, the origin of consciousness,[77] the foundation of morality, or the destiny of the human race. If so, reason can work on those problems over which it has authority while other sources of knowledge or opinion can have authority on the big questions.[78]

- **Tu quoque:** Philosophers Alasdair MacIntyre and Charles Taylor argue that those critics of traditional religion who are adherents of secular liberalism are also sometimes guilty of ignoring, suppressing, and forbidding some kinds of reasoning about subjects.[79][80] Similarly, philosophers of science such as Paul Feyaraband argue that scientists sometimes ignore or suppress evidence contrary to the dominant paradigm.

- **Unification:** Theologian Joseph Ratzinger, later Benedict XVI, asserted that "Christianity has understood itself as the religion of the Logos, as the religion according to reason," referring to John 1:Ἐν ἀρχῇ ἦν ὁ λόγος, usually translated as "In the beginning was the Word (Logos)." Thus, he said that the Christian faith is "open to all that is truly rational," and that the rationality of Western Enlightenment "is of Christian origin".[81]

Some commentators have claimed that Western civilization can be almost defined by its serious testing of the limits of tension between "unaided" reason and faith in "revealed" truths—figuratively summarized as Athens and Jerusalem, respectively.[82][83] Leo Strauss spoke of a "Greater West" that included all areas under the influence of the tension between Greek rationalism and Abrahamic revelation, including the Muslim lands. He was particularly influenced by the great Muslim philosopher Al-Farabi. To consider to what extent Eastern philosophy might have partaken of these important tensions, Strauss thought it best to consider whether dharma or tao may be equivalent to Nature (by which we mean *physis* in Greek). According to Strauss the beginning of philosophy involved the "discovery or invention of nature" and the "pre-philosophical equivalent of nature" was supplied by "such notions as 'custom' or 'ways'", which appear to be *really universal in all times and places*. The philosophical concept of nature or natures as a way of understanding *archai* (first principles of knowledge) brought about a peculiar tension between reasoning on the one hand, and tradition or faith on the other.[84]

Although there is this special history of debate concerning reason and faith in the Islamic, Christian and Jewish traditions, the pursuit of reason is sometimes argued to be compatible with the other practice of other religions of a different nature, such as Hinduism, because they do not define their tenets in such an absolute way.[85]

4.5 Reason in particular fields of study

4.5.1 Reason in political philosophy and ethics

Main articles: Political Philosophy, Ethics, and The Good

Aristotle famously described reason (with language) as a part of human nature, which means that it is best for humans to live "politically" meaning in communities of about the size and type of a small city state (*polis* in Greek). For example...

> It is clear, then, that a human being is more of a political [*politikon* = of the *polis*] animal [*zōion*] than is any bee or than any of those animals that live in herds. For nature, as we say, makes nothing in vain, and humans are the only animals who possess reasoned speech [*logos*]. Voice, of course, serves to indicate what is painful and pleasant; that is why it is also found in other animals, because their nature has reached the point where they can perceive what is painful and pleasant and express these to each other. But speech [*logos*] serves to make plain what is advantageous and harmful and so also what is just and unjust. For it is a peculiarity of humans, in contrast to the other animals, to have perception of good and bad, just and unjust, and the like; and the community in these things makes a household or city [*polis*]. [...] By nature, then, the drive for such a community exists in everyone, but the first to set one up is responsible for things of very great goodness. For as humans are the best of all animals when perfected, so they are the worst when divorced from law and right. The reason is that injustice is most difficult to deal with when furnished with weapons, and the weapons a human being has are meant by nature to go along with prudence and virtue, but it is only too possible to turn them to contrary uses. Consequently, if a human being lacks virtue, he is the most unholy and savage thing, and when it comes to sex and food, the worst. But justice is something political [to do with the *polis*], for right is the arrangement of the political community, and right is discrimination of what is just. (Aristotle's Politics 1253a 1.2. Peter Simpson's translation, with Greek terms inserted in square brackets.)

The concept of human nature being fixed in this way, implied, in other words, that we can define what type of community is always best for people. This argument has remained a central argument in all political, ethical and moral thinking since then, and has become especially controversial since firstly Rousseau's Second Discourse, and secondly, the Theory of Evolution. Already in Aristotle there was an awareness that the *polis* had not always existed and had needed to be invented or developed by humans themselves. The household came first, and the first villages and cities were just extensions of that, with the first cities being run as if they were still families with Kings acting like fathers.[86]

> Friendship [*philia*] seems to prevail [in] man and woman according to nature [*kata phusin*]; for people are by nature [*tēi phusei*] pairing [*sunduastikon*] more than political [*politikon* = of the *polis*], inasmuch as the household [*oikos*] is prior [*proteron* = earlier] and more necessary than the *polis* and making children is more common [*koinoteron*] with the animals. In the other animals, community [*koinōnia*] goes no further than this, but people live together [*sumoikousin*] not only for the sake of making children, but also for the things for life; for from the start the functions [*erga*] are divided, and are different [for] man and woman. Thus they supply each other, putting their own into the common [*eis to koinon*]. It is for these [reasons] that both utility [*chrēsimon*] and pleasure [*hēdu*] seem to be found in this kind of friendship. (Nicomachean Ethics, VIII.12.1162a. Rough literal translation with Greek terms shown in square brackets.)

Rousseau in his Second Discourse finally took the shocking step of claiming that this traditional account has things in reverse: with reason, language and rationally organized communities all having developed over a long period of time merely as a result of the fact that some habits of cooperation were found to solve certain types of problems, and that once such cooperation became more important, it forced people to develop increasingly complex cooperation—often only to defend themselves from each other.

In other words, according to Rousseau, reason, language and rational community did not arise because of any conscious decision or plan by humans or gods, nor because of any pre-existing human nature. As a result, he claimed, living together in rationally organized communities like modern humans is a development with many negative aspects compared to the original state of man as an ape. If anything is specifically human in this theory, it is the flexibility and adaptability of humans. This view of the animal origins of distinctive human characteristics later received support from Charles Darwin's Theory of Evolution.

The two competing theories concerning the origins of reason are relevant to political and ethical thought because, according to the Aristotelian theory, a best way of living together exists independently of historical circumstances. According to Rousseau, we should even doubt that reason, language and politics are a good thing, as opposed to being simply the best option given the particular course of events that lead to today. Rousseau's theory, that human nature is malleable rather than fixed, is often taken to imply, for example by Karl Marx, a wider range of possible ways of living together than traditionally known.

However, while Rousseau's initial impact encouraged bloody revolutions against traditional politics, including both the French Revolution and the Russian Revolution, his own conclusions about the best forms of community seem to have been remarkably classical, in favor of city-states such as Geneva, and rural living.

4.5.2 Psychology

Main article: Psychology of reasoning

Scientific research into reasoning is carried out within the fields of psychology and cognitive science. Psychologists attempt to determine whether or not people are capable of rational thought in a number of different circumstances.

Assessing how well someone engages in reasoning is the project of determining the extent to which the person is rational or acts rationally. It is a key research question in the psychology of reasoning. Rationality is often divided into its respective theoretical and practical counterparts.

Behavioral experiments on human reasoning

Experimental cognitive psychologists carry out research on reasoning behaviour. Such research may focus, for example, on how people perform on tests of reasoning such as intelligence or IQ tests, or on how well people's reasoning matches ideals set by logic (see, for example, the Wason test).[87] Experiments examine how people make inferences from conditionals e.g., *If A then B* and how they make inferences about alternatives, e.g., *A or else B*.[88] They test whether people can make valid deductions about spatial and temporal relations, e.g., *A is to the left of B*, or *A happens after B*, and about quantified assertions, e.g., *All the A are B*.[89] Experiments investigate how people make inferences about factual situations, hypothetical possibilities, probabilities, and counterfactual situations.[90]

Developmental studies of children's reasoning

Developmental psychologists investigate the development of reasoning from birth to adulthood. Piaget's theory of cognitive development was the first complete theory of reasoning development. Subsequently, several alternative theories were proposed, including the neo-Piagetian theories of cognitive development.[91]

Neuroscience of reasoning

The biological functioning of the brain is studied by neurophysiologists and neuropsychologists. Research in this area includes research into the structure and function of normally functioning brains, and of damaged or otherwise unusual brains. In addition to carrying out research into reasoning, some psychologists, for example, clinical psychologists and psychotherapists work to alter people's reasoning habits when they are unhelpful.

4.5.3 Computer science

Automated reasoning

Main articles: Automated reasoning and Computational logic

In artificial intelligence and computer science, scientists study and use automated reasoning for diverse applications including automated theorem proving the formal semantics of programming languages, and formal specification in software engineering.

Meta-reasoning

Main article: Metacognition

Meta-reasoning is reasoning about reasoning. In computer science, a system performs meta-reasoning when it is reasoning about its own operation.[92] This requires a programming language capable of reflection, the ability to observe and modify its own structure and behaviour.

4.5.4 Evolution of reason

Dan Sperber believes that reasoning in groups is more effective and promotes their evolutionary fitness.

A species could benefit greatly from better abilities to reason about, predict and understand the world. French social and cognitive scientist Dan Sperber, with his colleague Hugo Mercier, describes the idea that there could have been other forces driving the evolution of reason. Sperber points out that reasoning is very difficult for humans to do effectively, and that it is hard for individuals to doubt their own beliefs. Reasoning is most effective when it is done as a collective - as demonstrated by the success of projects like science. Sperber says this could suggest that there are not just individual, but group selection pressures at play. Any group that managed to find ways of reasoning effectively would reap benefits for all its members, increasing their fitness. This could also help explain why humans, according to Sperber, are not optimized to reason effectively alone.

They also claim that reason may have more to do with winning arguments than with the search for the truth.[93]

4.6 See also

- Confirmation bias
- Conformity
- Logic and rationality
- Outline of thought - topic tree that identifies many types of thoughts/thinking, types of reasoning, aspects of thought, related fields, and more.
- Outline of human intelligence - topic tree presenting the traits, capacities, models, and research fields of human intelligence, and more.

4.7 References

[1] "So We Need Something Else for Reason to Mean", *International Journal of Philosophical Studies* 8: 3, 271 — 295.

[2] Compare: MacIntyre, Alasdair (2013). *Dependent Rational Animals: Why Human Beings Need the Virtues.* The Paul Carus Lectures. Open Court. ISBN 9780812697056. Retrieved 2014-12-01. [...] the exercise of independent practical reasoning is one essential constituent to full human flourishing.

[3] Aristotle, Nicomachean Ethics 6 – The Intellectual Virtues

[4] Michel Foucault, "What is Enlightenment?" in *The Essential Foucault*, eds. Paul Rabinow and Nikolas Rose, New York: The New Press, 2003, 43-57. See also Nikolas Kompridis, "The Idea of a New Beginning: A Romantic Source of Normativity and Freedom," in *Philosophical Romanticism*, New York: Routledge, 2006, 32-59; "So We Need Something Else for Reason to Mean", *International Journal of Philosophical Studies* 8: 3, 271 — 295.

[5] Merriam-Webster.com Merriam-Webster Dictionary definition of reason

[6] Hintikka, J. "Philosophy of logic". *Encyclopædia Britannica*. Encyclopædia Britannica, Inc. Retrieved 12 November 2013.

[7] Liddell, Henry George; Scott, Robert, "logos", *A Greek-English Lexicon*. For etymology of English "logic" see any dictionary such as the Merriam Webster entry for logic.

[8] Lewis, Charlton; Short, Charles, "ratio", *A Latin Dictionary*

[9] See Merriam Webster "rational" and Merriam Webster "reasonable".

4.7. REFERENCES

[10] Habermas, Jürgen (1990). *The Philosophical Discourse of Modernity*. Cambridge, MA: MIT Press.

[11] Kirk; Raven; Schofield (1983), *The Presocratic Philosophers* (second ed.), Cambridge University Press. See pages 204 and 235.

[12] *Nicomachean Ethics* Book 1.

[13] Davidson, Herbert (1992), *Alfarabi, Avicenna, and Averroes, on Intellect*, Oxford University Press, page 3.

[14] Moore, Edward, "Plotinus", *Internet Encyclopedia of Philosophy*

[15] Dreyfus, Hubert. "Telepistemology: Descartes' Last Stand". socrates.berkeley.edu. Retrieved February 23, 2011.

[16] Descartes, "Second Meditation".

[17] Hobbes, Thomas, Molesworth, ed., *De Corpore*: "We must not therefore think that computation, that is, ratiocination, has place only in numbers, as if man were distinguished from other living creatures (which is said to have been the opinion of *Pythagoras*) by nothing but the faculty of numbering; for *magnitude, body, motion, time, degrees of quality, action, conception, proportion, speech and names* (in which all the kinds of philosophy consist) are capable of addition and substraction [*sic*]. Now such things as we add or substract, that is, which we put into an account, we are said to *consider*, in Greek λογίζεσθαι [*logizesthai*], in which language also συλλογίζεσθι [*syllogizesthai*] signifies to *compute, reason,* or *reckon.*"

[18] Hobbes, Thomas, "VII. Of the ends, or resolutions of discourse", *The English Works of Thomas Hobbes*, 3 (Leviathan) and Hobbes, Thomas, "IX. Of the several subjects of knowledge", *The English Works of Thomas Hobbes*, 3 (Leviathan)

[19] Locke, John (1824) [1689], "XXVII On Identity and Diversity", *An Essay concerning Human Understanding Part 1*, The Works of John Locke in Nine Volumes (12th ed.), Rivington

[20] Hume, David, "I.IV.VI. Of Personal Identity", *A Treatise of Human Nature*

[21] Hume, David, "II.III.III. Of the influencing motives of the will.", *A Treatise of Human Nature*

[22] Hume, David, "I.III.VII (footnote) Of the Nature of the Idea Or Belief", *A Treatise of Human Nature*

[23] Hume, David, "I.III.XVI. Of the reason of animals", *A Treatise of Human Nature*

[24] Immanuel Kant, *Critique of Pure Reason*; *Critique of Practical Reason.*

[25] Michael Sandel, *Justice: What's the Right Thing to Do?*, New York: Farrar, Straus and Giroux, 2009.

[26] Kant, Immanuel; translated by James W. Ellington [1785] (1993). *Grounding for the Metaphysics of Morals 3rd ed.* Hackett. p. 30. ISBN 0-87220-166-X.

[27] See Velkley, Richard (2002), "On Kant's Socratism", *Being After Rousseau*, University of Chicago Press and Kant's own first preface to *The Critique of Pure Reason*.

[28] Jürgen Habermas, *Moral Consciousness and Communicative Action*, Cambridge, MA: MIT Press, 1995.

[29] Jürgen Habermas, *The Theory of Communicative Action: Reason and the Rationalization of Society*, translated by Thomas McCarthy. Boston: Beacon Press, 1984.

[30] Nikolas Kompridis, *Critique and Disclosure: Critical Theory between Past and Future*, Cambridge, MA: MIT Press, 2006. See also Nikolas Kompridis, "So We Need Something Else for Reason to Mean", *International Journal of Philosophical Studies* 8:3, 271-295.

[31] Charles Taylor, *Philosophical Arguments* (Harvard University Press, 1997), 12; 15.

[32] Michel Foucault, "What is Enlightenment?", *The Essential Foucault*, New York: The New Press, 2003, 43-57.

[33] Douglas Hofstadter, *Gödel, Escher, Bach*, Vintage, 1979, ISBN 0-394-74502-7

[34] Aristotle, *Complete Works* (2 volumes), Princeton, 1995, ISBN 0-691-09950-2

[35] See this Perseus search, and compare English translations. and see LSJ dictionary entry for λογικός, section II.2.b.

[36] See the Treatise of Human Nature of David Hume, Book I, Part III, Sect. XVI.

[37] Locke, John (1824) [1689], "XVII Of Reason", *An Essay concerning Human Understanding Part 2 and Other Writings*, The Works of John Locke in Nine Volumes, **2** (12th ed.), Rivington

[38] Terrence Deacon, *The Symbolic Species: The Co-Evolution of Language and the Brain*, W.W. Norton & Company, 1998, ISBN 0-393-31754-4

[39] Leviathan Chapter IV Archived 2006-06-15 at the Wayback Machine.: "The Greeks have but one word, logos, for both speech and reason; not that they thought there was no speech without reason, but no reasoning without speech"

[40] *Posterior Analytics* II.19.

[41] See for example Ruth M.J. Byrne (2005). *The Rational Imagination: How People Create Counterfactual Alternatives to Reality.* Cambridge, MA: MIT Press.

[42] *De Anima* III.i-iii; *On Memory and Recollection, On Dreams*

[43] Mimesis in modern academic writing, starting with Erich Auerbach, is a technical word, which is not necessarily exactly the same in meaning as the original Greek. See Mimesis.

[44] Origins of the Modern Mind p.172

[45] Jacob Klein *A Commentary on the Meno* Ch.5

[46] Jacob Klein *A Commentary on the Meno* p.122

[47] Origins of the Modern Mind p.169

[48] "Introduction" to the translation of *Poetics* by Davis and Seth Benardete p. xvii, xxviii

[49] Davis is here using "poetic" in an unusual sense, questioning the contrast in Aristotle between action (*praxis*, the *praktikē*) and making (*poēsis*, the *poētikē*): "Human [peculiarly human] action is imitation of action because thinking is always rethinking. Aristotle can define human beings as at once rational animals, political animals, and imitative animals because in the end the three are the same."

[50] Aristotle On Memory 450a 15-16.

[51] Jacob Klein *A Commentary on the Meno* p.109

[52] Aristotle Hist. Anim. I.1.488b.25-26.

[53] Jacob Klein *A Commentary on the Meno* p. 112

[54] *The Origins of the Modern Mind* p.173 see also *A Mind So Rare* p.140-1

[55] Jeffrey, Richard. 1991. *Formal logic: its scope and limits*, (3rd ed.). New York: McGraw-Hill:1.

[56] Walton, Douglas N. (2014). "Argumentation schemes for argument from analogy". In Ribeiro, Henrique Jales. *Systematic approaches to argument by analogy*. Argumentation library. **25**. Cham; New York: Springer Verlag. pp. 23–40. ISBN 9783319063331. OCLC 884441074. doi:10.1007/978-3-319-06334-8_2.

[57] Vickers, John (2009). "The Problem of Induction". *The Stanford Encyclopedia of Philosophy*.

[58] Example: Aristotle *Metaphysics* 981b: τὴν ὀνομαζομένην σοφίαν περὶ τὰ πρῶτα αἴτια καὶ τὰς ἀρχὰς ὑπολαμβάνουσι πάντες· ὥστε, καθάπερ εἴρηται πρότερον, ὁ μὲν ἔμπειρος τῶν ὁποιανοῦν ἐχόντων αἴσθησιν εἶναι δοκεῖ σοφώτερος, ὁ δὲ τεχνίτης τῶν ἐμπείρων, χειροτέχνου δὲ ἀρχιτέκτων, αἱ δὲ θεωρητικαὶ τῶν ποιητικῶν μᾶλλον. English: "...what is called Wisdom is concerned with the primary causes and principles, so that, as has been already stated, the man of experience is held to be wiser than the mere possessors of any power of sensation, the artist than the man of experience, the master craftsman than the artisan; and the speculative sciences to be more learned than the productive."

[59] *Metaphysics* 1009b ποῖα οὖν τούτων ἀληθῆ ἢ ψευδῆ, ἄδηλον· οὐθὲν γὰρ μᾶλλον τάδε ἢ τάδε ἀληθῆ, ἀλλ' ὁμοίως. διὸ Δημόκριτός γέ φησιν ἤτοι οὐθὲν εἶναι ἀληθὲς ἢ ἡμῖν γ' ἄδηλον. English "Thus it is uncertain which of these impressions are true or false; for one kind is no more true than another, but equally so. And hence Democritus says that either there is no truth or we cannot discover it."

[60] For example Aristotle *Metaphysics* 983a: ἐπεὶ δὲ φανερὸν ὅτι τῶν ἐξ **ἀρχῆς** αἰτίων δεῖ λαβεῖν **ἐπιστήμην** (τότε γὰρ εἰδέναι φαμὲν ἕκαστον, ὅταν τὴν **πρώτην** αἰτίαν οἰώμεθα **γνωρίζειν**) English "It is clear that we must obtain knowledge of the **primary** causes, because it is when we think that we understand its **primary** cause that we claim to **know** each particular thing."

[61] Example: *Nicomachean Ethics* 1139b: ἀμφοτέρων δὴ τῶν **νοητικῶν** μορίων ἀλήθεια τὸ ἔργον. καθ' ἃς οὖν μάλιστα ἕξεις ἀληθεύσει ἑκάτερον, αὗται ἀρεταὶ ἀμφοῖν. English The attainment of truth is then the function of both the **intellectual** parts of the soul. Therefore their respective virtues are those dispositions that will best qualify them to attain truth.

[62] Example: Plato *Republic* 490b: μιγεὶς τῷ ὄντι ὄντως, γεννήσας νοῦν καὶ ἀλήθειαν, γνοίη English: "Consorting with reality really, he would beget intelligence and truth, attain to knowledge"

[63] "This quest for the beginnings proceeds through sense perception, reasoning, and what they call *noesis*, which is literally translated by "understanding" or intellect," and which we can perhaps translate a little bit more cautiously by "awareness," an awareness of the mind's eye as distinguished from sensible awareness." "Progress or Return" in An Introduction to Political Philosophy: Ten Essays by Leo Strauss. (Expanded version of Political Philosophy: Six Essays by Leo Strauss, 1975.) Ed. Hilail Gilden. Detroit: Wayne State UP, 1989.

[64] However, the empiricism of Aristotle must certainly be doubted. For example in *Metaphysics* 1009b, cited above, he criticizes people who think knowledge might not be possible because, "They say that the impression given through sense-perception is necessarily true; for it is on these grounds that both Empedocles and Democritus and practically all the rest have become obsessed by such opinions as these."

[65] G.W.F. Hegel *The Philosophy of History*, p. 9, Dover Publications Inc., ISBN 0-486-20112-0; 1st ed. 1899

[66] Velkley, Richard (2002), "Speech. Imagination, Origins: Rousseau and the Political Animal", *Being after Rousseau: Philosophy and Culture in Question*, University of Chicago Press

[67] Rousseau (1997), "Preface", in Gourevitch, *Discourse on the Origin and Foundations of Inequality Among Men or Second Discourse*, Cambridge University Press

[68] Velkley, Richard (2002), "Freedom, Teleology, and Justification of Reason", *Being after Rousseau: Philosophy and Culture in Question*, University of Chicago Press

[69] Plattner, Marc (1997), "Rousseau and the Origins of Nationalism", *The Legacy of Rousseau*, University of Chicago Press

[70] Dawkins, Richard (2008-01-16). *The God Delusion* (Reprint ed.). Mariner Books. ISBN 9780618918249. Scientists... see the fight for evolution as only one battle in a larger war: a looming war between supernaturalism on the one side and rationality on the other.

[71] Strauss, Leo, "Progress or Return", *An Introduction to Political Philosophy*

[72] Locke, John (1824) [1689], "XVIII Of Faith and Reason, and their distinct Provinces.", *An Essay concerning Human Understanding Part 2 and Other Writings*, The Works of John Locke in Nine Volumes, **2** (An Essay concerning Human Understanding Part 2 and Other Writings) (12th ed.), Rivington

[73] Plantinga, Alvin (2011-12-09). *Where the Conflict Really Lies: Science, Religion, and Naturalism* (1 ed.). Oxford University Press. ISBN 9780199812097.

[74] *Natural Signs and Knowledge of God: A New Look at Theistic Arguments* (Reprint ed.). Oxford: Oxford University Press. 2012-12-15. ISBN 9780199661077.

[75] "Stephen Jay Gould, "Nonoverlapping Magisteria," 1997". *www.stephenjaygould.org*. Retrieved 2016-04-06. To say it for all my colleagues and for the umpteenth millionth time (from college bull sessions to learned treatises): science simply cannot (by its legitimate methods) adjudicate the issue of God's possible superintendence of nature. We neither affirm nor deny it; we simply can't comment on it as scientists.

[76] Dawkins, Richard (2008-01-16). "4". *The God Delusion* (Reprint ed.). Mariner Books. ISBN 9780618918249. This sounds terrific, right up until you give it a moment's thought. You then realize that the presence of a creative deity in the universe is clearly a scientific hypothesis. Indeed, it is hard to imagine a more momentous hypothesis in all of science. A universe with a god would be a completely different kind of universe from one without, and it would be a scientific difference. God could clinch the matter in his favour at any moment by staging a spectacular demonstration of his powers, one that would satisfy the exacting standards of science. Even the infamous Templeton Foundation recognized that God is a scientific hypothesis — by funding double-blind trials to test whether remote prayer would speed the recovery of heart patients. It didn't, of course, although a control group who knew they had been prayed for tended to get worse (how about a class action suit against the Templeton Foundation?) Despite such well-financed efforts, no evidence for God's existence has yet appeared.

[77] Moreland, J.P. "Consciousness and the Existence of God: A Theistic Argument". Routledge. Retrieved 2016-04-06.

[78] "The Meaning of Life as Narrative: A New Proposal for Interpreting Philosophy's 'Primary' Question - Joshua W. Seachris - Philo (Philosophy Documentation Center)". *www.pdcnet.org*. Retrieved 2016-04-06.

[79] *Three Rival Versions of Moral Enquiry: Encyclopaedia, Genealogy, and Tradition* (60067th ed.). University of Notre Dame Press. 1991-08-31. ISBN 9780268018771.

[80] Taylor, Charles (2007-09-20). *A Secular Age* (1st ed.). The Belknap Press of Harvard University Press. ISBN 9780674026766.

[81] http://www.catholiceducation.org/en/culture/catholic-contributions/cardinal-ratzinger-on-europe-s-crisis-of-culture.html

[82] *When Athens Met Jerusalem: An Introduction to Classical and Christian Thought* (58760th ed.). IVP Academic. 2009-05-21. ISBN 9780830829231.

[83] Shestov, Lev (1968-01-01). *Athens and Jerusalem*. Simon and Schuster.

[84] "Progress or Return" in An Introduction to Political Philosophy: Ten Essays by Leo Strauss. (Expanded version of Political Philosophy: Six Essays by Leo Strauss, 1975.) Ed. Hilail Gilden. Detroit: Wayne State UP, 1989.

[85] *Bhagavad Gita*, Sarvepalli Radhakrishnan: "Hinduism is not just a faith. It is the union of reason and intuition that can not be defined but is only to be experienced."

[86] Politics I.2.1252b15

[87] Manktelow, K.I. 1999. *Reasoning and Thinking (Cognitive Psychology: Modular Course.)*. Hove, Sussex:Psychology Press

[88] Johnson-Laird, P.N. & Byrne, R.M.J. (1991). *Deduction*. Hillsdale: Erlbaum

[89] Johnson-Laird, P.N. (2006). *How we reason*. Oxford: Oxford University Press

[90] Byrne, R.M.J. (2005). *The Rational Imagination: How People Create Counterfactual Alternatives to Reality*. Cambridge, MA: MIT Press

[91] Demetriou, A. (1998). Cognitive development. In A. Demetriou, W. Doise, K.F.M. van Lieshout (Eds.), Life-span developmental psychology (pp. 179-269). London: Wiley.

[92] Costantini, Stefania (2002), "Meta-reasoning: A Survey", *Lecture Notes in Computer Science*, 2408/2002 (65), doi:10.1007/3-540-45632-5_11

[93] "Dan Sperber on the Enigma of Reason". *Philosophy Bites*. 25 September 2011.

4.8 Further reading

- Reason at PhilPapers

- Beer, Francis A., "Words of Reason", *Political Communication* 11 (Summer, 1994): 185-201.

- Gilovich, Thomas (1991), *How We Know What Isn't So: The Fallibility of Human Reason in Everyday Life*, New York: The Free Press, ISBN 0-02-911705-4

- Tripurari, Swami, *On Faith and Reason*, *The Harmonist*, May 27, 2009.

Chapter 5

Ethics

Ethics or **moral philosophy** is a branch of philosophy that involves systematizing, defending, and recommending concepts of right and wrong conduct.[1] The term *ethics* derives from Ancient Greek ἠθικός *(ethikos)*, from ἦθος *(ethos)*, meaning 'habit, custom'. The branch of philosophy axiology comprises the sub-branches of ethics and aesthetics, each concerned with values.[2]

Ethics seeks to resolve questions of human morality by defining concepts such as good and evil, right and wrong, virtue and vice, justice and crime. As a field of intellectual enquiry, moral philosophy also is related to the fields of moral psychology, descriptive ethics, and value theory.

Three major areas of study within ethics recognised today are:[1]

1. Meta-ethics, concerning the theoretical meaning and reference of moral propositions, and how their truth values (if any) can be determined

2. Normative ethics, concerning the practical means of determining a moral course of action

3. Applied ethics, concerning what a person is obligated (or permitted) to do in a specific situation or a particular domain of action[1]

5.1 Defining ethics

Rushworth Kidder states that "standard definitions of *ethics* have typically included such phrases as 'the science of the ideal human character' or 'the science of moral duty'".[3] Richard William Paul and Linda Elder define ethics as "a set of concepts and principles that guide us in determining what behavior helps or harms sentient creatures".[4] The *Cambridge Dictionary of Philosophy* states that the word ethics is "commonly used interchangeably with 'morality' ... and sometimes it is used more narrowly to mean the moral principles of a particular tradition, group or individual."[5] Paul and Elder state that most people confuse ethics with behaving in accordance with social conventions, religious beliefs and the law and don't treat ethics as a stand-alone concept.[4]

The word *ethics* in English refers to several things.[6] It can refer to philosophical ethics or moral philosophy—a project that attempts to use reason to answer various kinds of ethical questions. As the English philosopher Bernard Williams writes, attempting to explain moral philosophy: "What makes an inquiry a philosophical one is reflective generality and a style of argument that claims to be rationally persuasive."[7] And Williams describes the content of this area of inquiry as addressing the very broad question, "how one should live"[8] Ethics can also refer to a common human ability to think about ethical problems that is not particular to philosophy. As bioethicist Larry Churchill has written: "Ethics, understood as the capacity to think critically about moral values and direct our actions in terms of such values, is a generic human capacity."[9] Ethics can also be used to describe a particular person's own idiosyncratic principles or habits.[10] For example: "Joe has strange ethics."

The English word ethics is derived from an Ancient Greek word *êthikos*, which means "relating to one's character". The Ancient Greek adjective *êthikos* is itself derived from another Greek word, the noun *êthos* meaning "character, disposition".[11]

5.2 Meta-ethics

Main article: Meta-ethics

Meta-ethics asks how we understand, know about, and what we mean when we talk about what is right and what is wrong.[12] An ethical question fixed on some particular practical question—such as, "Should I eat this particular piece of chocolate cake?"—cannot be a meta-ethical question. A meta-ethical question is abstract and relates to a wide range of more specific practical questions. For example, "Is it ever possible to have secure knowledge of what is right and wrong?" would be a meta-ethical question.

Meta-ethics has always accompanied philosophical ethics. For example, Aristotle implies that less precise knowledge is possible in ethics than in other spheres of inquiry, and he regards ethical knowledge as depending upon habit and acculturation in a way that makes it distinctive from other kinds of knowledge. Meta-ethics is also important in G.E. Moore's *Principia Ethica* from 1903. In it he first wrote about what he called *the naturalistic fallacy*. Moore was seen to reject naturalism in ethics, in his Open Question Argument. This made thinkers look again at second order questions about ethics. Earlier, the Scottish philosopher David Hume had put forward a similar view on the difference between facts and values.

Studies of how we know in ethics divide into cognitivism and non-cognitivism; this is similar to the contrast between descriptivists and non-descriptivists. Non-cognitivism is the claim that when we judge something as right or wrong, this is neither true nor false. We may for example be only expressing our emotional feelings about these things.[13] Cognitivism can then be seen as the claim that when we talk about right and wrong, we are talking about matters of fact.

The ontology of ethics is about value-bearing things or properties, i.e. the kind of things or stuff referred to by ethical propositions. Non-descriptivists and non-cognitivists believe that ethics does not need a specific ontology, since ethical propositions do not refer. This is known as an anti-realist position. Realists on the other hand must explain what kind of entities, properties or states are relevant for ethics, how they have value, and why they guide and motivate our actions.[14]

5.3 Normative ethics

Main article: Normative ethics

Normative ethics is the study of ethical action. It is the branch of ethics that investigates the set of questions that arise when considering how one ought to act, morally speaking. Normative ethics is distinct from meta-ethics because it examines standards for the rightness and wrongness of actions, while meta-ethics studies the meaning of moral language and the metaphysics of moral facts.[12] Normative ethics is also distinct from descriptive ethics, as the latter is an empirical investigation of people's moral beliefs. To put it another way, descriptive ethics would be concerned with determining what proportion of people believe that killing is always wrong, while normative ethics is concerned with whether it is correct to hold such a belief. Hence, normative ethics is sometimes called prescriptive, rather than descriptive. However, on certain versions of the meta-ethical view called moral realism, moral facts are both descriptive and prescriptive at the same time.[15]

Traditionally, normative ethics (also known as moral theory) was the study of what makes actions right and wrong. These theories offered an overarching moral principle one could appeal to in resolving difficult moral decisions.

At the turn of the 20th century, moral theories became more complex and are no longer concerned solely with rightness and wrongness, but are interested in many different kinds of moral status. During the middle of the century, the study of normative ethics declined as meta-ethics grew in prominence. This focus on meta-ethics was in part caused by an intense linguistic focus in analytic philosophy and by the popularity of logical positivism.

In 1971 John Rawls published *A Theory of Justice*, noteworthy in its pursuit of moral arguments and eschewing of meta-ethics.

5.3.1 Virtue ethics

Main article: Virtue ethics

Virtue ethics describes the character of a moral agent as a driving force for ethical behavior, and is used to describe the ethics of Socrates, Aristotle, and other early Greek philosophers. Socrates (469–399 BC) was one of the first Greek philosophers to encourage both scholars and the common citizen to turn their attention from the outside world to the condition of humankind. In this view, knowledge bearing on human life was placed highest, while all other knowledge were secondary. Self-knowledge was considered necessary for success and inherently an essential good. A self-aware person will act completely within his capabilities to his pinnacle, while an ignorant person will flounder and encounter difficulty. To Socrates, a person must become aware of every fact (and its context) relevant to his existence, if he wishes to attain self-knowledge. He posited that people will naturally do what is good, if they know what is right. Evil or bad actions are the result of ignorance. If a criminal was truly aware of the intellectual and spiritual consequences of his actions, he would neither commit nor even consider committing those actions. Any person who knows what is truly right will automatically do it, according to Socrates. While he correlated knowledge with virtue, he similarly equated virtue with joy. The truly wise man will know what is right, do what is good, and therefore be happy.[16]:32–33

5.3. NORMATIVE ETHICS

Socrates

Aristotle (384–323 BC) posited an ethical system that may be termed "self-realizationism". In Aristotle's view, when a person acts in accordance with his nature and realizes his full potential, he will do good and be content. At birth, a baby is not a person, but a potential person. To become a "real" person, the child's inherent potential must be realized. Unhappiness and frustration are caused by the unrealized potential of a person, leading to failed goals and a poor life. Aristotle said, "Nature does nothing in vain." Therefore, it is imperative for people to act in accordance with their nature and develop their latent talents in order to be content and complete. Happiness was held to be the ultimate goal. All other things, such as civic life or wealth, are merely means to the end. Self-realization, the awareness of one's nature and the development of one's talents, is the surest path to happiness.[16]:33–35

Aristotle asserted that man had three natures: body (physical/metabolism), animal (emotional/appetite) and rational (mental/conceptual). Physical nature can be assuaged through exercise and care, emotional nature through indulgence of instinct and urges, and mental through human reason and developed potential. Rational development was considered the most important, as essential to philosophical self-awareness and as uniquely human. Moderation was encouraged, with the extremes seen as degraded and immoral. For example, courage is the moderate virtue between the extremes of cowardice and recklessness. Man should not simply live, but live well with conduct governed by moderate virtue. This is regarded as difficult, as virtue denotes doing the right thing, to the right person, at the right time, to the proper extent, in the correct fashion, for the right reason.[16]:35–37

Stoicism

Main article: Stoicism

The Stoic philosopher Epictetus posited that the greatest

Epictetus

good was contentment and serenity. Peace of mind, or Apatheia, was of the highest value; self-mastery over one's desires and emotions leads to spiritual peace. The "unconquerable will" is central to this philosophy. The individual's will should be independent and inviolate. Allowing a person to disturb the mental equilibrium is in essence offering yourself in slavery. If a person is free to anger you at will, you have no control over your internal world, and therefore no freedom. Freedom from material attachments is also necessary. If a thing breaks, the person should not be upset, but realize it was a thing that could break. Similarly, if someone should die, those close to them should hold to their serenity because the loved one was made of flesh and blood destined to death. Stoic philosophy says to accept things that cannot be changed, resigning oneself to existence and enduring in a rational fashion. Death is not feared. People do not "lose" their life, but instead "return", for they are returning to God (who initially gave what the person is as a person). Epictetus said difficult problems in life should not be avoided, but rather embraced. They are spiritual exercises needed for the health of the spirit, just as physical exercise is required for the health of the body. He also stated that sex and sexual desire are to be avoided as the greatest threat to the integrity and equilibrium of a man's mind. Abstinence is highly desirable. Epictetus said remaining abstinent in the face of temptation was a victory for which a man could be proud.[16]:38–41

Contemporary virtue ethics

Modern virtue ethics was popularized during the late 20th century in large part as a response to G. E. M. Anscombe's "Modern Moral Philosophy". Anscombe argues that consequentialist and deontological ethics are only feasible as universal theories if the two schools ground themselves in divine law. As a deeply devoted Christian herself, Anscombe proposed that either those who do not give ethical credence to notions of divine law take up virtue ethics, which does not necessitate universal laws as agents themselves are investigated for virtue or vice and held up to "universal standards", or that those who wish to be utilitarian or consequentialist ground their theories in religious conviction.[17] Alasdair MacIntyre, who wrote the book *After Virtue*, was a key contributor and proponent of modern virtue ethics, although MacIntyre supports a relativistic account of virtue based on cultural norms, not objective standards.[17] Martha Nussbaum, a contemporary virtue ethicist, objects to MacIntyre's relativism, among that of others, and responds to relativist objections to form an objective account in her work "Non-Relative Virtues: An Aristotelian Approach".[18] *Complete Conduct Principles for the 21st Century*[19] blended the Eastern virtue ethics and the Western virtue ethics, with some modifications to suit the 21st Century, and formed a part of contemporary virtue ethics.[19]

5.3.2 Hedonism

Main article: Hedonism

Hedonism posits that the principal ethic is maximizing pleasure and minimizing pain. There are several schools of Hedonist thought ranging from those advocating the indulgence of even momentary desires to those teaching a pursuit of spiritual bliss. In their consideration of consequences, they range from those advocating self-gratification regardless of the pain and expense to others, to those stating that the most ethical pursuit maximizes pleasure and happiness for the most people.[16]:37

Cyrenaic hedonism

Founded by Aristippus of Cyrene, Cyrenaics supported immediate gratification or pleasure. "Eat, drink and be merry, for tomorrow we die." Even fleeting desires should be indulged, for fear the opportunity should be forever lost. There was little to no concern with the future, the present dominating in the pursuit for immediate pleasure. Cyrenaic hedonism encouraged the pursuit of enjoyment and indulgence without hesitation, believing pleasure to be the only good.[16]:37

Epicureanism

Main article: Epicureanism

Epicurean ethics is a hedonist form of virtue ethics. Epicurus "...presented a sustained argument that pleasure, correctly understood, will coincide with virtue."[20] He rejected the extremism of the Cyrenaics, believing some pleasures and indulgences to be detrimental to human beings. Epicureans observed that indiscriminate indulgence sometimes resulted in negative consequences. Some experiences were therefore rejected out of hand, and some unpleasant experiences endured in the present to ensure a better life in the future. To Epicurus the *summum bonum*, or greatest good, was prudence, exercised through moderation and caution. Excessive indulgence can be destructive to pleasure and can even lead to pain. For example, eating one food too often makes a person lose taste for it. Eating too much food at once leads to discomfort and ill-health. Pain and fear were to be avoided. Living was essentially good, barring pain and illness. Death was not to be feared. Fear was considered the source of most unhappiness. Conquering the fear of death would naturally lead to a happier life. Epicurus reasoned if there was an afterlife and immortality, the fear of death was irrational. If there was no life after death, then the person would not be alive to suffer, fear or worry; he would be non-existent in death. It is irrational to fret over circumstances that do not exist, such as one's state in death in the absence of an afterlife.[16]:37–38

5.3.3 State consequentialism

Main article: State consequentialism

State consequentialism, also known as Mohist consequentialism,[21] is an ethical theory that evaluates the moral worth of an action based on how much it contributes to the basic goods of a state.[21] The *Stanford Encyclopedia of Philosophy* describes Mohist consequentialism, dating back to the 5th century BC, as "a remarkably sophisticated version based on a plurality of intrinsic goods taken as constitutive of human welfare".[22] Unlike utilitarianism, which views pleasure as a moral good, "the basic goods in Mohist consequentialist thinking are ... order, material wealth, and increase in population".[23] During Mozi's era, war and famines were common, and population growth was seen as a moral necessity for a harmonious society. The "material wealth" of Mohist consequentialism refers to basic needs like shelter and clothing, and the "order" of Mohist consequentialism refers to Mozi's stance against warfare and violence, which he viewed as pointless and a threat to social stability.[24]

Stanford sinologist David Shepherd Nivison, in *The Cambridge History of Ancient China*, writes that the moral goods of Mohism "are interrelated: more basic wealth, then more reproduction; more people, then more production and wealth ... if people have plenty, they would be good, filial, kind, and so on unproblematically."[23] The Mohists believed that morality is based on "promoting the benefit of all under heaven and eliminating harm to all under heaven". In contrast to Bentham's views, state consequentialism is not utilitarian because it is not hedonistic or individualistic. The importance of outcomes that are good for the community outweigh the importance of individual pleasure and pain.[25]

5.3.4 Consequentialism/Teleology

Main article: Consequentialism
See also: Ethical egoism

Consequentialism refers to moral theories that hold that the consequences of a particular action form the basis for any valid moral judgment about that action (or create a structure for judgment, see rule consequentialism). Thus, from a consequentialist standpoint, a morally right action is one that produces a good outcome, or consequence. This view is often expressed as the aphorism *"The ends justify the means"*.

The term "consequentialism" was coined by G. E. M. Anscombe in her essay "Modern Moral Philosophy" in 1958, to describe what she saw as the central error of certain moral theories, such as those propounded by Mill and Sidgwick.[26] Since then, the term has become common in English-language ethical theory.

The defining feature of consequentialist moral theories is the weight given to the consequences in evaluating the rightness and wrongness of actions.[27] In consequentialist theories, the consequences of an action or rule generally outweigh other considerations. Apart from this basic outline, there is little else that can be unequivocally said about consequentialism as such. However, there are some questions that many consequentialist theories address:

- What sort of consequences count as good consequences?
- Who is the primary beneficiary of moral action?
- How are the consequences judged and who judges them?

One way to divide various consequentialisms is by the types of consequences that are taken to matter most, that is, which consequences count as good states of affairs. According to utilitarianism, a good action is one that results in an increase in a positive effect, and the best action is one that results in that effect for the greatest number. Closely related is eudaimonic consequentialism, according to which a full, flourishing life, which may or may not be the same as enjoying a great deal of pleasure, is the ultimate aim. Similarly, one might adopt an aesthetic consequentialism, in which the ultimate aim is to produce beauty. However, one might fix on non-psychological goods as the relevant effect. Thus, one might pursue an increase in material equality or political liberty instead of something like the more ephemeral "pleasure". Other theories adopt a package of several goods, all to be promoted equally. Whether a particular consequentialist theory focuses on a single good or many, conflicts and tensions between different good states of affairs are to be expected and must be adjudicated.

Utilitarianism

Main article: Utilitarianism

Jeremy Bentham

John Stuart Mill

Utilitarianism is an ethical theory that argues the proper course of action is one that maximizes a positive effect, such as "happiness", "welfare", or the ability to live according to personal preferences.[28] Jeremy Bentham and John Stuart Mill are influential proponents of this school of thought. In *A Fragment on Government* Bentham says 'it is the greatest happiness of the greatest number that is the measure of right and wrong' and describes this as a fundamental axiom. In *An Introduction to the Principles of Morals and Legislation* he talks of 'the principle of utility' but later prefers "the greatest happiness principle".[29][30]

Utilitarianism is the paradigmatic example of a consequentialist moral theory. This form of utilitarianism holds that the morally correct action is the one that produces the best outcome for all people affected by the action. John Stuart Mill, in his exposition of utilitarianism, proposed a hierarchy of pleasures, meaning that the pursuit of certain kinds of pleasure is more highly valued than the pursuit of other pleasures.[31] Other noteworthy proponents of utilitarianism are neuroscientist Sam Harris, author of The Moral Landscape, and moral philosopher Peter Singer, author of, amongst other works, Practical Ethics.

There are two types of utilitarianism, *act utilitarianism* and *rule utilitarianism*. In act utilitarianism, the principle of utility applies directly to each alternative act in a situation of choice. The right act is the one that brings about the best results (or the least amount of bad results). In rule utilitarianism, the principle of utility determines the validity of rules of conduct (moral principles). A rule like promise-keeping is established by looking at the consequences of a world in which people break promises at will, and a world in which promises are binding. Right and wrong are the following or breaking of rules that are sanctioned by their utilitarian value.[32]

5.3.5 Deontology

Main article: Deontological ethics

Deontological ethics or deontology (from Greek δέον,

Immanuel Kant

deon, "obligation, duty"; and -λογία, -*logia*) is an approach to ethics that determines goodness or rightness from examining acts, or the rules and duties that the person doing the act strove to fulfill.[33] This is in contrast to consequentialism, in which rightness is based on the consequences of an act, and not the act by itself. In deontology, an act may be considered right even if the act produces a bad consequence,[34] if it follows the *rule* that "one should do unto others as they would have done unto them",[35] and even if the person who does the act lacks virtue and had a bad intention in doing the act. According to deontology, people have a *duty* to act in a way that does those things that are inherently good as acts ("truth-telling" for example), or follow an objectively obligatory rule (as in rule utilitarianism). For deontologists, the ends or consequences of people's actions are not important in and of themselves, and people's intentions are not important in and of themselves.

Immanuel Kant's theory of ethics is considered deontological for several different reasons.[36][37] First, Kant argues that to act in the morally right way, people must act from duty (*deon*).[38] Second, Kant argued that it was not the consequences of actions that make them right or wrong but the motives (*maxime*) of the person who carries out the action. Kant's argument that to act in the morally right way, one must act from duty, begins with an argument that the highest good must be both good in itself, and good without qualification.[39] Something is 'good in itself' when it is intrinsically good, and 'good without qualification' when the addition of that thing never makes a situation ethically worse. Kant then argues that those things that are usually thought to be good, such as intelligence, perseverance and pleasure, fail to be either intrinsically good or good without qualification. Pleasure, for example, appears to not be good without qualification, because when people take pleasure in watching someone suffer, they make the situation ethically worse. He concludes that there is only one thing that is truly good:

> Nothing in the world—indeed nothing even beyond the world—can possibly be conceived which could be called good without qualification except a *good will*.[39]

5.3.6 Pragmatic ethics

Main article: Pragmatic ethics

Associated with the pragmatists, Charles Sanders Peirce, William James, and especially John Dewey, pragmatic ethics holds that moral correctness evolves similarly to scientific knowledge: socially over the course of many lifetimes. Thus, we should prioritize social reform over attempts to account for consequences, individual virtue or duty (although these may be worthwhile attempts, if social reform is provided for).[40]

5.3.7 Ethics of care

Main article: Ethics of care

Care ethics contrasts with more well-known ethical models, such as consequentialist theories (e.g. utilitarianism) and deontological theories (e.g., Kantian ethics) in that it seeks to incorporate traditionally feminized virtues and values that—proponents of care ethics contend—are absent in such traditional models of ethics. These values include the importance of empathetic relationships and compassion.

Care-focused feminism is a branch of feminist thought, informed primarily by ethics of care as developed by Carol Gilligan and Nel Noddings.[41] This body of theory is critical of how caring is socially assigned to women, and consequently devalued. They write, "Care-focused feminists regard women's capacity for care as a human strength," that should be taught to and expected of men as well as women. Noddings proposes that ethical caring has the potential to be a more concrete evaluative model of moral dilemma, than an ethic of justice.[42] Noddings' care-focused feminism requires practical application of relational ethics, predicated on an ethic of care.[43]

5.3.8 Role ethics

Main article: Role ethics

Role ethics is an ethical theory based on family roles.[44] Unlike virtue ethics, role ethics is not individualistic. Morality is derived from a person's relationship with their community.[45] Confucian ethics is an example of role ethics[44] though this is not straightforwardly uncontested.[46] Confucian roles center around the concept of filial piety or *xiao*, a respect for family members.[47] According to Roger Ames and Henry Rosemont, "Confucian normativity is defined by living one's family roles to maximum effect." Morality is determined through a person's fulfillment of a role, such as that of a parent or a child. Confucian roles are not rational, and originate through the *xin*, or human emotions.[45]

5.3.9 Anarchist ethics

Main article: Anarchism

Anarchist ethics is an ethical theory based on the studies of anarchist thinkers. The biggest contributor to the anarchist ethics is the Russian zoologist, geographer, economist, and political activist Peter Kropotkin. The anarchist ethics is a large, vague field that can depend on different historical situations and different anarchist thinkers—but as Peter Kropotkin explains, "any "bourgeois" or "proletarian" ethics rests, after all, on the common basis, on the common ethnological foundation, which at times exerts a very strong influence on the principles of the class or group morality." Still, most of the anarchist ethics schools are based on three fundamental ideas, which are: "solidarity, equality and justice". Kropotkin argues that Ethics is evolutionary and is inherited as a sort of a social instinct through History, and by so, he rejects any religious and transcendental explanation of ethics.[48] Kropotkin suggests that the principle of equality at the core of anarchism is the same as the Golden rule:

> This principle of treating others as one wishes to be treated oneself, what is it but the very same

principle as equality, the fundamental principle of anarchism? And how can any one manage to believe himself an anarchist unless he practices it? We do not wish to be ruled. And by this very fact, do we not declare that we ourselves wish to rule nobody? We do not wish to be deceived, we wish always to be told nothing but the truth. And by this very fact, do we not de- clare that we ourselves do not wish to deceive anybody, that we promise to always tell the truth, nothing but the truth, the whole truth? We do not wish to have the fruits of our labor stolen from us. And by that very fact, do we not declare that we respect the fruits of others' labor? By what right indeed can we demand that we should be treated in one fashion, reserving it to ourselves to treat others in a fashion entirely different? Our sense of equality revolts at such an idea.[49]

5.3.10 Postmodern ethics

Main article: Postmodernism

The 20th century saw a remarkable expansion and evolution of critical theory, following on earlier Marxist Theory efforts to locate individuals within larger structural frameworks of ideology and action.

Antihumanists such as Louis Althusser and Michel Foucault and structuralists such as Roland Barthes challenged the possibilities of individual agency and the coherence of the notion of the 'individual' itself. As critical theory developed in the later 20th century, post-structuralism sought to problematize human relationships to knowledge and 'objective' reality. Jacques Derrida argued that access to meaning and the 'real' was always deferred, and sought to demonstrate via recourse to the linguistic realm that "there is no outside-text/non-text" ("*il n'y a pas de hors-texte*" is often mistranslated as "there is nothing outside the text"); at the same time, Jean Baudrillard theorised that signs and symbols or simulacra mask reality (and eventually the absence of reality itself), particularly in the consumer world.

Post-structuralism and postmodernism argue that ethics must study the complex and relational conditions of actions. A simple alignment of ideas of right and particular acts is not possible. There will always be an ethical remainder that cannot be taken into account or often even recognized. Such theorists find narrative (or, following Nietzsche and Foucault, genealogy) to be a helpful tool for understanding ethics because narrative is always about particular lived experiences in all their complexity rather than the assignment of an idea or norm to separate and individual actions.

Zygmunt Bauman says postmodernity is best described as modernity without illusion, the illusion being the belief that humanity can be repaired by some ethic principle. Postmodernity can be seen in this light as accepting the messy nature of humanity as unchangeable.

David Couzens Hoy states that Emmanuel Levinas's writings on the face of the Other and Derrida's meditations on the relevance of death to ethics are signs of the "ethical turn" in Continental philosophy that occurred in the 1980s and 1990s. Hoy describes post-critique ethics as the "obligations that present themselves as necessarily to be fulfilled but are neither forced on one or are enforceable" (2004, p. 103).

Hoy's post-critique model uses the term *ethical resistance*. Examples of this would be an individual's resistance to consumerism in a retreat to a simpler but perhaps harder lifestyle, or an individual's resistance to a terminal illness. Hoy describes Levinas's account as "not the attempt to use power against itself, or to mobilize sectors of the population to exert their political power; the ethical resistance is instead the resistance of the powerless"(2004, p. 8).

Hoy concludes that

> The ethical resistance of the powerless others to our capacity to exert power over them is therefore what imposes unenforceable obligations on us. The obligations are unenforceable precisely because of the other's lack of power. That actions are at once obligatory and at the same time unenforceable is what put them in the category of the ethical. Obligations that were enforced would, by the virtue of the force behind them, not be freely undertaken and would not be in the realm of the ethical. (2004, p.184)

In present-day terms the powerless may include the unborn, the terminally sick, the aged, the insane, and non-human animals. It is in these areas that ethical action in Hoy's sense apply. Until legislation or the state apparatus enforces a moral order that addresses the causes of resistance these issues remain in the ethical realm. For example, should animal experimentation become illegal in a society, it is longer be an ethical issue by Hoy's definition. Likewise, one hundred and fifty years ago, not having a black slave in America would have been an ethical choice. This later issue has been absorbed into the fabric of an enforceable social order, and is therefore no longer an ethical issue in Hoy's sense.

5.4 Applied ethics

Main article: Applied ethics

5.4. APPLIED ETHICS

Applied ethics is a discipline of philosophy that attempts to apply ethical theory to real-life situations. The discipline has many specialized fields, such as engineering ethics, bioethics, geoethics, public service ethics and business ethics.

5.4.1 Specific questions

Applied ethics is used in some aspects of determining public policy, as well as by individuals facing difficult decisions. The sort of questions addressed by applied ethics include: "Is getting an abortion immoral?" "Is euthanasia immoral?" "Is affirmative action right or wrong?" "What are human rights, and how do we determine them?" "Do animals have rights as well?" and "Do individuals have the right of self determination?"[12]

A more specific question could be: "If someone else can make better out of his/her life than I can, is it then moral to sacrifice myself for them if needed?" Without these questions there is no clear fulcrum on which to balance law, politics, and the practice of arbitration—in fact, no common assumptions of all participants—so the ability to formulate the questions are prior to rights balancing. But not all questions studied in applied ethics concern public policy. For example, making ethical judgments regarding questions such as, "Is lying always wrong?" and, "If not, when is it permissible?" is prior to any etiquette.

People in general are more comfortable with dichotomies (two opposites). However, in ethics the issues are most often multifaceted and the best proposed actions address many different areas concurrently. In ethical decisions the answer is almost never a "yes or no", "right or wrong" statement. Many buttons are pushed so that the overall condition is improved and not to the benefit of any particular faction.

5.4.2 Particular fields of application

Bioethics

Main article: Bioethics

Bioethics is the study of controversial ethics brought about by advances in biology and medicine. Bioethicists are concerned with the ethical questions that arise in the relationships among life sciences, biotechnology, medicine, politics, law, and philosophy. It also includes the study of the more commonplace questions of values ("the ethics of the ordinary") that arise in primary care and other branches of medicine.

Bioethics also needs to address emerging biotechnologies that affect basic biology and future humans. These developments include cloning, gene therapy, human genetic engineering, astroethics and life in space,[50] and manipulation of basic biology through altered DNA, RNA and proteins, e.g.- "three parent baby, where baby is born from genetically modified embryos, would have DNA from a mother, a father and from a female donor.[51] Correspondingly, new bioethics also need to address life at its core. For example, biotic ethics value organic gene/protein life itself and seek to propagate it.[52] With such life-centered principles, ethics may secure a cosmological future for life.[53]

Business ethics

Main article: Business ethics

Business ethics (also corporate ethics) is a form of applied ethics or professional ethics that examines ethical principles and moral or ethical problems that arise in a business environment, including fields like medical ethics. It applies to all aspects of business conduct and is relevant to the conduct of individuals and entire organizations.

Business ethics has both normative and descriptive dimensions. As a corporate practice and a career specialization, the field is primarily normative. Academics attempting to understand business behavior employ descriptive methods. The range and quantity of business ethical issues reflects the interaction of profit-maximizing behavior with non-economic concerns. Interest in business ethics accelerated dramatically during the 1980s and 1990s, both within major corporations and within academia. For example, today most major corporations promote their commitment to non-economic values under headings such as ethics codes and social responsibility charters. Adam Smith said, "People of the same trade seldom meet together, even for merriment and diversion, but the conversation ends in a conspiracy against the public, or in some contrivance to raise prices."[54] Governments use laws and regulations to point business behavior in what they perceive to be beneficial directions. Ethics implicitly regulates areas and details of behavior that lie beyond governmental control.[55] The emergence of large corporations with limited relationships and sensitivity to the communities in which they operate accelerated the development of formal ethics regimes.[56]

Machine ethics

Main article: Machine ethics

In *Moral Machines: Teaching Robots Right from Wrong*, Wendell Wallach and Colin Allen conclude that issues in machine ethics will likely drive advancement in understanding of human ethics by forcing us to address gaps in modern normative theory and by providing a platform for experimental investigation.[57] The effort to actually program a machine or artificial agent to behave as though instilled with a sense of ethics requires new specificity in our normative theories, especially regarding aspects customarily considered common-sense. For example, machines, unlike humans, can support a wide selection of learning algorithms, and controversy has arisen over the relative ethical merits of these options. This may reopen classic debates of normative ethics framed in new (highly technical) terms.

Military ethics

See also: Geneva Conventions and Nuremberg Principles

Military ethics are concerned with questions regarding the application of force and the ethos of the soldier and are often understood as applied professional ethics.[58] Just war theory is generally seen to set the background terms of military ethics. However individual countries and traditions have different fields of attention.[59]

Military ethics involves multiple subareas, including the following among others:

1. what, if any, should be the laws of war.
2. justification for the initiation of military force.
3. decisions about who may be targeted in warfare.
4. decisions on choice of weaponry, and what collateral effects such weaponry may have.
5. standards for handling military prisoners.
6. methods of dealing with violations of the laws of war.

Political ethics

Main article: Political ethics

Political ethics (also known as political morality or public ethics) is the practice of making moral judgements about political action and political agents.[60]

Public sector ethics

Main article: Public sector ethics

Public sector ethics is a set of principles that guide public officials in their service to their constituents, including their decision-making on behalf of their constituents. Fundamental to the concept of public sector ethics is the notion that decisions and actions are based on what best serves the public's interests, as opposed to the official's personal interests (including financial interests) or self-serving political interests.[61]

Publication ethics

Publication ethics is the set of principles that guide the writing and publishing process for all professional publications. To follow these principles, authors must verify that the publication does not contain plagiarism or publication bias.[62] As a way to avoid misconduct in research these principles can also apply to experiments that are referenced or analyzed in publications by ensuring the data is recorded honestly and accurately.[63]

Plagiarism is the failure to give credit to another author's work or ideas, when it is used in the publication.[64] It is the obligation of the editor of the journal to ensure the article does not contain any plagiarism before it is published.[65] If a publication that has already been published is proven to contain plagiarism, the editor of the journal can retract the article.[66]

Publication bias occurs when the publication is one-sided or "prejudiced against results".[67] In best practice, an author should try to include information from all parties involved, or affected by the topic. If an author is prejudiced against certain results, than it can "lead to erroneous conclusions being drawn".[68]

Misconduct in research can occur when an experimenter falsifies results.[69] Falsely recorded information occurs when the researcher "fakes" information or data, which was not used when conducting the actual experiment.[69] By faking the data, the researcher can alter the results from the experiment to better fit the hypothesis they originally predicted. When conducting medical research, it is important to honor the healthcare rights of a patient by protecting their anonymity in the publication.[62] *Respect for autonomy* is the principle that decision-making should allow individuals to be autonomous; they should be able to make decisions that apply to their own lives. This means that individuals should have control of their lives. *Justice* is the principle that decision makers must focus on actions that are fair to those affected. Ethical decisions need to be consistent with the ethical theory. There are cases where the management has made decisions that seem to be unfair to the employees, shareholders, and other stakeholders (Solomon, 1992, pp49). Such decisions are unethical.

Relational ethics

Relational ethics are related to an ethics of care.[70]:62–63 They are used in qualitative research, especially ethnography and autoethnography. Researchers who employ relational ethics value and respect the connection between themselves and the people they study, and "...between researchers and the communities in which they live and work." (Ellis, 2007, p. 4).[71] Relational ethics also help researchers understand difficult issues such as conducting research on intimate others that have died and developing friendships with their participants.[72][73] Relational ethics in close personal relationships form a central concept of contextual therapy.

Animal ethics

Main article: Animal ethics

Animal ethics is a term used in academia to describe human-animal relationships and how animals ought to be treated. The subject matter includes animal rights, animal welfare, animal law, speciesism, animal cognition, wildlife conservation, the moral status of nonhuman animals, the concept of nonhuman personhood, human exceptionalism, the history of animal use, and theories of justice.

5.5 Moral psychology

Main article: Moral psychology

Moral psychology is a field of study that began as an issue in philosophy and that is now properly considered part of the discipline of psychology. Some use the term "moral psychology" relatively narrowly to refer to the study of moral development.[74] However, others tend to use the term more broadly to include any topics at the intersection of ethics and psychology (and philosophy of mind).[75] Such topics are ones that involve the mind and are relevant to moral issues. Some of the main topics of the field are moral responsibility, moral development, moral character (especially as related to virtue ethics), altruism, psychological egoism, moral luck, and moral disagreement.[76]

5.5.1 Evolutionary ethics

Main article: Evolutionary ethics
See also: Evolution of morality

Evolutionary ethics concerns approaches to ethics (morality) based on the role of evolution in shaping human psychology and behavior. Such approaches may be based in scientific fields such as evolutionary psychology or sociobiology, with a focus on understanding and explaining observed ethical preferences and choices.[77]

5.6 Descriptive ethics

Main article: Descriptive ethics

Descriptive ethics is on the less philosophical end of the spectrum, since it seeks to gather particular information about how people live and draw general conclusions based on observed patterns. Abstract and theoretical questions that are more clearly philosophical—such as, "Is ethical knowledge possible?"—are not central to descriptive ethics. Descriptive ethics offers a value-free approach to ethics, which defines it as a social science rather than a humanity. Its examination of ethics doesn't start with a preconceived theory, but rather investigates observations of actual choices made by moral agents in practice. Some philosophers rely on descriptive ethics and choices made and unchallenged by a society or culture to derive categories, which typically vary by context. This can lead to situational ethics and situated ethics. These philosophers often view aesthetics, etiquette, and arbitration as more fundamental, percolating "bottom up" to imply the existence of, rather than explicitly prescribe, theories of value or of conduct. The study of descriptive ethics may include examinations of the following:

- Ethical codes applied by various groups. Some consider aesthetics itself the basis of ethics—and a personal moral core developed through art and storytelling as very influential in one's later ethical choices.

- Informal theories of etiquette that tend to be less rigorous and more situational. Some consider etiquette a simple negative ethics, i.e., where can one evade an uncomfortable truth without doing wrong? One notable advocate of this view is Judith Martin ("Miss Manners"). According to this view, ethics is more a summary of common sense social decisions.

- Practices in arbitration and law, e.g., the claim that ethics itself is a matter of balancing "right versus right", i.e., putting priorities on two things that are both right, but that must be traded off carefully in each situation.

- Observed choices made by ordinary people, without expert aid or advice, who vote, buy, and decide what is worth valuing. This is a major concern of sociology, political science, and economics.

5.7 See also

- Contemporary ethics
- Corporate social responsibility
- Declaration of Geneva
- Declaration of Helsinki
- Deductive reasoning
- Descriptive ethics
- Dharma
- Ethical movement
- Ethics paper
- Index of ethics articles—alphabetical list of ethics-related articles
- Moral psychology
- Outline of ethics—list of ethics-related articles, arranged by sub-topic
- Practical philosophy
- Science of morality
- Theory of justification

5.8 Notes

[1] *Internet Encyclopedia of Philosophy* "Ethics"

[2] *Random House Unabridged Dictionary*: Entry on Axiology.

[3] Kidder, Rushworth (2003). *How Good People Make Tough Choices: Resolving the Dilemmas of Ethical Living*. New York: Harper Collins. p. 63. ISBN 0-688-17590-2.

[4] Paul, Richard; Elder, Linda (2006). *The Miniature Guide to Understanding the Foundations of Ethical Reasoning*. United States: Foundation for Critical Thinking Free Press. p. np. ISBN 0-944583-17-2.

[5] John Deigh in Robert Audi (ed), *The Cambridge Dictionary of Philosophy*, 1995.

[6] "Definition of ethic by Merriam Webster". Merriam Webster. Retrieved October 4, 2015.

[7] Williams, Bernard. *Ethics and the Limits of Philosophy*. p. 2.

[8] Williams, Bernard. *Ethics and the Limits of Philosophy*. p. 1.

[9] "Are We Professionals? A Critical Look at the Social Role of Bioethicists.". *Daedalus*. 1999. pp. 253–274.

[10] David Tanguay (January 24, 2014). "Buddha and Socrates share Common ground". Soul of Wit. Archived from the original on July 22, 2014. Retrieved July 22, 2014.

[11] *An Intermediate Greek-English Lexicon*. 1889.

[12] "What is ethics?". BBC. Archived from the original on October 28, 2013. Retrieved July 22, 2014.

[13] http://www.iep.utm.edu/non-cogn/

[14] Miller, C. (2009). The Conditions of Moral Realism. The Journal of Philosophical Research, 34, 123-155.

[15] Cavalier, Robert. "Meta-ethics, Normative Ethics, and Applied Ethics". *Online Guide to Ethics and Moral Philosophy*. Archived from the original on November 12, 2013. Retrieved February 26, 2014.

[16] William S. Sahakian; Mabel Lewis Sahakian (1966). *Ideas of the Great Philosophers*. Barnes & Noble. ISBN 978-1-56619-271-2.

[17] Professor Michiel S. S. De De Vries; Professor Pan Suk Kim (October 28, 2011). *Value and Virtue in Public Administration: A Comparative Perspective*. Palgrave Macmillan. p. 42. ISBN 978-0-230-35709-9.

[18] Nussbaum, Martha (1987). *Non-Relative Virtues: An Aristotelian Approach*.

[19] John Newton, Ph.D., *Complete Conduct Principles for the 21st Century* (2000). ISBN 0-9673705-7-4.

[20] Ancient Ethical Theory, *Stanford Encyclopedia of Philosophy*.

[21] Ivanhoe, P.J.; Van Norden, Bryan William (2005). *Readings in classical Chinese philosophy*. Hackett Publishing. p. 60. ISBN 978-0-87220-780-6. he advocated a form of state consequentialism, which sought to maximize three basic goods: the wealth, order, and population of the state

[22] Fraser, Chris, "Mohism", *The Stanford Encyclopedia of Philosophy*, Edward N. Zalta.

[23] Loewe, Michael; Shaughnessy, Edward L. (1999). *The Cambridge History of Ancient China*. Cambridge University Press. p. 761. ISBN 978-0-521-47030-8.

[24] Van Norden, Bryan W. (2011). *Introduction to Classical Chinese Philosophy*. Hackett Publishing. p. 52. ISBN 978-1-60384-468-0.

[25] Jay L. Garfield; William Edelglass (June 9, 2011). *The Oxford Handbook of World Philosophy*. Oxford University Press. p. 62. ISBN 978-0-19-532899-8. The goods that serve as criteria of morality are collective or public, in contrast, for instance, to individual happiness or well-being

[26] Anscombe, G. E. M. (1958). "Modern Moral Philosophy". *Philosophy*. 1958. **33** (124): 1–19. doi:10.1017/S0031819100037943.

[27] Mackie, J. L. (1990) [1977]. *Ethics: Inventing Right and Wrong*. London: Penguin. ISBN 0-14-013558-8.

[28] Baqgini, Julian; Fosl, Peter S. (2007). *The Ethics Toolkit: A Compendium of Ethical Concepts and Methods*. Malden: Blackwell. pp. 57–58. ISBN 978-1-4051-3230-5.

[29] Bentham, Jeremy (2001). *The Works of Jeremy Bentham: Published under the Superintendence of His Executor, John Bowring. Volume 1*. Adamant Media Corporation. p. 18. ISBN 978-1-4021-6393-7.

[30] Mill, John Stuart, Utilitarianism (Project Gutenberg online edition)

[31] Mill, John Stuart (1998). *Utilitarianism*. Oxford: Oxford University Press. ISBN 978-0-19-875163-2.

[32] Department of Philosophy, CMU

[33] Stanford.edu

[34] Olson, Robert G. 1967. 'Deontological Ethics'. In Paul Edwards (ed.) *The Encyclopedia of Philosophy*. London: Collier Macmillan: 343.

[35] http://plato.stanford.edu/entries/ethics-virtue/

[36] Orend, Brian. 2000. *War and International Justice: A Kantian Perspective*. West Waterloo, Ontario: Wilfrid Laurier University Press: 19.

[37] Kelly, Eugene. 2006. *The Basics of Western Philosophy*. Greenwood Press: 160.

[38] Kant, Immanuel. 1780. 'Preface'. In *The Metaphysical Elements of Ethics*. Translated by Thomas Kingsmill Abbott

[39] Kant, Immanuel. 1785. 'First Section: Transition from the Common Rational Knowledge of Morals to the Philosophical', Groundwork of the Metaphysic of Morals.

[40] Lafollette, Hugh, ed. (February 2000). *The Blackwell Guide to Ethical Theory*. Blackwell Philosophy Guides (1 ed.). Wiley-Blackwell. ISBN 978-0-631-20119-9.

[41] Tong, Rosemarie; Williams, Nancy (May 4, 2009). "Feminist Ethics". *Stanford Encyclopedia of Philosophy*. The Metaphysics Research Lab. Retrieved January 6, 2017.

[42] Noddings, Nel: Caring: A Feminine Approach to Ethics and Moral Education, page 3-4. University of California Press, Berkeley, 1984.

[43] Noddings, Nel: Women and Evil, page 222. University of California Press, Berkeley, 1989.

[44] Roger T. Ames (April 30, 2011). *Confucian Role Ethics: A Vocabulary*. University of Hawai'i Press. ISBN 978-0-8248-3576-7.

[45] Chris Fraser; Dan Robins; Timothy O'Leary (May 1, 2011). *Ethics in Early China: An Anthology*. Hong Kong University Press. pp. 17–35. ISBN 978-988-8028-93-1.

[46] Sim, May, 2015, "Why Confucius' Ethics is a Virtue Ethics", in Besser-Jones and Slote (2015), pp. 63–76

[47] Wonsuk Chang; Leah Kalmanson (November 8, 2010). *Confucianism in Context: Classic Philosophy and Contemporary Issues, East Asia and Beyond*. SUNY Press. p. 68. ISBN 978-1-4384-3191-8.

[48] "Ethics: Origin and Development" by Pëtr Kropotkin

[49] "Anarchist morality", chapter VI, Pëtr Kropotkin

[50] "Astroethics". Archived from the original on October 23, 2013. Retrieved December 21, 2005.

[51] Freemont, P. F.; Kitney, R. I. (2012). *Synthetic Biology*. New Jersey: World Scientific. ISBN 978-1-84816-862-6.

[52] Mautner, Michael N. (2009). "Life-centered ethics, and the human future in space" (PDF). *Bioethics*. **23**: 433–440. PMID 19077128. doi:10.1111/j.1467-8519.2008.00688.x.

[53] Mautner, Michael N. (2000). *Seeding the Universe with Life: Securing Our Cosmological Future* (PDF). Washington D. C.: Legacy Books (www.amazon.com). ISBN 0-476-00330-X.

[54] Smith, A (1776/1952). *An Inquiry Into the Nature and Causes of the Wealth of Nations*. Chicago, Illinois: University of Chicago Press, p. 55.

[55] Berle, A. A., & Means, G. C. (1932). The Modern Corporation and Private Property. New Jersey: Transaction Publishers. In this book, Berle and Means observe, "Corporations have ceased to be merely legal devices through which the private business transactions of individuals may be carried on. Though still much used for this purpose, the corporate form has acquired a much larger significance. The corporation has, in fact, become both a method of property tenure and a means of organizing economic life. Grown to tremendous proportions, there may be said to have evolved a 'corporate system'—as there once was a feudal system—which has attracted to itself a combination of attributes and powers, and has attained a degree of prominence entitling it to be dealt with as a major social institution. ... We are examining this institution probably before it has attained its zenith. Spectacular as its rise has been, every indication seems to be that the system will move forward to proportions which stagger imagination today ... They [management] have placed the community in a position to demand that the modern corporation serve not only the owners ... but all society." p. 1.

[56] Jones, Parker & et al. 2005, p. 17

[57] Wallach, Wendell; Allen, Colin (November 2008). *Moral Machines: Teaching Robots Right from Wrong*. USA: Oxford University Press. ISBN 978-0-19-537404-9.

[58] Cook, Martin L.; Syse, Henrik (2010). "What Should We Mean by 'Military Ethics'?". *Journal of Military Ethics*. **9** (2). p. 122.

[59] Goffi, Emmanuel (2011). *Les Armée Françaises Face à la Morale* [*The French Army Facing Morale*] (in French). France: L'Harmattan. ISBN 978-2296542495.

[60] Thompson, Dennis F. "Political Ethics". *International Encyclopedia of Ethics*, ed. Hugh LaFollette (Blackwell Publishing, 2012).

[61] See, for example, work of Institute for Local Government, at www.ca-ilg.org/trust.

[62] Morton, Neil (October 2009). "Publication ethics". *Pediatric Anesthesia*. **19** (10): 1011–1013. doi:10.1111/j.1460-9592.2009.03086.x.

[63] Wager, E; Fiack, S; Graf, C; Robinson, A; Rowlands, I (31 March 2009). "Science journal editors' views on publication ethics: results of an international survey". *Journal of Medical Ethics*. **35**: 348–353. doi:10.1136/jme.2008.028324.

[64] Scollon, Ron (June 1999). "Plagiarism". *Journal of Linguistic Anthropology*. **9**: 188–190. JSTOR 43102462. doi:10.1525/jlin.1999.9.1-2.188.

[65] Wager, Elizabeth; Williams, Peter (September 2011). "Why and how do journals retract articles? An analysis of Medline retractions 1988—2008". *Journal of Medical Ethics*. **37** (9): 567–570. JSTOR 23034717. doi:10.1136/jme.2010.040964.

[66] Sanjeev, Handa (2008). "Plagiarism and publication ethics: Dos and don'ts". *Indian Journal of Dermatology Venereology and Leprology*. **74** (4): 301–303.

[67] Sigelman, Lee (2000). "Publication Bias Reconsidered". *Political Analysis*. **8** (2): 201–210. JSTOR 25791607. doi:10.1093/oxfordjournals.pan.a029813.

[68] Peters, Jamie L.; Sutton, Alex J.; Jones, David R.; Abrams, Keith R.; Rushton, Lesley; Moreno, Santiago G. (July 2010). "Assessing publication bias in meta-analysis in the presence of between-study heterogeneity". *Journal of the Royal Statistical Society. Series A (Statistics in Society)*. **173** (3): 575–591. doi:10.1111/j.1467-985x.2009.00629.x.

[69] Smith, Richard (July 26, 1997). "Misconduct in Research: Editors Respond: The Committee on Publication Ethics (COPE) Is Formed". *British Medical Journal*. **315** (7102): 201–202. JSTOR 25175246. doi:10.1136/bmj.315.7102.201.

[70] Carol GILLIGAN (June 30, 2009). *IN A DIFFERENT VOICE*. Harvard University Press. ISBN 978-0-674-03761-8.

[71] Ellis, C. (2007) Telling secrets, revealing lives: Relational ethics in research with intimate others. *Qualitative Inquiry*, 13, 3-29.

[72] Ellis, C. (1986). *Fisher folk. Two communities on Chesapeake Bay*. Lexington: University Press of Kentucky.

[73] Ellis, C. (1995).*Final negotiations: A story of love, loss, and chronic illness*. Philadelphia: Temple University Press.

[74] See, for example, Lapsley (2006) and "moral psychology" (2007).

[75] See, for example, Doris & Stich (2008) and Wallace (2007). Wallace writes: "Moral psychology is the study of morality in its psychological dimensions" (p. 86).

[76] See Doris & Stich (2008), §1.

[77] Doris Schroeder. "Evolutionary Ethics". Archived from the original on October 7, 2013. Retrieved January 5, 2010.

5.9 References

- Hoy, D. (2005). *Critical Resistance from Poststructuralism to Postcritique*. Massachusetts Institute of Technology, Cambridge, Massachusetts.

- Lyon, D. (1999). *Postmodernity* (2nd ed.). Open University Press, Buckingham.

- Singer, P. (2000). *Writings on an Ethical Life*. Harper Collins Publishers, London.

5.10 Further reading

- Aristotle, *Nicomachean Ethics*

- The London Philosophy Study Guide offers many suggestions on what to read, depending on the student's familiarity with the subject: Ethics

- *Encyclopedia of Ethics*. Lawrence C. Becker and Charlotte B. Becker, editors. Second edition in three volumes. New York: Routledge, 2002. A scholarly encyclopedia with over 500 signed, peer-reviewed articles, mostly on topics and figures of, or of special interest in, Western philosophy.

- Azurmendi, J. 1998: "The violence and the search for new values" in *Euskal Herria krisian*, (Elkar, 1999), pp. 11–116. ISBN 84-8331-572-6

- Blackburn, S. (2001). *Being good: A short introduction to ethics*. Oxford: Oxford University Press.

- De Finance, Joseph, *An Ethical Inquiry*, Rome, Editrice Pontificia Università Gregoriana, 1991.

- De La Torre, Miguel A., "Doing Christian Ethics from the Margins", Orbis Books, 2004.

- Derrida, J. 1995, *The Gift of Death*, translated by David Wills, University of Chicago Press, Chicago.

- Fagothey, Austin, *Right and Reason*, Tan Books & Publishers, Rockford, Illinois, 2000.

- Levinas, E. 1969, *Totality and infinity, an essay on exteriority*, translated by Alphonso Lingis, Duquesne University Press, Pittsburgh.

- Perle, Stephen (March 11, 2004). "Morality and Ethics: An Introduction". Retrieved February 13, 2007., Butchvarov, Panayot. Skepticism in Ethics (1989).

- Jadranka Skorin-Kapov, *The Intertwining of Aesthetics and Ethics: Exceeding of Expectations, Ecstasy, Sublimity*. Lexington Books, 2016. ISBN 978-1-4985-2456-8

- Solomon, R.C., *Morality and the Good Life: An Introduction to Ethics Through Classical Sources*, New York: McGraw-Hill Book Company, 1984.

- Vendemiati, Aldo, *In the First Person, An Outline of General Ethics*, Rome, Urbaniana University Press, 2004.

- John Paul II, Encyclical Letter *Veritatis Splendor*, August 6, 1993.

- D'Urance, Michel, *Jalons pour une éthique rebelle*, Alétheia, Paris, 2005.

- John Newton, Ph.D. *Complete Conduct Principles for the 21st Century*, 2000. ISBN 0-9673705-7-4.

- Guy Cools & Pascal Gielen, The Ethics of Art. Valiz: Amsterdam, 2014.

- Lafollette, Hugh [ed.]: *Ethics in Practice: An Anthology*. Wiley Blackwell, 4th edition, Oxford 2014. ISBN 978-0470671832

- An entire issue of *Pacific Island Studies* devoted to studying "Constructing Moral Communities" in Pacific islands, 2002, vol. 25: Link

- Paul R. Ehrlich (May 2016), Conference on population, environment, ethics: where we stand now (video, 93 min), *University of Lausanne*

5.11 External links

- Meta-Ethics at PhilPapers

- Normative Ethics at PhilPapers

- Applied Ethics at PhilPapers

- Ethics at the Indiana Philosophy Ontology Project

- "Ethics". *Internet Encyclopedia of Philosophy*.

- An Introduction to Ethics by Paul Newall, aimed at beginners.

- *Ethics, 2d ed., 1973. by William Frankena*

- Ethics Bites, Open University podcast series podcast exploring ethical dilemmas in everyday life.

- National Reference Center for Bioethics Literature World's largest library for ethical issues in medicine and biomedical research

- Ethics entry in Encyclopædia Britannica by Peter Singer

- The Philosophy of Ethics on Philosophy Archive

- Kirby Laing Institute for Christian Ethics Resources, events, and research on a range of ethical subjects from a Christian perspective.

- Basic principle of ethics summary talk

- International Association for Geoethics (IAGETH)

- International Association for Promoting Geoethics (IAPG)

- Markkula Center for Applied Ethics at Santa Clara University Resources for analyzing real-world ethical issues and tools to address them.

Chapter 6

Social justice

For the early-20th-century periodical, see Social Justice (periodical).

Social justice is the fair and just relation between the

College students live in tents for a week to call attention to perceived social injustice

individual and society. This is measured by the explicit and tacit terms for the distribution of wealth, opportunities for personal activity and social privileges. In Western as well as in older Asian cultures, the concept of social justice has often referred to the process of ensuring that individuals fulfill their societal roles and receive what was their due from society.[1][2][3] In the current global grassroots movements for social justice, the emphasis has been on the breaking of barriers for social mobility, the creation of safety nets and economic justice.[4][5][6][7][8]

Social justice assigns rights and duties in the institutions of society, which enables people to receive the basic benefits and burdens of cooperation. The relevant institutions often include taxation, social insurance, public health, public school, public services, labour law and regulation of markets, to ensure fair distribution of wealth, and equal opportunity.[9]

Interpretations that relate justice to a reciprocal relationship to society are mediated by differences in cultural traditions, some of which emphasize the individual responsibility toward society and others the equilibrium between access to power and its responsible use.[10] Hence, social justice is invoked today while reinterpreting historical figures such as Bartolomé de las Casas, in philosophical debates about differences among human beings, in efforts for gender, racial and social equality, for advocating justice for migrants, prisoners, the environment, and the physically and mentally disabled.[11][12][13][14]

While the concept of social justice can be traced through the theology of Augustine of Hippo and the philosophy of Thomas Paine, the term "social justice" became used explicitly from the 1840s. A Jesuit priest named Luigi Taparelli is typically credited with coining the term, and it spread during the revolutions of 1848 with the work of Antonio Rosmini-Serbati.[2][15][16] In the late industrial revolution, progressive American legal scholars began to use the term more, particularly Louis Brandeis and Roscoe Pound. From the early 20th century it was also embedded in international law and institutions; the preamble to establish the International Labour Organization recalled that "universal and lasting peace can be established only if it is based upon social justice." In the later 20th century, social justice was made central to the philosophy of the social contract, primarily by John Rawls in *A Theory of Justice* (1971). In 1993, the Vienna Declaration and Programme of Action treats social justice as a purpose of human rights education.[17][18]

6.1 History

Main articles: Social contract, Justice, Corrective justice, and Distributive justice

The different concepts of justice, as discussed in ancient Western philosophy, were typically centered upon the community.

> Plato wrote in *The Republic* that it would be an ideal state that "every member of the community must be assigned to the class for which he finds himself best fitted."[19] In an article for J.N.V University, author D.R. Bhandari says, "Justice is, for Plato, at once a part of human virtue and the bond, which joins man together in society. It is the identical quality that makes

Plato

Aristotle

good and social. Justice is an order and duty of the parts of the soul, it is to the soul as health is to the body. Plato says that justice is not mere strength, but it is a harmonious strength. Justice is not the right of the stronger but the effective harmony of the whole. All moral conceptions revolve about the good of the whole-individual as well as social".[20]

Aristotle believed rights existed only between free people, and the law should take "account in the first instance of relations of inequality in which individuals are treated in proportion to their worth and only secondarily of relations of equality." Reflecting this time when slavery and subjugation of women was typical, ancient views of justice tended to reflect the rigid class systems that still prevailed. On the other hand, for the privileged groups, strong concepts of fairness and the community existed. Distributive justice was said by Aristotle to require that people were distributed goods and assets according to their merit.[21]

Socrates (through Plato's dialogue *Crito*) is attributed with developing the idea of a social contract, whereby people ought to follow the rules of a society, and accept its burdens because they have accepted its benefits.[22] During the Middle Ages, religious scholars particularly, such as Thomas Aquinas continued discussion of justice in various ways, but ultimately connected being a good citizen to the purpose of serving God.

After the Renaissance and Reformation, the modern concept of social justice, as developing human potential, began to emerge through the work of a series of authors. Baruch Spinoza in *On the Improvement of the Understanding* (1677) contended that the one true aim of life should be to acquire "a human character much more stable than [one's] own", and to achieve this "pitch of perfection... The chief good is that he should arrive, together with other individuals if possible, at the possession of the aforesaid character."[23] During the enlightenment and responding to the French and American Revolutions, Thomas Paine similarly wrote in *The Rights of Man* (1792) society should give "genius a fair and universal chance" and so "the construction of government ought to be such as to bring forward... all that extent of capacity which never fails to appear in revolutions."[24]

The first modern usage of the specific term "social justice" is typically attributed to Catholic thinkers from the 1840s, including the Jesuit Luigi Taparelli in *Civiltà Cattolica*, based on the work of St. Thomas Aquinas. He argued that rival capitalist and socialist theories, based on subjective Cartesian thinking, undermined the unity of society present in Thomistic metaphysics as neither were sufficiently concerned with moral philosophy. Writing in 1861, the influential British philosopher and economist, John Stu-

Socrates

"Social justice" was coined by Jesuit priest Luigi Taparelli in the 1840s.

art Mill stated in *Utilitarianism* his view that "Society should treat all equally well who have deserved equally well of it, that is, who have deserved equally well absolutely. This is the highest abstract standard of social and distributive justice; towards which all institutions, and the efforts of all virtuous citizens, should be made in the utmost degree to converge."[25]

In the later 19th and early 20th century, social justice became an important theme in American political and legal philosophy, particularly in the work of John Dewey, Roscoe Pound and Louis Brandeis. One of the prime concerns was the *Lochner era* decisions of the US Supreme Court to strike down legislation passed by state governments and the Federal government for social and economic improvement, such as the eight-hour day or the right to join a trade union. After the First World War, the founding document of the International Labour Organization took up the same terminology in its preamble, stating that "peace can be established only if it is based on social justice". From this point, the discussion of social justice entered into mainstream legal and academic discourse. In the late 20th century, a number of liberal and conservative thinkers, notably Friedrich von Hayek rejected the concept by stating that it did not mean anything, or meant too many things.[26] However the concept remained highly influential, particularly with its promotion by philosophers such as John Rawls.

6.2 Contemporary theory

6.2.1 Philosophical perspectives

Cosmic values

Hunter Lewis' work promoting natural healthcare and sustainable economies advocates for conservation as a key premise in social justice. His manifesto on sustainability ties the continued thriving of human life to real conditions, the environment supporting that life, and associates injustice with the detrimental effects of unintended consequences of human actions. Quoting classical Greek thinkers like Epicurus on the good of pursuing happiness, Hunter also cites ornithologist, naturalist, and philosopher

Alexander Skutch in his book Moral Foundations:

> The common feature which unites the activities most consistently forbidden by the moral codes of civilized peoples is that by their very nature they cannot be both habitual and enduring, because they tend to destroy the conditions which make them possible.[27]

Pope Benedict XVI cites Teilhard de Chardin in a vision of the cosmos as a 'living host'[28] embracing an understanding of ecology that includes humanity's relationship to others, that pollution affects not just the natural world but interpersonal relations as well. Cosmic harmony, justice and peace are closely interrelated:

> If you want to cultivate peace, protect creation.[29]

John Rawls

Main article: John Rawls

Political philosopher John Rawls draws on the utilitarian insights of Bentham and Mill, the social contract ideas of John Locke, and the categorical imperative ideas of Kant. His first statement of principle was made in *A Theory of Justice* where he proposed that, "Each person possesses an inviolability founded on justice that even the welfare of society as a whole cannot override. For this reason justice denies that the loss of freedom for some is made right by a greater good shared by others."[30] A deontological proposition that echoes Kant in framing the moral good of justice in absolutist terms. His views are definitively restated in *Political Liberalism* where society is seen "as a fair system of co-operation over time, from one generation to the next".[31]

All societies have a basic structure of social, economic, and political institutions, both formal and informal. In testing how well these elements fit and work together, Rawls based a key test of legitimacy on the theories of social contract. To determine whether any particular system of collectively enforced social arrangements is legitimate, he argued that one must look for agreement by the people who are subject to it, but not necessarily to an objective notion of justice based on coherent ideological grounding. Obviously, not every citizen can be asked to participate in a poll to determine his or her consent to every proposal in which some degree of coercion is involved, so one has to assume that all citizens are reasonable. Rawls constructed an argument for a two-stage process to determine a citizen's hypothetical agreement:

- The citizen agrees to be represented by X for certain purposes, and, to that extent, X holds these powers as a trustee for the citizen.
- X agrees that enforcement in a particular social context is legitimate. The citizen, therefore, is bound by this decision because it is the function of the trustee to represent the citizen in this way.

This applies to one person who represents a small group (e.g., the organiser of a social event setting a dress code) as equally as it does to national governments, which are ultimate trustees, holding representative powers for the benefit of all citizens within their territorial boundaries. Governments that fail to provide for welfare of their citizens according to the principles of justice are not legitimate. To emphasise the general principle that justice should rise from the people and not be dictated by the law-making powers of governments, Rawls asserted that, "There is ... a general presumption against imposing legal and other restrictions on conduct without sufficient reason. But this presumption creates no special priority for any particular liberty."[32] This is support for an unranked set of liberties that reasonable citizens in all states should respect and uphold — to some extent, the list proposed by Rawls matches the normative human rights that have international recognition and direct enforcement in some nation states where the citizens need encouragement to act in a way that fixes a greater degree of equality of outcome. According to Rawls, the basic liberties that every good society should guarantee are,

- Freedom of thought;
- Liberty of conscience as it affects social relationships on the grounds of religion, philosophy, and morality;
- Political liberties (e.g., representative democratic institutions, freedom of speech and the press, and freedom of assembly);
- Freedom of association;
- Freedoms necessary for the liberty and integrity of the person (namely: freedom from slavery, freedom of movement and a reasonable degree of freedom to choose one's occupation); and
- Rights and liberties covered by the rule of law.

Thomas Pogge

Thomas Pogge's arguments pertain to a standard of social justice that creates human rights deficits. He assigns responsibility to those who actively cooperate in designing or imposing the social institution, that the order is foreseeable as harming the global poor and is reasonably avoidable. Pogge argues that social institutions have a negative duty to not harm the poor.[33][34]

Pogge speaks of "institutional cosmopolitanism" and assigns responsibility to institutional schemes[35] for deficits of human rights. An example given is slavery and third parties. A third party should not recognize or enforce slavery. The institutional order should be held responsible only for deprivations of human rights that it establishes or authorizes. The current institutional design, he says, systematically harms developing economies by enabling corporate tax evasion,[36] illicit financial flows, corruption, trafficking of people and weapons. Joshua Cohen disputes his claims based on the fact that some poor countries have done well with the current institutional design.[37] Elizabeth Kahn argues that some of these responsibilities should apply globally.[38]

United Nations

The United Nations' 2006 document *Social Justice in an Open World: The Role of the United Nations*, states that "Social justice may be broadly understood as the fair and compassionate distribution of the fruits of economic growth..."[39]:16

The term "social justice" was seen by the U.N. "as a substitute for the protection of human rights [and] first appeared in United Nations texts during the second half of the 1960s. At the initiative of the Soviet Union, and with the support of developing countries, the term was used in the Declaration on Social Progress and Development, adopted in 1969."[39]:52

The same document reports, "From the comprehensive global perspective shaped by the United Nations Charter and the Universal Declaration of Human Rights, neglect of the pursuit of social justice in all its dimensions translates into de facto acceptance of a future marred by violence, repression and chaos."[39]:6 The report concludes, "Social justice is not possible without strong and coherent redistributive policies conceived and implemented by public agencies."[39]:16

The same UN document offers a concise history: "[T]he notion of social justice is relatively new. None of history's great philosophers—not Plato or Aristotle, or Confucius or Averroes, or even Rousseau or Kant—saw the need to consider justice or the redress of injustices from a social perspective. The concept first surfaced in Western thought and political language in the wake of the industrial revolution and the parallel development of the socialist doctrine. It emerged as an expression of protest against what was perceived as the capitalist exploitation of labour and as a focal point for the development of measures to improve the human condition. It was born as a revolutionary slogan embodying the ideals of progress and fraternity. Following the revolutions that shook Europe in the mid-1800s, social justice became a rallying cry for progressive thinkers and political activists.... By the mid-twentieth century, the concept of social justice had become central to the ideologies and programmes of virtually all the leftist and centrist political parties around the world..."[39]:11–12

6.3 Religious perspectives

6.3.1 Hinduism

The present-day Jāti hierarchy is undergoing changes for a variety of reasons including 'social justice', which is a politically popular stance in democratic India. Institutionalized affirmative action has promoted this. The disparity and wide inequalities in social behaviour of the jātis – exclusive, endogamous communities centred on traditional occupations – has led to various reform movements in Hinduism. While legally outlawed, the caste system remains strong in practice.[40]

6.3.2 Islam

The Quran contains numerous references to elements of social justice. For example, one of Islam's Five Pillars is Zakāt, or alms-giving. Charity and assistance to the poor – concepts central to social justice – are and have historically been important parts of the Islamic faith.

In Muslim history, Islamic governance has often been associated with social justice. Establishment of social justice was one of the motivating factors of the Abbasid revolt against the Umayyads.[41] The Shi'a believe that the return of the *Mahdi* will herald in "the messianic age of justice" and the Mahdi along with the Isa (Jesus) will end plunder, torture, oppression and discrimination.[42]

For the Muslim Brotherhood the implementation of social justice would require the rejection of consumerism and communism. The Brotherhood strongly affirmed the right to private property as well as differences in personal wealth due to factors such as hard work. However, the Brotherhood held Muslims had an obligation to assist those Muslims in need. It held that *zakat* (alms-giving) was not voluntary charity, but rather the poor had the right to assistance from the more fortunate.[43] Most Islamic governments therefore enforce the *zakat* through taxes.

Though monetary donations are the most practiced way of zakat, Islam is deeply rooted in the tenets of volunteerism and social activism. Areas of one's communities which require assistance and beneficiaries must be a Muslim's foci if need be, rather than strictly her or his personal or superficial wants. For example, the ecological well-being of the planet (i.e.: animal rights, global warming, natural resources degradation); locally, nationally, globally, is a campaign to which every Muslim must adhere. Many Muslims practice this today by ensuring that they produce minimal waste, give to charity what they no longer need, and spend time in prayer and meditation upon the bounties of nature so as to more mindfully approach all that is provided by nature, and ultimately, Allah.[44] Other areas of society in need may be the safety and security of minority populations, i.e.: women or persons of color, children, the elderly, the developmentally or physically disabled, animals, et al.[44]

6.3.3 Judaism

Main article: Tikkun olam

In *To Heal a Fractured World: The Ethics of Responsibility*, Rabbi Jonathan Sacks states that social justice has a central place in Judaism. One of Judaism's most distinctive and challenging ideas is its ethics of responsibility reflected in the concepts of simcha ("gladness" or "joy"), tzedakah ("the religious obligation to perform charity and philanthropic acts"), chesed ("deeds of kindness"), and tikkun olam ("repairing the world").

6.3.4 Christianity

Methodism

From its founding, Methodism was a Christian social justice movement. Under John Wesley's direction, Methodists became leaders in many social justice issues of the day, including the prison reform and abolition movements. Wesley himself was among the first to preach for slaves rights attracting significant opposition.[45][46][47]

Today, social justice plays a major role in the United Methodist Church. The *Book of Discipline of the United Methodist Church* says, "We hold governments responsible for the protection of the rights of the people to free and fair elections and to the freedoms of speech, religion, assembly, communications media, and petition for redress of grievances without fear of reprisal; to the right to privacy; and to the guarantee of the rights to adequate food, clothing, shelter, education, and health care."[48] The United Methodist Church also teaches population control as part of its doctrine.[49]

Catholicism

Main article: Catholic social teaching

Catholic social teaching consists of those aspects of Roman Catholic doctrine which relate to matters dealing with the respect of the individual human life. A distinctive feature of Catholic social doctrine is its concern for the poorest and most vulnerable members of society. Two of the seven key areas[50] of "Catholic social teaching" are pertinent to social justice:

- Life and dignity of the human person: The foundational principle of all "Catholic Social Teaching" is the sanctity of all human life and the inherent dignity of every human person, from conception to natural death. Human life must be valued above all material possessions.

- Preferential option for the poor and vulnerable: Catholics believe Jesus taught that on the Day of Judgement God will ask what each person did to help the poor and needy: "Amen, I say to you, whatever you did for one of these least brothers of mine, you did for me."[51] The Catholic Church believes that through words, prayers and deeds one must show solidarity with, and compassion for, the poor. The moral test of any society is "how it treats its most vulnerable members. The poor have the most urgent moral claim on the conscience of the nation. People are called to look at public policy decisions in terms of how they affect the poor."[52]

Even before it was propounded in the Catholic social doctrine, social justice appeared regularly in the history of the Catholic Church:

- Pope Leo XIII, who studied under Taparelli, published in 1891 the encyclical *Rerum novarum* (On the Condition of the Working Classes; lit. "On new things"), rejecting both socialism and capitalism, while defending labor unions and private property. He stated that society should be based on cooperation and not class conflict and competition. In this document, Leo set out the Catholic Church's response to the social instability and labor conflict that had arisen in the wake of industrialization and had led to the rise of socialism. The Pope advocated that the role of the State was to promote social justice through the protection of rights, while the Church must speak out on social issues in order to teach correct social principles and ensure class harmony.

- The encyclical *Quadragesimo anno* (On Reconstruction of the Social Order, literally "in the fortieth year") of 1931 by Pope Pius XI, encourages a living wage,[53] subsidiarity, and advocates that social justice is a personal virtue as well as an attribute of the social order, saying that society can be just only if individuals and institutions are just.

- Pope John Paul II added much to the corpus of the Catholic social teaching, penning three encyclicals which focus on issues such as economics, politics, geopolitical situations, ownership of the means of production, private property and the "social mortgage", and private property. The encyclicals *Laborem exercens*, *Sollicitudo rei socialis*, and *Centesimus annus* are just a small portion of his overall contribution to Catholic social justice. Pope John Paul II was a strong advocate of justice and human rights, and spoke forcefully for the poor. He addresses issues such as the problems that technology can present should it be misused, and admits a fear that the "progress" of the world is not true progress at all, if it should denigrate the value of the human person. He argued in *Centesimus annus* that private property, markets, and honest labor were the keys to alleviating the miseries of the poor and to enabling a life that can express the fullness of the human person.

- Pope Benedict XVI's encyclical *Deus caritas est* ("God is Love") of 2006 claims that justice is the defining concern of the state and the central concern of politics, and not of the church, which has charity as its central social concern. It said that the laity has the specific responsibility of pursuing social justice in civil society and that the church's active role in social justice should be to inform the debate, using reason and natural law, and also by providing moral and spiritual formation for those involved in politics.

- The official Catholic doctrine on **social justice** can be found in the book *Compendium of the Social Doctrine of the Church*, published in 2004 and updated in 2006, by the Pontifical Council *Iustitia et Pax*.

The Catechism (§1928–1948) contain more detail of the Church's view of social justice.[54]

6.4 Criticism

Many authors criticize the idea that there exists an objective standard of social justice. Moral relativists deny that there is any kind of objective standard for justice in general. Non-cognitivists, moral skeptics, moral nihilists, and most logical positivists deny the epistemic possibility of objective notions of justice. Political realists believe that any ideal of social justice is ultimately a mere justification for the status quo.

Many other people accept some of the basic principles of social justice, such as the idea that all human beings have a basic level of value, but disagree with the elaborate conclusions that may or may not follow from this. One example is the statement by H. G. Wells that all people are "equally entitled to the respect of their fellowmen."[55]

On the other hand, some scholars reject the very idea of social justice as meaningless, religious, self-contradictory, and ideological, believing that to realize any degree of social justice is unfeasible, and that the attempt to do so must destroy all liberty. Perhaps the most complete rejection of the concept of social justice comes from Friedrich Hayek of the Austrian School of economics:

> There can be no test by which we can discover what is 'socially unjust' because there is no subject by which such an injustice can be committed, and there are no rules of individual conduct the observance of which in the market order would secure to the individuals and groups the position which as such (as distinguished from the procedure by which it is determined) would appear just to us. [Social justice] does not belong to the category of error but to that of nonsense, like the term 'a moral stone'.[56]

Ben O'Neill of the University of New South Wales argues that, for proponents of "social justice":[57]

> the notion of "rights" is a mere term of entitlement, indicative of a claim for any possible desirable good, no matter how important or trivial, abstract or tangible, recent or ancient. It is merely an assertion of desire, and a declaration of intention to use the language of rights to acquire said desire.
>
> In fact, since the program of social justice inevitably involves claims for government provision of goods, paid for through the efforts of others, the term actually refers to an intention to use *force* to acquire one's desires. Not to earn desirable goods by rational thought and action, production and voluntary exchange, but to go in there and forcibly take goods from those who can supply them!

Janusz Korwin-Mikke states, "Either 'social justice' has the same meaning as 'justice' – or not. If so – why use the additional word 'social?' We lose time, we destroy trees to obtain paper necessary to print this word. If not, if 'social justice' means something different from 'justice' – then 'something different from justice' is by definition 'injustice.'"[58]

Sociologist Carl L. Bankston has argued that a secular, leftist view of social justice entails viewing the redistribution of goods and resources as based on the rights of disadvantaged categories of people, rather than on compassion or national interest. Bankston maintains that this secular version of social justice became widely accepted due to the rise of demand-side economics and to the moral influence of the civil rights movement.[59]

6.5 Social justice movements

Social justice is also a concept that is used to describe the movement towards a socially just world, e.g., the Global Justice Movement. In this context, social justice is based on the concepts of human rights and equality, and can be defined as *"the way in which human rights are manifested in the everyday lives of people at every level of society"*.[60]

A number of movements are working to achieve social justice in society. These movements are working towards the realization of a world where all members of a society, regardless of background or procedural justice, have basic human rights and equal access to the benefits of their society.[61]

6.5.1 Liberation theology

Main article: Liberation theology

Liberation theology[62] is a movement in Christian theology which conveys the teachings of Jesus Christ in terms of a liberation from unjust economic, political, or social conditions. It has been described by proponents as "an interpretation of Christian faith through the poor's suffering, their struggle and hope, and a critique of society and the Catholic faith and Christianity through the eyes of the poor",[63] and by detractors as Christianity perverted by Marxism and Communism.[64]

Although liberation theology has grown into an international and inter-denominational movement, it began as a movement within the Catholic Church in Latin America in the 1950s–1960s. It arose principally as a moral reaction to the poverty caused by social injustice in that region.[65] It achieved prominence in the 1970s and 1980s. The term was coined by the Peruvian priest, Gustavo Gutiérrez, who wrote one of the movement's most famous books, *A Theology of Liberation* (1971). According to Sarah Kleeb, "Marx would surely take issue," she writes, "with the appropriation of his works in a religious context...there is no way to reconcile Marx's views of religion with those of Gutierrez, they are simply incompatible. Despite this, in terms of their understanding of the necessity of a just and righteous world, and the nearly inevitable obstructions along such a path, the two have much in common; and, particularly in the first edition of [A Theology of Liberation], the use of Marxian theory is quite evident."[66]

Other noted exponents are Leonardo Boff of Brazil, Carlos Mugica of Argentina, Jon Sobrino of El Salvador, and Juan Luis Segundo of Uruguay.[67][68]

6.5.2 Health care

Social justice has more recently made its way into the field of bioethics. Discussion involves topics such as affordable access to health care, especially for low income households and families. The discussion also raises questions such as whether society should bear healthcare costs for low income families, and whether the global marketplace is the best way to distribute healthcare. Ruth Faden of the Johns Hopkins Berman Institute of Bioethics and Madison Powers of Georgetown University focus their analysis of social justice on which inequalities matter the most. They develop a social justice theory that answers some of these questions in concrete settings.

Social injustices occur when there is a preventable difference in health states among a population of people. These social injustices take the form of health inequities when negative health states such as malnourishment, and infectious diseases are more prevalent in impoverished nations.[69] These negative health states can often be prevented by providing social and economic structures such as primary healthcare which ensures the general population has equal access to health care services regardless of income level, gender, education or any other stratifying factors. Integrating social justice with health inherently reflects the social determinants of health model without discounting the role of the bio-medical model.[70]

6.5.3 Human rights education

Main article: Human rights education

The Vienna Declaration and Programme of Action affirm that "Human rights education should include peace, democracy, development and social justice, as set forth in international and regional human rights instruments, in order to achieve common understanding and awareness with a view to strengthening universal commitment to human rights."[71]

6.6 See also

- "Beyond Vietnam: A Time to Break Silence", an anti-Vietnam war and pro-social justice speech delivered by Martin Luther King, Jr. in 1967
- Counterculture of the 1960s
- Climate justice
- Environmental justice
- Environmental racism
- Essentially contested concept
- Labour law and labour rights
- Left-wing politics
- Resource justice
- Right to education
- Right to health
- Right to housing
- Right to social security
- Social justice art
- Social justice warrior
- Social law
- Social work
- Solidarity
- World Day of Social Justice
- All pages beginning with "Social justice"
- All pages with a title containing *Social justice*

6.7 Notes

[1] Aristotle, *The Politics* (ca 350 BC)

[2] Clark, Mary T. (2015). *"Augustine on Justice," a Chapter in Augustine and Social Justice*. Lexington Books. pp. 3–10. ISBN 9781498509183.

[3] Banai, Ayelet; Ronzoni, Miriam; Schemmel, Christian (2011). *Social Justice, Global Dynamics : Theoretical and Empirical Perspectives*. Florence: Taylor and Francis. ISBN 9780203819296.

[4] Kitching, G. N. (2001). *Seeking Social Justice Through Globalization Escaping a Nationalist Perspective*. University Park, Pa: Pennsylvania State University Press. pp. 3–10. ISBN 0271023775.

[5] Hillman, Arye L. (2008). "Globalization and Social Justice". *The Singapore Economic Review*. **53** (2): 173–189.

[6] Agartan, Kaan (2014). "Globalization and the Question of Social Justice". *Sociology Compass*. **8** (6): 903–915. doi:10.1111/soc4.12162.

[7] El Khoury, Ann (2015). *Globalization Development and Social Justice : A propositional political approach*. Florence: Taylor and Francis. pp. 1–20. ISBN 9781317504801.

[8] Lawrence, Cecile & Natalie Churn (2012). *Movements in Time Revolution, Social Justice, and Times of Change*. Newcastle upon Tyne, UK:: Cambridge Scholars Pub. pp. xi–xv. ISBN 1443845523.

[9] John Rawls, *A Theory of Justice* (1971) 4, "the principles of social justice: they provide a way of assigning rights and duties in the basic institutions of society and they define the appropriate distribution of benefits and burdens of social co-operation."

[10] Zhang, Aiqing, Feifei Xia and Chengwei Li (2007). "THE ANTECEDENTS OF HELP GIVING IN CHINESE CULTURe: ATTRIBUTION, JUDGMENT OF RESPONSIBILITY, EXPECTATION CHANGE AND THE REACTION OF AFFECT". *Social Behavior and Personality*. **35** (1): 135–142.

[11] Jalata, Asafa (2013). "Indigenous Peoples and the Capitalist World System: Researching, Knowing, and Promoting Social Justice". *Sociology Mind*. **3** (2): 156–178.

[12] Smith, Justin E. H. (2015). *Nature, Human Nature, and Human Difference : Race in Early Modern Philosophy*. Princeton University Press. p. 17. ISBN 9781400866311.

[13] Trường, Thanh-Đạm (2013). *Migration, Gender and Social Justice: Perspectives on Human Insecurity*. Springer. pp. 3–26. ISBN 9783642280122.

[14] Teklu, Abebe Abay (2010). "We Cannot Clap with One Hand: Global Socio–Political Differences in Social Support for People with Visual Impairment". *International Journal of Ethiopian Studies*. **5** (1): 93–105.

[15] Paine, Thomas. *Agrarian Justice*.

[16] J. Zajda, S. Majhanovich, V. Rust, *Education and Social Justice*, 2006, ISBN 1-4020-4721-5

[17] The Preamble of ILO Constitution

[18] Vienna Declaration and Programme of Action, Part II, D.

[19] Plato, *The Republic* (ca 380BC)

[20] "20th WCP: Plato's Concept Of Justice: An Analysis".

6.7. NOTES

[21] *Nicomachean Ethics* V.3

[22] Plato, *Crito* (ca 380 BC)

[23] B Spinoza, *On the Improvement of the Understanding* (1677) para 13

[24] T Paine, *Rights of Man* (1792) 197

[25] JS Mill, *Utilitarianism* (1863)

[26] FA Hayek, *Law, Legislation and Liberty* (1973) vol II, ch 3

[27] Hunter Lewis (14 October 2009). "Sustainability, The Complete Concept, Environment, Healthcare, and Economy" (PDF). ChangeThis.

[28] John Allen Jr. (28 July 2009). "Ecology – The first stirring of an 'evolutionary leap' in late Jesuit's official standing?". National Catholic Reporter. Archived from the original on 24 August 2012.

[29] Sandro Magister (11 January 2010). "Benedict XVI to the Diplomats: Three Levers for Lifting Up the World". www.chiesa, Rome.

[30] John Rawls, A Theory of Justice (2005 reissue), Chapter 1, "Justice as Fairness" – 1. The Role of Justice, pp. 3–4

[31] John Rawls, *Political Liberalism* 15 (Columbia University Press 2003)

[32] John Rawls, Political Liberalism 291–92 (Columbia University Press 2003)

[33] James, Nickel. "Human Rights". *stanford.edu*. The Stanford Encyclopedia of Philosophy. Retrieved 10 February 2015.

[34] Pogge, Thomas Pogge. "World Poverty and Human Rights". *thomaspogge.com*.

[35] North, James. "The Resource Privilege". *thenation.com*. Retrieved 10 February 2015.

[36] Pogge, Thomas. "Human Rights and Just Taxation – Global Financial Transparency".

[37] Jaggar, edited by Alison M. (2010). *Thomas Pogge and His Critics*. (1. publ. ed.). Cambridge: Polity Press. ISBN 978-0-7456-4258-1.

[38] Kahn, Elizabeth (June–December 2012). "Global Economic Justice: A Structural Approach". *Public Reason*. **4** (1–2): 48–67.

[39] "Social Justice in an Open World: The Role of the United Nations", The International Forum for Social Development, Department of Economic and Social Affairs, Division for Social Policy and Development, ST/ESA/305" (PDF). New York: United Nations. 2006.

[40] Patil, Vijaykumar. "Caste system hindering the goal of social justice: Siddaramaiah".

[41] John L. Esposito (1998). *Islam and Politics*. Syracuse University Press. p. 17.

[42] John L. Esposito (1998). *Islam and Politics*. Syracuse University Press. p. 205.

[43] John L. Esposito (1998). *Islam and Politics*. Syracuse University Press. pp. 147–8.

[44] "The Eco Muslim".

[45] S. R. Valentine, John Bennet & the Origins of Methodism and the Evangelical revival in England, Scarecrow Press, Lanham, 1997.

[46] Carey, Brycchan. "John Wesley (1703–1791)." The British Abolitionists. Brycchan Carey, 11 July 2008. 5 October 2009. Brycchancarey.com

[47] Wesley John, "Thoughts Upon Slavery," John Wesley: Holiness of Heart and Life. Charles Yrigoyen, 1996. 5 October 2009. Gbgm-umc.org Archived 16 October 2014 at the Wayback Machine.

[48] The Book of Discipline of The United Methodist Church – 2012 ¶164 V, umc.org Archived 6 December 2013 at the Wayback Machine.

[49] The Book of Discipline of The United Methodist Church – 2008 ¶ 162 K, umc.org Archived 6 December 2013 at the Wayback Machine.

[50] "Seven Key Themes of Catholic Social Teaching". Web.archive.org. Archived from the original on 8 June 2007. Retrieved 29 March 2014.

[51] Matthew 25:40.

[52] Option for the Poor, Major themes from Catholic Social Teaching Archived 16 February 2006 at the Wayback Machine., Office for Social Justice, Archdiocese of St. Paul and Minneapolis.

[53] Popularised by John A. Ryan, although see Sidney Webb and Beatrice Webb, *Industrial Democracy* (1897)

[54] "Catechism of the Catholic Church – Social justice". Vatican.va. Retrieved 29 March 2014.

[55] "The Rights of Man", *Daily Herald*, London, February 1940

[56] Hayek, F.A. (1982). *Law, Legislation and Liberty, Vol. 2*. Routledge. p. 78.

[57] O'Neill, Ben (16 March 2011) The Injustice of Social Justice, *Mises Institute*

[58] Suresh Murugan (1 December 2013). *Social Problems And Social Legislation*. Social work department, PSGCAS. p. 22. GGKEY:KUBPEL1FG8Q.

[59] Social Justice: Cultural Origins of a Theory and a Perspective By Carl L. Bankston III, Independent Review vol. 15 no. 2, pp. 165–178, 2010

[60] Just Comment – Volume 3 Number 1, 2000

[61] Capeheart, Loretta; Milovanovic, Dragan. *Social Justice: Theories, Issues, and Movements.*

[62] In the mass media, 'Liberation Theology' can sometimes be used loosely, to refer to a wide variety of activist Christian thought. This article uses the term in the narrow sense outlined here.

[63] Berryman, Phillip, *Liberation Theology: essential facts about the revolutionary movement in Latin America and beyond*(1987)

[64] "[David] Horowitz first describes liberation theology as 'a form of Marxised Christianity,' which has validity despite the awkward phrasing, but then he calls it a form of 'Marxist-Leninist ideology,' which is simply not true for most liberation theology..." Robert Shaffer, "Acceptable Bounds of Academic Discourse Archived 4 September 2013 at the Wayback Machine.," Organization of American Historians Newsletter 35, November 2007. URL retrieved 12 July 2010.

[65] *Liberation Theology and Its Role in Latin America.* Elisabeth Erin Williams. Monitor: Journal of International Studies. The College of William and Mary.

[66] Sarah Kleeb, "Envisioning Emancipation: Karl Marx, Gustavo Gutierrez, and the Struggle of Liberation Theology"; Presented at the Annual Meeting of the Canadian Society for the Study of Religion (CSSR), Toronto, 2006. Retrieved 22 October 2012.

[67] Richard P. McBrien, *Catholicism* (Harper Collins, 1994), chapter IV.

[68] Gustavo Gutierrez, *A Theology of Liberation*, First (Spanish) edition published in Lima, Peru, 1971; first English edition published by Orbis Books (Maryknoll, New York), 1973.

[69] Farmer, Paul E., Bruce Nizeye, Sara Stulac, and Salmaan Keshavjee. 2006. Structural Violence and Clinical Medicine. PLoS Medicine, 1686–1691

[70] Cueto, Marcos. 2004. The ORIGINS of Primary Health Care and SELECTIVE Primary Health Care. Am J Public Health 94 (11):1868

[71] Vienna Declaration and Programme of Action, Part II, paragraph 80

6.8 References

6.8.1 Articles

- LD Brandeis, 'The Living Law' (1915–1916) 10 Illinois Law Review 461

- A Etzioni, 'The Fair Society, Uniting America: Restoring the Vital Center to American Democracy' in N Garfinkle and D Yankelovich (eds) (Yale University Press 2005) pp. 211–223

- Otto von Gierke, *The Social Role of Private Law* (2016) translated and introduced by E McGaughey, originally *Die soziale Aufgabe des Privatrechts*

- M Novak, 'Defining Social Justice' (2000) First Things

- B O'Neill, 'The Injustice of Social Justice' (Mises Institute)

- R Pound, 'Social Justice and Legal Justice' (1912) 75 Central Law Journal 455

- M Powers and R Faden, 'Inequalities in health, inequalities in health care: four generations of discussion about justice and cost-effectiveness analysis' (2000) 10(2) Kennedy Inst Ethics Journal 109–127

- M Powers and R Faden, 'Racial and Ethnic Disparities in Health Care: An Ethical Analysis of When and How They Matter,' in *Unequal Treatment: Confronting Racial and Ethnic Disparities in Health Care* (National Academy of Sciences, Institute of Medicine, 2002) 722–38

- United Nations, Department of Economic and Social Affairs, 'Social Justice in an Open World: The Role of the United Nations' (2006) ST/ESA/305

6.8.2 Books

- AB Atkinson, *Social Justice and Public Policy* (1982) previews

- Gad Barzilai, *Communities and Law: Politics and Cultures of Legal Identities* (University of Michigan Press) analysis of justice for non-ruling communities

- TN Carver, *Essays in Social Justice* (1915) Chapter links.

- C Quigley *The Evolution Of Civilizations: An Introduction to Historical Analysis* (1961) 2nd edition 1979

- P Corning, *The Fair Society: The Science of Human Nature and the Pursuit of Social Justice* (Chicago UP 2011)

- R Faden and M Powers, *Social Justice: The Moral Foundations of Public Health and Health Policy* (OUP 2006)

- J Franklin (ed), *Life to the Full: Rights and Social Justice in Australia* (Connor Court 2007)

- FA Hayek, *Law, Legislation and Liberty* (1973) vol II, ch 3

- G Kitching, *Seeking Social Justice through Globalization: Escaping a Nationalist Perspective* (2003)

- JS Mill, *Utilitarianism* (1863)

- John Rawls, *A Theory of Justice* (Harvard University Press 1971)

- John Rawls, *Political Liberalism* (Columbia University Press 1993)

- C Philomena, B Hoose and G Mannion (eds), *Social Justice: Theological and Practical Explorations* (2007)

- A Swift, *Political Philosophy* (3rd edn 2013) ch 1

- Michael J. Thompson, *The Limits of Liberalism: A Republican Theory of Social Justice* (International Journal of Ethics: vol. 7, no. 3 (2011)

6.9 External links

- Leading Social Justice Organizations in the United States

- Social Justice Month

- The Forward covers the emergent Jewish Social Justice Roundtable

- Social Justice Movements in the US

- Interfaith Social Assistance Reform Coalition, Ontario, Canada

- Social Justice Now – Global Social Justice News

- Social Justice Solutions – Social Justice News, Topics & Issues

- - Centre for Social Justice and Wellbeing in Education, UK

Chapter 7

Naturalism (philosophy)

This article is about the term that is used in philosophy. For other uses, see Naturalism (disambiguation).

In philosophy, **naturalism** is the "idea or belief that only natural (as opposed to supernatural or spiritual) laws and forces operate in the world."[1] Adherents of naturalism (i.e., naturalists) assert that natural laws are the rules that govern the structure and behavior of the natural universe, that the changing universe at every stage is a product of these laws.[2]

"Naturalism can intuitively be separated into an ontological and a methodological component."[3] "Ontological" refers to the philosophical study of the nature of reality. Some philosophers equate naturalism with materialism. For example, philosopher Paul Kurtz argues that nature is best accounted for by reference to material principles. These principles include mass, energy, and other physical and chemical properties accepted by the scientific community. Further, this sense of naturalism holds that spirits, deities, and ghosts are not real and that there is no "purpose" in nature. Such an absolute belief in naturalism is commonly referred to as *metaphysical naturalism*.[4]

Assuming naturalism in working methods as the current paradigm, without the unfounded consideration of naturalism as an absolute truth with philosophical entailment, is called *methodological naturalism*.[5] The subject matter here is a philosophy of acquiring knowledge based on an assumed paradigm.

With the exception of pantheists—who believe that Nature and God are one and the same thing—theists challenge the idea that nature contains all of reality. According to some theists, natural laws may be viewed as so-called secondary causes of God(s).

In the 20th century, Willard Van Orman Quine, George Santayana, and other philosophers argued that the success of naturalism in science meant that scientific methods should also be used in philosophy. Science and philosophy are said to form a continuum, according to this view.

7.1 Origins and history

The ideas and assumptions of philosophical naturalism were first seen in the works of the Ionian School pre-Socratic philosophers. One such was Thales, considered to be the father of science, as he was the first to give explanations of natural events without the use of supernatural causes. These early philosophers subscribed to principles of empirical investigation that strikingly anticipate naturalism.[6]

Naturalism, in classical Indian philosophies, was the foundation of two (Vaisheshika, Nyaya) of six orthodox schools and one (Carvaka) heterodox school of Hinduism.[7][8] The Vaisheshika school is traced to 2nd century BCE.[9][10]

The modern emphasis in methodological naturalism primarily originated in the ideas of medieval scholastic thinkers during the Renaissance of the 12th century:

> By the late Middle Ages the search for natural causes had come to typify the work of Christian natural philosophers. Although characteristically leaving the door open for the possibility of direct divine intervention, they frequently expressed contempt for contemporaries who invoked miracles rather than searching for natural explanations. The University of Paris cleric Jean Buridan (a. 1295-ca. 1358), described as "perhaps the most brilliant arts master of the Middle Ages," contrasted the philosopher's search for "appropriate natural causes" with the common folk's habit of attributing unusual astronomical phenomena to the supernatural. In the fourteenth century the natural philosopher Nicole Oresme (ca. 1320–82), who went on to become a Roman Catholic bishop, admonished that, in discussing various marvels of nature, "there is no reason to take recourse to the heavens, the last refuge of the weak, or demons, or to our glorious God as if He would produce these effects directly, more so than those effects whose causes we believe are well known to us."

Enthusiasm for the naturalistic study of nature picked up in the sixteenth and seventeenth centuries as more and more Christians turned their attention to discovering the so-called secondary causes that God employed in operating the world. The Italian Catholic Galileo Galilei (1564–1642), one of the foremost promoters of the new philosophy, insisted that nature "never violates the terms of the laws imposed upon her."[11]

During the Enlightenment, a number of philosophers including Francis Bacon and Voltaire outlined the philosophical justifications for removing appeal to supernatural forces from investigation of the natural world. Subsequent scientific revolutions would offer modes of explanation not inherently theistic for biology, geology, physics, and other natural sciences.

Pierre Simon de Laplace, when asked about the lack of mention of intervention by God in his work on celestial mechanics, is said to have replied, "I had no need of that hypothesis."[12]

According to Steven Schafersman, president of Texas Citizens for Science, an advocacy group opposing creationism in public schools,[13] the progressive adoption of methodological naturalism—and later of metaphysical naturalism—followed the advances of science and the increase of its explanatory power.[14] These advances also caused the diffusion of positions associated with metaphysical naturalism, such as existentialism.[15]

The current usage of the term naturalism "derives from debates in America in the first half of the last century. The self-proclaimed 'naturalists' from that period included John Dewey, Ernest Nagel, Sidney Hook and Roy Wood Sellars." For them nature is the only reality. There is no such thing as 'supernatural'. The scientific method is to be used to investigate all reality, including the human spirit: "So understood, 'naturalism' is not a particularly informative term... The great majority of contemporary philosophers would happily... reject 'supernatural' entities, and allow that science is a possible route (if not necessarily the only one) to important truths about the 'human spirit'."[16]

7.1.1 Etymology

The term "methodological naturalism" for this approach is much more recent. According to Ronald Numbers, it was coined in 1983 by Paul de Vries, a Wheaton College philosopher. De Vries distinguished between what he called "methodological naturalism," a disciplinary method that says nothing about God's existence, and "metaphysical naturalism," which "denies the existence of a transcendent God."[17] The term "methodological naturalism" had been used in 1937 by Edgar S. Brightman in an article in *The Philosophical Review* as a contrast to "naturalism" in general, but there the idea was not really developed to its more recent distinctions.[18]

7.2 Metaphysical naturalism

Main article: Metaphysical naturalism

Metaphysical naturalism, also called "ontological naturalism" and "philosophical naturalism", is a philosophical worldview and belief system that holds that there is nothing but natural elements, principles, and relations of the kind studied by the natural sciences, i.e., those required to understand our physical environment by mathematical modeling. Methodological naturalism, on the other hand, refers exclusively to the methodology of science, for which metaphysical naturalism provides only one possible ontological foundation.

Metaphysical naturalism holds that all properties related to consciousness and the mind are reducible to, or supervene upon, nature. Broadly, the corresponding theological perspective is religious naturalism or spiritual naturalism. More specifically, metaphysical naturalism rejects the supernatural concepts and explanations that are part of many religions.

7.3 Methodological naturalism

Further information: Alternatives to natural selection

Methodological naturalism does not concern itself with claims about what exists, but with methods of learning what nature is. It attempts to explain and test scientific endeavors, hypotheses, and events with reference to natural causes and events. This second sense of the term "naturalism" seeks to provide a framework within which to conduct the scientific study of the laws of nature. Methodological naturalism is a way of acquiring knowledge. It is a distinct system of thought concerned with a cognitive approach to reality, and is thus a philosophy of knowledge. Studies by sociologist Elaine Ecklund suggest that religious scientists in practice apply methodological naturalism. They report that their religious beliefs affect the way they think about the implications - often moral - of their work, but not the way they practice science.[19][20]

In a series of articles and books from 1996 onward, Robert T. Pennock wrote using the term "methodological naturalism" to clarify that the scientific method confines itself to natural explanations without assuming the existence or non-existence of the supernatural, and is not based on dogmatic metaphysical naturalism (as claimed by creationists and proponents of intelligent design, in particular by Phillip E. Johnson). Pennock's testimony as an expert witness[21] at the Kitzmiller v. Dover Area School District trial was cited by the Judge in his *Memorandum Opinion* concluding that "Methodological naturalism is a 'ground rule' of science today".[22]

> Expert testimony reveals that since the scientific revolution of the 16th and 17th centuries, science has been limited to the search for natural causes to explain natural phenomena.... While supernatural explanations may be important and have merit, they are not part of science." Methodological naturalism is thus "a paradigm of science." It is a "ground rule" that "requires scientists to seek explanations in the world around us based upon what we can observe, test, replicate, and verify.[23]

7.4 Views

7.4.1 Alvin Plantinga

Alvin Plantinga, Professor Emeritus of Philosophy at Notre Dame,[24] and a Christian, has become a well-known critic of naturalism.[25] He suggests, in his evolutionary argument against naturalism, that the probability that evolution has produced humans with reliable true beliefs, is low or inscrutable, unless their evolution was guided (for example, by God). According to David Kahan of the University of Glasgow, in order to understand how beliefs are warranted, a justification must be found in the context of supernatural theism, as in Plantinga's epistemology.[26][27][28] *(See also supernormal stimuli).*

Plantinga argues that together, naturalism and evolution provide an insurmountable "*defeater* for the belief that our cognitive faculties are reliable", i.e., a skeptical argument along the lines of Descartes' Evil demon or Brain in a vat.[29]

> Take *philosophical naturalism* to be the belief that there aren't any supernatural entities - no such person as God, for example, but also no other supernatural entities, and nothing at all like God. My claim was that naturalism and contemporary evolutionary theory are at serious odds with one another - and this despite the fact that the latter is ordinarily thought to be one of the main pillars supporting the edifice of the former. (Of course I am *not* attacking the theory of evolution, or anything in that neighborhood; I am instead attacking the conjunction of *naturalism* with the view that human beings have evolved in that way. I see no similar problems with the conjunction of *theism* and the idea that human beings have evolved in the way contemporary evolutionary science suggests.) More particularly, I argued that the conjunction of naturalism with the belief that we human beings have evolved in conformity with current evolutionary doctrine... is in a certain interesting way self-defeating or self-referentially incoherent.
> — Alvin Plantinga, Naturalism Defeated?: Essays on Plantinga's Evolutionary Argument Against Naturalism, "Introduction"[29]

7.4.2 Robert T. Pennock

Robert T. Pennock contends[30] that as supernatural agents and powers "are above and beyond the natural world and its agents and powers" and "are not constrained by natural laws", only logical impossibilities constrain what a supernatural agent could not do. He states: "If we could apply natural knowledge to understand supernatural powers, then, by definition, they would not be supernatural". As the supernatural is necessarily a mystery to us, it can provide no grounds on which to judge scientific models. "Experimentation requires observation and control of the variables.... But by definition we have no control over supernatural entities or forces." Science does not deal with meanings; the closed system of scientific reasoning cannot be used to define itself. Allowing science to appeal to untestable supernatural powers would make the scientist's task meaningless, undermine the discipline that allows science to make progress, and "would be as profoundly unsatisfying as the ancient Greek playwright's reliance upon the *deus ex machina* to extract his hero from a difficult predicament."

Naturalism of this sort says nothing about the existence or nonexistence of the supernatural, which by this definition is beyond natural testing. As a practical consideration, the rejection of supernatural explanations would merely be pragmatic, thus it would nonetheless be possible, for an ontological supernaturalist to espouse and practice methodological naturalism. For example, scientists may believe in God while practicing methodological naturalism in their scientific work. This position does not preclude knowledge that is somehow connected to the supernatural. Generally however, anything that can be scientifically examined and explained would not be supernatural, simply by definition.

7.4.3 W. V. O. Quine

Main article: Naturalized epistemology

W. V. O. Quine describes naturalism as the position that there is no higher tribunal for truth than natural science itself. In his view, there is no better method than the scientific method for judging the claims of science, and there is neither any need nor any place for a "first philosophy", such as (abstract) metaphysics or epistemology, that could stand behind and justify science or the scientific method.

Therefore, philosophy should feel free to make use of the findings of scientists in its own pursuit, while also feeling free to offer criticism when those claims are ungrounded, confused, or inconsistent. In Quine's view, philosophy is "continuous with" science and *both* are empirical.[31] Naturalism is not a dogmatic belief that the modern view of science is entirely correct. Instead, it simply holds that science is the best way to explore the processes of the universe and that those processes are what modern science is striving to understand. However, this Quinean Replacement Naturalism finds relatively few supporters among philosophers.[32]

7.4.4 Karl Popper

Karl Popper equated naturalism with inductive theory of science. He rejected it based on his general critique of induction (see problem of induction), yet acknowledged its utility as means for inventing conjectures.

> A naturalistic methodology (sometimes called an "inductive theory of science") has its value, no doubt.... I reject the naturalistic view: It is uncritical. Its upholders fail to notice that whenever they believe to have discovered a fact, they have only proposed a convention. Hence the convention is liable to turn into a dogma. This criticism of the naturalistic view applies not only to its criterion of meaning, but also to its idea of science, and consequently to its idea of empirical method.
> — Karl R. Popper, The Logic of Scientific Discovery, (Routledge, 2002), pp. 52–53, ISBN 0-415-27844-9.

Popper instead proposed that science should adopt a methodology based on falsifiability for demarcation, because no number of experiments can ever prove a theory, but a single experiment can contradict one. Popper holds that scientific theories are characterized by falsifiability.

7.5 See also

- Daoism
- Deism
- Empiricism
- Epicureanism
- Hylomorphism
- Liberal naturalism
- Materialism
- Metaphysical naturalism
- Naturalistic pantheism
- Physicalism
- Religious naturalism
- Scientism
- Sociological naturalism
- Supernaturalism
- Alfred North Whitehead
- Vaisheshika
- Carvaka

7.6 Notes

[1] Oxford English Dictionary Online naturalism

[2] "CATHOLIC ENCYCLOPEDIA: Naturalism". 21 November 2009. Retrieved 6 March 2012. Naturalism is not so much a special system as a point of view or tendency common to a number of philosophical and religious systems; not so much a well-defined set of positive and negative doctrines as an attitude or spirit pervading and influencing many doctrines. As the name implies, this tendency consists essentially in looking upon nature as the one original and fundamental source of all that exists, and in attempting to explain everything in terms of nature. Either the limits of nature are also the limits of existing reality, or at least the first cause, if its existence is found necessary, has nothing to do with the working of natural agencies. All events, therefore, find their adequate explanation within nature itself. But, as the terms nature and natural are themselves used in more than one sense, the term naturalism is also far from having one fixed meaning.

[3] Papineau, David (22 February 2007). "Naturalism". *Stanford Encyclopedia of Philosophy*.

[4] Kurtz, Paul (Spring 1998). "Darwin Re-Crucified: Why Are So Many Afraid of Naturalism?". *Free Inquiry.* **18** (2).

[5] Schafersman, Steven D. (1996). "Naturalism is Today An Essential Part of Science". Methodological naturalism is the adoption or assumption of naturalism in scientific belief and practice without really believing in naturalism.

[6] Jonathan Barnes's introduction to Early Greek Philosophy (Penguin)

[7] A Chatterjee (2012), Naturalism in Classical Indian Philosophy, The Stanford Encyclopedia of Philosophy (Fall 2014 Edition), Edward N. Zalta (ed.)

[8] Dale Riepe (1996), Naturalistic Tradition in Indian Thought, Motilal Banarsidass, ISBN 978-8120812932, pages 227-246

[9] Oliver Leaman (1999), *Key Concepts in Eastern Philosophy.* Routledge, ISBN 978-0415173629, page 269

[10] J Ganeri (2012), The Self: Naturalism, Consciousness, and the First-Person Stance, Oxford University Press, ISBN 978-0199652365

[11] Ronald L. Numbers (2003). "Science without God: Natural Laws and Christian Beliefs." In: When Science and Christianity Meet, edited by David C. Lindberg, Ronald L. Numbers. Chicago: University Of Chicago Press, p. 267.

[12] Rouse Ball, W. W. [1908] (2003) "Pierre Simon Laplace (1749–1827)", in *A Short Account of the History of Mathematics*, 4th ed., Dover, ISBN 0-486-20630-0

[13] Williams, Sally (July 4, 2007). "The God curriculum". London: The Telegraph. Retrieved 2008-12-26.

[14] Schafersman, Steven D. (1996). "Naturalism is Today An Essential Part of Science". Section "The Origin of Naturalism and Its Relation to Science". Naturalism did not exist as a philosophy before the nineteenth century, but only as an occasionally adopted and non-rigorous method among natural philosophers. It is a unique philosophy in that it is not ancient or prior to science, and that it developed largely due to the influence of science.

[15] Schafersman, Steven D. (1996). "Naturalism is Today An Essential Part of Science". Section "The Origin of Naturalism and Its Relation to Science". Naturalism is almost unique in that it would not exist as a philosophy without the prior existence of science. It shares this status, in my view, with the philosophy of existentialism.

[16] Papineau, David "Naturalism", in "The Stanford Encyclopedia of Philosophy"

[17] Nick Matzke: On the Origins of Methodological Naturalism. *The Pandas Thumb* (March 20, 2006)

[18] ASA March 2006 – Re: Methodological Naturalism

[19] Belief Net, "What do scientists say"

[20] Elaine Ecklund's book "Science versus Religion: What do scientists really think"

[21] Kitzmiller trial: testimony of Robert T. Pennock

[22] Kitzmiller v. Dover: Whether ID is Science

[23] Judge John E. Jones, III Decision of the Court Expert witnesses were John F. Haught, Robert T. Pennock, and Kenneth R. Miller. Links in the original to specific testimony records have been deleted here.

[24] http://philosophy.nd.edu/people/alvin-plantinga/

[25] Beilby, J.K. (2002). *Naturalism Defeated?: Essays on Plantinga's Evolutionary Argument Against Naturalism.* G - Reference, Information and Interdisciplinary Subjects Series. Cornell University Press. p. 9. ISBN 9780801487637. LCCN 2001006111.

[26] "Gifford Lecture Series - Warrant and Proper Function 1987-1988".

[27] Plantinga, Alvin (11 April 2010). "Evolution, Shibboleths, and Philosophers — Letters to the Editor". The Chronicle of Higher Education. ...I do indeed think that evolution functions as a contemporary shibboleth by which to distinguish the ignorant fundamentalist goats from the informed and scientifically literate sheep.
According to Richard Dawkins, 'It is absolutely safe to say that, if you meet somebody who claims not to believe in evolution, that person is ignorant, stupid, or insane (or wicked, but I'd rather not consider that).' Daniel Dennett goes Dawkins one (or two) further: 'Anyone today who doubts that the variety of life on this planet was produced by a process of evolution is simply ignorant—inexcusably ignorant.' You wake up in the middle of the night; you think, can that whole Darwinian story really be true? Wham! You are inexcusably ignorant.
I do think that evolution has become a modern idol of the tribe. But of course it doesn't even begin to follow that I think the scientific theory of evolution is false. And I don't.

[28] Plantinga, Alvin (1993). *Warrant and Proper Function.* Oxford: Oxford University Press. Chap. 11. ISBN 0-19-507863-2.

[29] Beilby, J.K., ed. (2002). "Introduction by Alvin Plantinga". *Naturalism Defeated?: Essays on Plantinga's Evolutionary Argument Against Naturalism.* Reference, Information and Interdisciplinary Subjects Series. Ithaca: Cornell University Press. pp. 1–2, 10. ISBN 978-0-8014-8763-7. LCCN 2001006111.

[30] Robert T. Pennock, Supernaturalist Explanations and the Prospects for a Theistic Science or "How do you know it was the lettuce?"

[31] Lynne Rudder (2013). *Naturalism and the First-Person Perspective.* Oxford University Press. p. 5. ISBN 0199914745.

[32] Feldman, Richard (2012). "Naturalized Epistemology". In Zalta, Edward N. *The Stanford Encyclopedia of Philosophy* (Summer 2012 ed.). Retrieved 2014-06-04. Quinean Replacement Naturalism finds relatively few supporters.

7.7 References

- Audi, Robert (1996). "Naturalism". In Borchert, Donald M. *The Encyclopedia of Philosophy Supplement*. USA: Macmillan Reference. pp. 372–374.

- Danto, Arthur C. (1967). "Naturalism". In Edwords, Paul. *The Encyclopedia of Philosophy*. New York: The Macmillan Co. and The Free Press. pp. 448–450.

- Kurtz, Paul (1990). *Philosophical Essays in Pragmatic Naturalism*. Prometheus Books.

- Lacey, Alan R. (1995). "Naturalism". In Honderich, Ted. *The Oxford Companion to Philosophy*. Oxford University Press. pp. 604–606.

- Post, John F. (1995). "Naturalism". In Audi, Robert. *The Cambridge Dictionary of Philosophy*. Cambridge University Press. pp. 517–518.

- Sagan, Carl (2002). *Cosmos*. Random House. ISBN 978-0-375-50832-5.

7.8 Further reading

- Mario De Caro and David Macarthur (eds) *Naturalism in Question*. Cambridge, Mass: Harvard University Press, 2004.

- Mario De Caro and David Macarthur (eds) *Naturalism and Normativity*. New York: Columbia University Press, 2010.

- Friedrich Albert Lange, *The History of Materialism*, London: Kegan Paul, Trench, Trubner & Co Ltd, 1925, ISBN 0-415-22525-6

- David Macarthur, "Quinean Naturalism in Question," Philo. vol 11, no. 1 (2008).

7.9 External links

7.9.1 Supportive

- naturalism.org

- Naturalist Newsletter
- Center for Naturalism
- Naturalism: The Naturalistic Worldview
- Naturalism David Papineau, The Stanford Encyclopedia of Philosophy
- The Brights Illuminating and elevating the naturalistic worldview

7.9.2 Neutral

- Naturalism at PhilPapers
- Naturalism entry in the *Stanford Encyclopedia of Philosophy*
- Naturalism at the Indiana Philosophy Ontology Project
- "Naturalism". *Internet Encyclopedia of Philosophy*.
- The Craig-Taylor Debate: Is The Basis Of Morality Natural Or Supernatural? William Lane Craig and Richard Taylor October 1993, Union College (Schenectady, New York)

7.9.3 Critical

- biologos.org
- "Naturalism" article in *The Catholic Encyclopedia*
- Alvin Plantinga (1994). "Naturalism Defeated" (PDF). (pdf)
- A shorter version of C. S. Lewis' Dangerous Idea
- Philip Johnson's Evolution as Dogma: The Establishment of Naturalism from *First Things*
- Robert A. Delfino's (2007) Replacing Methodological Naturalism Metanexus Institute. Archived from the original.
- Robert A. Delfino's (2011) Scientific Naturalism and the Need for a Neutral Metaphysical Framework

Chapter 8

Utilitarianism

This article discusses utilitarian ethical theory. For a discussion of John Stuart Mill's book *Utilitarianism*, see Utilitarianism (book). For the architectural theory, see Utilitarianism (architecture).

Utilitarianism is an ethical theory which states that the best action is the one that maximizes utility. "Utility" is defined in various ways, usually in terms of the well-being of sentient entities. Jeremy Bentham, the founder of utilitarianism, described utility as the sum of all pleasure that results from an action, minus the suffering of anyone involved in the action. Utilitarianism is a version of consequentialism, which states that the consequences of any action are the only standard of right and wrong. Unlike other forms of consequentialism, such as egoism, utilitarianism considers all interests equally.

Proponents of utilitarianism have disagreed on a number of points, such as whether actions should be chosen based on their likely results (act utilitarianism) or whether agents should conform to rules that maximize utility (rule utilitarianism). There is also disagreement as to whether total (total utilitarianism) or average (average utilitarianism) utility should be maximized.

Though the seeds of the theory can be found in the hedonists Aristippus and Epicurus, who viewed happiness as the only good, the tradition of utilitarianism properly began with Bentham, and has included John Stuart Mill, Henry Sidgwick, R. M. Hare, David Braybrooke, and Peter Singer. It has been applied to social welfare economics, the crisis of global poverty, the ethics of raising animals for food and the importance of avoiding existential risks to humanity.

8.1 Etymology

Benthamism, the utilitarian philosophy founded by Jeremy Bentham, was substantially modified by his successor John Stuart Mill, who popularized the word 'Utilitarianism'.[1] In 1861, Mill acknowledged in a footnote that, though "believing himself to be the first person who brought the word 'utilitarian' into use, he did not invent it. Rather, he adopted it from a passing expression in" John Galt's 1821 novel *Annals of the Parish*.[2] Mill seems to have been unaware that Bentham had used the term 'utilitarian' in his 1781 letter to George Wilson and his 1802 letter to Étienne Dumont.[1]

8.2 Historical background

8.2.1 Chinese philosophy

In Chinese philosophy the Mohists and their successors the "Chinese Legalists" might be considered utilitarians, or at least the "earliest form of consequentialism". Of particular concern for them, the fourth century witnessed the emergence of discussions polarizing the concepts of self and private, commonly used in conjunction with profit and associated with fragmentation, division, partiality, and one-sidelines, with that of the state and "public", represented by the duke and referring to what is official or royal, that is, the ruler himself, associated with unity, wholeness, objectivity, and universality. The later denotes the "universal Way".[3]

However, the Mohists did not focus on emotional happiness, but promoted objective public goods: material wealth, a large population or family, and social order.[4] On the other hand, the "Legalist" Han Fei "is motivated almost totally from the ruler's point of view."[5]

8.2.2 Western philosophy

See also: Hedonism

The importance of happiness as an end for humans has long been recognized. Forms of hedonism were put forward by Aristippus and Epicurus; Aristotle argued that eudaimonia is the highest human good and Augustine wrote that "all men agree in desiring the last end, which is happiness." Happiness was also explored in depth by

8.2. HISTORICAL BACKGROUND

Aquinas.[6][7][8][9][10] Different varieties of consequentialism also existed in the ancient and medieval world, like the state consequentialism of Mohism or the political philosophy of Niccolò Machiavelli. Mohist consequentialism advocated communitarian moral goods including political stability, population growth, and wealth, but did not support the utilitarian notion of maximizing individual happiness.[11] Machiavelli was also an exponent of consequentialism. He believed that the actions of a state, however cruel or ruthless they may be, must contribute towards the common good of a society.[12] Utilitarianism as a distinct ethical position only emerged in the eighteenth century.

Although utilitarianism is usually thought to start with Jeremy Bentham, there were earlier writers who presented theories that were strikingly similar. In *An Enquiry Concerning the Principles of Morals*, David Hume writes:[13]

> In all determinations of morality, this circumstance of public utility is ever principally in view; and wherever disputes arise, either in philosophy or common life, concerning the bounds of duty, the question cannot, by any means, be decided with greater certainty, than by ascertaining, on any side, the true interests of mankind. If any false opinion, embraced from appearances, has been found to prevail; as soon as farther experience and sounder reasoning have given us juster notions of human affairs, we retract our first sentiment, and adjust anew the boundaries of moral good and evil.

Hume studied the works of, and corresponded with, Francis Hutcheson, and it was he who first introduced a key utilitarian phrase. In *An Inquiry into the Original of Our Ideas of Beauty and Virtue* (1725), Hutcheson says[14] when choosing the most moral action, virtue is in proportion to the number of people a particular action brings happiness to. In the same way, moral evil, or vice, is proportionate to the number of people made to suffer. The best action is the one that procures the greatest happiness of the greatest numbers—and the worst is the one that causes the most misery.

In the first three editions of the book, Hutcheson included various mathematical algorithms "...to compute the Morality of any Actions." In this, he pre-figured the hedonic calculus of Bentham.

Some claim that John Gay developed the first systematic theory of utilitarian ethics.[15] In *Concerning the Fundamental Principle of Virtue or Morality* (1731), Gay argues that:[16]

> happiness, private happiness, is the proper or ultimate end of all our actions... each particular action may be said to have its proper and peculiar end...(but).... they still tend or ought to tend to something farther; as is evident from hence, viz. that a man may ask and expect a reason why either of them are pursued: now to ask the reason of any action or pursuit, is only to enquire into the end of it: but to expect a reason, i.e. an end, to be assigned for an ultimate end, is absurd. To ask why I pursue happiness, will admit of no other answer than an explanation of the terms.

This pursuit of happiness is given a theological basis:[17]

> Now it is evident from the nature of God, viz. his being infinitely happy in himself from all eternity, and from his goodness manifested in his works, that he could have no other design in creating mankind than their happiness; and therefore he wills their happiness; therefore the means of their happiness: therefore that my behaviour, as far as it may be a means of the happiness of mankind, should be such...thus the will of God is the immediate criterion of Virtue, and the happiness of mankind the criterion of the wilt of God; and therefore the happiness of mankind may be said to be the criterion of virtue, but once removed...(and)... I am to do whatever lies in my power towards promoting the happiness of mankind.

Gay's theological utilitarianism was developed and popularized by William Paley. It has been claimed that Paley was not a very original thinker and that the philosophical part of his treatise on ethics is "an assemblage of ideas developed by others and is presented to be learned by students rather than debated by colleagues."[18] Nevertheless, his book *The Principles of Moral and Political Philosophy* (1785) was a required text at Cambridge[18] and Smith says that Paley's writings were "once as well known in American colleges as were the readers and spellers of William McGuffey and Noah Webster in the elementary schools."[19] Although now largely missing from the philosophical canon, Schneewind writes that "utilitarianism first became widely known in England through the work of William Paley."[20] The now forgotten significance of Paley can be judged from the title of Thomas Rawson Birks's 1874 work *Modern Utilitarianism or the Systems of Paley, Bentham and Mill Examined and Compared*.

Apart from restating that happiness as an end is grounded in the nature of God, Paley also discusses the place of rules. He writes:[21]

> ...actions are to be estimated by their tendency. Whatever is expedient, is right. It is the

MODERN UTILITARIANISM,

OR THE

SYSTEMS OF PALEY, BENTHAM, AND MILL EXAMINED AND COMPARED.

BY

THOMAS RAWSON BIRKS,
KNIGHTBRIDGE PROFESSOR OF MORAL PHILOSOPHY.

London:
MACMILLAN AND CO.
1874.
[All Rights reserved.]

Modern Utilitarianism *by Thomas Rawson Birks 1874*

utility of any moral rule alone, which constitutes the obligation of it.

But to all this there seems a plain objection, viz. that many actions are useful, which no man in his senses will allow to be right. There are occasions, in which the hand of the assassin would be very useful... The true answer is this; that these actions, after all, are not useful, and for that reason, and that alone, are not right.

To see this point perfectly, it must be observed that the bad consequences of actions are twofold, particular and general. The particular bad consequence of an action, is the mischief which that single action directly and immediately occasions. The general bad consequence is, the violation of some necessary or useful general rule...

You cannot permit one action and forbid another, without showing a difference between them. Consequently, the same sort of actions must be generally permitted or generally forbidden. Where, therefore, the general permission of them would be pernicious, it becomes necessary to lay down and support the rule which generally forbids them.

8.3 Classical utilitarianism

8.3.1 Jeremy Bentham

Main article: Jeremy Bentham
Bentham's book *An Introduction to the Principles of Morals*

Jeremy Bentham.

and Legislation was printed in 1780 but not published until 1789. It is possible that Bentham was spurred on to publish after he saw the success of Paley's *The Principles of Moral and Political Philosophy*.[22] Bentham's book was not an immediate success[23] but his ideas were spread further when Pierre Étienne Louis Dumont translated edited selections from a variety of Bentham's manuscripts into French. *Traité de legislation civile et pénale* was published in 1802 and then later retranslated back into English by Hildreth as *The Theory of Legislation*, although by this time significant portions of Dumont's work had already been retranslated and incorporated into Sir John Bowring's edition of Bentham's works, which was issued in parts between 1838 and 1843.

Bentham's work opens with a statement of the principle of utility:[24]

> Nature has placed mankind under the governance of two sovereign masters, pain and pleasure. It is for them alone to point out what we ought to do... By the principle of utility is meant that principle which approves or disapproves of every action whatsoever according to the tendency it appears to have to augment or diminish the happiness of the party whose interest is in question: or, what is the same thing in other words to promote or to oppose that happiness. I say of every action whatsoever, and therefore not only of every action of a private individual, but of every measure of government.

In Chapter IV, Bentham introduces a method of calculating the value of pleasures and pains, which has come to be known as the hedonic calculus. Bentham says that the value of a pleasure or pain, considered by itself, can be measured according to its intensity, duration, certainty/uncertainty and propinquity/remoteness. In addition, it is necessary to consider "the tendency of any act by which it is produced" and, therefore, to take account of the act's fecundity, or the chance it has of being followed by sensations of the same kind and its purity, or the chance it has of not being followed by sensations of the opposite kind. Finally, it is necessary to consider the extent, or the number of people affected by the action.

Perhaps aware that Hutcheson eventually removed his algorithms for calculating the greatest happiness because they "appear'd useless, and were disagreeable to some readers",[25] Bentham contends that there is nothing novel or unwarranted about his method, for "in all this there is nothing but what the practice of mankind, wheresoever they have a clear view of their own interest, is perfectly conformable to."

Rosen warns that descriptions of utilitarianism can bear "little resemblance historically to utilitarians like Bentham and J. S. Mill" and can be more "a crude version of act utilitarianism conceived in the twentieth century as a straw man to be attacked and rejected."[26] It is a mistake to think that Bentham is not concerned with rules. His seminal work is concerned with the principles of legislation and the hedonic calculus is introduced with the words "Pleasures then, and the avoidance of pains, are the ends that the legislator has in view." In Chapter VII, Bentham says: "The business of government is to promote the happiness of the society, by punishing and rewarding... In proportion as an act tends to disturb that happiness, in proportion as the tendency of it is pernicious, will be the demand it creates for punishment."

The question then arises as to when, if at all, it might be legitimate to break the law. This is considered in *The Theory of Legislation*, where Bentham distinguishes between evils of the first and second orders. Those of the first order are the more immediate consequences; those of the second are when the consequences spread through the community causing "alarm" and "danger".

> It is true there are cases in which, if we confine ourselves to the effects of the first order, the good will have an incontestable preponderance over the evil. Were the offence considered only under this point of view, it would not be easy to assign any good reasons to justify the rigour of the laws. Every thing depends upon the evil of the second order; it is this which gives to such actions the character of crime, and which makes punishment necessary. Let us take, for example, the physical desire of satisfying hunger. Let a beggar, pressed by hunger, steal from a rich man's house a loaf, which perhaps saves him from starving, can it be possible to compare the good which the thief acquires for himself, with the evil which the rich man suffers? ... It is not on account of the evil of the first order that it is necessary to erect these actions into offences, but on account of the evil of the second order.[27]

8.3.2 John Stuart Mill

Main article: John Stuart Mill

Mill was brought up as a Benthamite with the explicit intention that he would carry on the cause of utilitarianism.[28] Mill's book *Utilitarianism* first appeared as a series of three articles published in *Fraser's Magazine* in 1861 and was reprinted as a single book in 1863.[29]

Higher and lower pleasures

Mill rejects a purely quantitative measurement of utility and says:[30]

> It is quite compatible with the principle of utility to recognize the fact, that some kinds of pleasure are more desirable and more valuable than others. It would be absurd that while, in estimating all other things, quality is considered as well as quantity, the estimation of pleasures should be supposed to depend on quantity alone.

The word utility is used to mean general well-being or happiness, and Mill's view is that utility is the consequence of a good action. Utility, within the context of utilitarianism, refers to people performing actions for social utility. With social utility, he means the well-being of many people. Mill's explanation of the concept of utility in his work, Utilitarianism, is that people really do desire happiness, and since each individual desires their own happiness, it must follow that all of us desire the happiness of everyone, contributing to a larger social utility. Thus, an action that results in the greatest pleasure for the utility of society is the best action, or as Jeremy Bentham, the founder of early Utilitarianism put it, as the greatest happiness of the greatest number.

Mill not only viewed actions as a core part of utility, but as the directive rule of moral human conduct. The rule being that we should only be committing actions that provide pleasure to society. This view of pleasure was hedonistic, as it pursued the thought that pleasure is the highest good in life. This concept was adopted by Jeremy Bentham, the founder of Utilitarianism, and can be seen in his works. According to Mill, good actions result in pleasure, and that there is no higher end than pleasure. Mill says that good actions lead to pleasure and define good character. Better put, the justification of character, and whether an action is good or not, is based on how the person contributes to the concept of social utility. In the long run the best proof of a good character is good actions; and resolutely refuse to consider any mental disposition as good, of which the predominant tendency is to produce bad conduct. In the last chapter of Utilitarianism, Mill concludes that justice, as a classifying factor of our actions (being just or unjust) is one of the certain moral requirements, and when the requirements are all regarded collectively, they are viewed as greater according to this scale of "social utility" as Mill puts it.

He also notes that, contrary to what its critics might say, there is "no known Epicurean theory of life which does not assign to the pleasures of the intellect... a much higher value as pleasures than to those of mere sensation." However, he accepts that this is usually because the intellectual pleasures are thought to have circumstantial advantages, i.e. "greater permanency, safety, uncostliness, &c." Instead, Mill will argue that some pleasures are intrinsically better than others.

The accusation that hedonism is "doctrine worthy only of swine" has a long history. In Nicomachean Ethics (Book 1 Chapter 5), Aristotle says that identifying the good with pleasure is to prefer a life suitable for beasts. The theological utilitarians had the option of grounding their pursuit of happiness in the will of God; the hedonistic utilitarians needed a different defence. Mill's approach is to argue that the pleasures of the intellect are intrinsically superior to physical pleasures.

Few human creatures would consent to be changed into any of the lower animals, for a promise of the fullest allowance of a beast's pleasures; no intelligent human being would consent to be a fool, no instructed person would be an ignoramus, no person of feeling and conscience would be selfish and base, even though they should be persuaded that the fool, the dunce, or the rascal is better satisfied with his lot than they are with theirs... A being of higher faculties requires more to make him happy, is capable probably of more acute suffering, an certainly accessible to it at more points, than one of an inferior type; but in spite of these liabilities, he can never really wish to sink into what he feels to be a lower grade of existence... It is better to be a human being dissatisfied than a pig satisfied; better to be Socrates dissatisfied than a fool satisfied. And if the fool, or the pig, are of a different opinion, it is because they only know their own side of the question...[31]

Mill argues that if people who are "competently acquainted" with two pleasures show a decided preference for one even if it be accompanied by more discontent and "would not resign it for any quantity of the other", then it is legitimate to regard that pleasure as being superior in quality. Mill recognizes that these "competent judges" will not always agree, and states that, in cases of disagreement, the judgment of the majority is to be accepted as final. Mill also acknowledges that "many who are capable of the higher pleasures, occasionally, under the influence of temptation, postpone them to the lower. But this is quite compatible with a full appreciation of the intrinsic superiority of the higher." Mill says that this appeal to those who have experienced the relevant pleasures is no different from what must happen when assessing the quantity of pleasure, for there is no other way of measuring "the acutest of two pains, or the intensest of two pleasurable sensations." "It is indisputable that the being whose capacities of enjoyment are low, has the greatest chance of having them fully satisfied; and a highly-endowed being will always feel that any happiness which he can look for, as the world is constitute, is imperfect."[32]

Mill's 'proof' of the principle of utility

In Chapter Four of *Utilitarianism*, Mill considers what proof can be given for the principle of utility. He says:[33]

> The only proof capable of being given that an object is visible, is that people actually see it. The only proof that a sound is audible, is that people

hear it... In like manner, I apprehend, the sole evidence it is possible to produce that anything is desirable, is that people do actually desire it... No reason can be given why the general happiness is desirable, except that each person, so far as he believes it to be attainable, desires his own happiness... we have not only all the proof which the case admits of, but all which it is possible to require, that happiness is a good: that each person's happiness is a good to that person, and the general happiness, therefore, a good to the aggregate of all persons.

It is usual[34] to say that Mill is committing a number of fallacies. He is accused of committing the naturalistic fallacy, because he is trying to deduce what people ought to do from what they in fact do; the fallacy of equivocation, because he moves from the fact that (1) something is desirable, i.e. is capable of being desired, to the claim that (2) it is desirable, i.e. that it ought to be desired; and the fallacy of composition, because the fact that people desire their own happiness does not imply that the aggregate of all persons will desire the general happiness.

Such allegations began to emerge in Mill's lifetime, shortly after the publication of *Utilitarianism*, and persisted for well over a century, though the tide has been turning in recent discussions.

A defence of Mill against all three charges, with a chapter devoted to each, can be found in Necip Fikri Alican's *Mill's Principle of Utility: A Defense of John Stuart Mill's Notorious Proof* (1994). This is the first, and remains the only, book-length treatment of the subject matter. Yet the alleged fallacies in the proof continue to attract scholarly attention in journal articles and book chapters.

Hall[35] and Popkin[36] defend Mill against this accusation pointing out that he begins Chapter Four by asserting that "questions of ultimate ends do not admit of proof, in the ordinary acceptation of the term" and that this is "common to all first principles." According to Hall and Popkin, therefore, Mill does not attempt to "establish that what people do desire is desirable but merely attempts to make the principles acceptable."[34] The type of "proof" Mill is offering "consists only of some considerations which, Mill thought, might induce an honest and reasonable man to accept utilitarianism."[34]

Having claimed that people do, in fact, desire happiness, Mill now has to show that it is the *only* thing they desire. Mill anticipates the objection that people desire other things such as virtue. He argues that whilst people might start desiring virtue as a *means* to happiness, eventually, it becomes part of someone's happiness and is then desired as an end in itself.

The principle of utility does not mean that any given pleasure, as music, for instance, or any given exemption from pain, as for example health, are to be looked upon as means to a collective something termed happiness, and to be desired on that account. They are desired and desirable in and for themselves; besides being means, they are a part of the end. Virtue, according to the utilitarian doctrine, is not naturally and originally part of the end, but it is capable of becoming so; and in those who love it disinterestedly it has become so, and is desired and cherished, not as a means to happiness, but as a part of their happiness.[37]

We may give what explanation we please of this unwillingness; we may attribute it to pride, a name which is given indiscriminately to some of the most and to some of the least estimable feelings of which is mankind are capable; we may refer it to the love of liberty and personal independence, an appeal to which was with the Stoics one of the most effective means for the inculcation of it; to the love of power, or the love of excitement, both of which do really enter into and contribute to it: but its most appropriate appellation is a sense of dignity, which all humans beings possess in one form or other, and in some, though by no means in exact, proportion to their higher faculties, and which is so essential a part of the happiness of those in whom it is strong, that nothing which conflicts with it could be, otherwise than momentarily, an object of desire to them.[38]

8.4 Twentieth-century developments

8.4.1 Ideal utilitarianism

The description of ideal utilitarianism was first used by Hastings Rashdall in *The Theory of Good and Evil* (1907), but it is more often associated with G. E. Moore. In *Ethics* (1912), Moore rejected a purely hedonistic utilitarianism and argued that there is a range of values that might be maximized. Moore's strategy was to show that it is intuitively implausible that pleasure is the sole measure of what is good. He says that such an assumption:[39]

involves our saying, for instance, that a world in which absolutely nothing except pleasure existed—no knowledge, no love, no enjoyment of beauty, no moral qualities—must yet be intrinsically better—better worth creating—provided only the total quantity of pleasure in it were the least bit greater, than one in which all these things existed as well as pleasure.

It involves our saying that, even if the total quantity of pleasure in each was exactly equal, yet the fact that all the beings in the one possessed in addition knowledge of many different kinds and a full appreciation of all that was beautiful or worthy of love in their world, whereas none of the beings in the other possessed any of these things, would give us no reason whatever for preferring the former to the latter.

Moore admits that it is impossible to prove the case either way, but he believed that it was intuitively obvious that even if the amount of pleasure stayed the same a world that contained such things as beauty and love would be a better world. He adds that, if a person was to take the contrary view, then "I think it is self-evident that he would be wrong."[39]

8.4.2 Act and rule utilitarianism

In the mid-twentieth century a number of philosophers focused on the place of rules in utilitarian thinking.[40] It was already accepted that it is necessary to use rules to help you choose the right action because the problems of calculating the consequences on each and every occasion would almost certainly result in you frequently choosing something less than the best course of action. Paley had justified the use of rules and Mill says:[41]

It is truly a whimsical supposition that, if mankind were agreed in considering utility to be the test of morality, they would remain without any agreement as to what is useful, and would take no measures for having their notions on the subject taught to the young, and enforced by law and opinion... to consider the rules of morality as improvable, is one thing; to pass over the intermediate generalisations entirely, and endeavour to test each individual action directly by the first principle, is another... The proposition that happiness is the end and aim of morality, does not mean that no road ought to be laid down to that goal... Nobody argues that the art of navigation is not founded on astronomy, because sailors cannot wait to calculate the Nautical Almanack. Being rational creatures, they go to sea with it ready calculated; and all rational creatures go out upon the sea of life with their minds made up on the common questions of right and wrong.

However, rule utilitarianism proposes a more central role for rules that was thought to rescue the theory from some of its more devastating criticisms, particularly problems to do with justice and promise keeping. Throughout the 1950s and 1960s, articles were published both for and against the new form of utilitarianism, and through this debate the theory we now call rule utilitarianism was created. In an introduction to an anthology of these articles, the editor was able to say: "The development of this theory was a dialectical process of formulation, criticism, reply and reformulation; the record of this process well illustrates the co-operative development of a philosophical theory."[42]

Smart[43] and McCloskey[44] initially used the terms 'extreme' and 'restricted' utilitarianism but eventually everyone settled on the terms 'act' and 'rule' utilitarianism.

The essential difference is in what determines whether or not an action is the right action. Act utilitarianism maintains that an action is right if it maximizes utility; rule utilitarianism maintains that an action is right if it conforms to a rule that maximizes utility.

In 1956, Urmson published an influential article[45] arguing that Mill justified rules on utilitarian principles. From then on, articles have debated this interpretation of Mill. In all probability, it was not a distinction that Mill was particularly trying to make and so the evidence in his writing is inevitably mixed. A collection of Mill's writing published in 1977 includes a letter in which he says:[46]

I agree with you that the right way of testing actions by their consequences, is to test them by the natural consequences of the particular action, and not by those which would follow if everyone did the same. But, for the most part, the consideration of what would happen if everyone did the same, is the only means we have of discovering the tendency of the act in the particular case.

This seems to tip the balance in favour of saying that Mill is best classified as an act utilitarian.

Some school level textbooks and at least one UK examination board[47] make a further distinction between strong and weak rule utilitarianism. However, it is not clear that this distinction is made in the academic literature.

It has been argued that rule utilitarianism collapses into act utilitarianism, because for any given rule, in the case where breaking the rule produces more utility, the rule can be refined by the addition of a sub-rule that handles cases like the exception.[48] This process holds for all cases of exceptions, and so the "rules" have as many "sub-rules" as there are exceptional cases, which, in the end, makes an agent seek out whatever outcome produces the maximum utility.[49]

8.4.3 Two-level utilitarianism

Main article: Two-level utilitarianism

In *Principles* (1973),[50] R. M. Hare accepts that rule utilitarianism collapses into act utilitarianism but claims that this is a result of allowing the rules to be "as specific and un-general as we please." He argues that one of the main reasons for introducing rule utilitarianism was to do justice to the general rules that people need for moral education and character development and he proposes that "a difference between act-utilitarianism and rule-utilitarianism can be introduced by limiting the specificity of the rules, i.e., by increasing their generality."[50]:14 This distinction between a "specific rule utilitarianism" (which collapses into act utilitarianism) and "general rule utilitarianism" forms the basis of Hare's two-level utilitarianism.

When we are "playing God or the ideal observer", we use the specific form, and we will need to do this when we are deciding what general principles to teach and follow. When we are "inculcating" or in situations where the biases of our human nature are likely to prevent us doing the calculations properly, then we should use the more general rule utilitarianism.

Hare argues that in practice, most of the time, we should be following the general principles:[50]:17

> One ought to abide by the general principles whose general inculcation is for the best; harm is more likely to come, in actual moral situations, from questioning these rules than from sticking to them, unless the situations are very extra-ordinary; the results of sophisticated felicific calculations are not likely, human nature and human ignorance being what they are, to lead to the greatest utility.

In *Moral Thinking* (1981), Hare illustrated the two extremes. The "archangel" is the hypothetical person who has perfect knowledge of the situation and no personal biases or weaknesses and always uses critical moral thinking to decide the right thing to do; the "prole" is the hypothetical person who is completely incapable of critical thinking and uses nothing but intuitive moral thinking and, of necessity, has to follow the general moral rules they have been taught or learned through imitation.[51] It is not that some people are archangels and others proles, but rather that "we all share the characteristics of both to limited and varying degrees and at different times."[51]

Hare does not specify when we should think more like an "archangel" and more like a "prole" as this will, in any case, vary from person to person. However, the critical moral thinking underpins and informs the more intuitive moral thinking. It is responsible for formulating and, if necessary, reformulating the general moral rules. We also switch to critical thinking when trying to deal with unusual situations or in cases where the intuitive moral rules give conflicting advice.

8.4.4 Preference utilitarianism

Main article: Preference utilitarianism

Preference utilitarianism was first put forward in 1977 by John Harsanyi in *Morality and the theory of rational behaviour*,[52] but preference utilitarianism is more commonly associated with R. M. Hare,[51] Peter Singer[53] and Richard Brandt.[54]

Harsanyi claimed that his theory is indebted to Adam Smith, who equated the moral point of view with that of an impartial but sympathetic observer; to Kant, who insisted on the criterion of universality, which may also be described as a criterion of reciprocity; to the classical utilitarians who made maximizing social utility the basic criterion of morality; and to "the modern theory of rational behaviour under risk and uncertainty, usually described as Bayesian decision theory".[52]:42

Harsanyi rejects hedonistic utilitarianism as being dependent on an outdated psychology saying that it is far from obvious that everything we do is motivated by a desire to maximize pleasure and minimize pain. He also rejects ideal utilitarianism because "it is certainly not true as an empirical observation that people's only purpose in life is to have 'mental states of intrinsic worth'."[52]:54

According to Harsanyi, "preference utilitarianism is the only form of utilitarianism consistent with the important philosophical principle of preference autonomy. By this I mean the principle that, in deciding what is good and what is bad for a given individual, the ultimate criterion can only be his own wants and his own preferences."[52]:55

Harsanyi adds two caveats. People sometimes have irrational preferences. To deal with this, Harsanyi distinguishes between "manifest" preferences and "true" preferences. The former are those "manifested by his observed behaviour, including preferences possibly based on erroneous factual beliefs, or on careless logical analysis, or on strong emotions that at the moment greatly hinder rational choice" whereas the latter are "the preferences he would have if he had all the relevant factual information, always reasoned with the greatest possible care, and were in a state of mind most conducive to rational choice."[52]:55 It is the latter that preference utilitarianism tries to satisfy.

The second caveat is that antisocial preferences, such

as sadism, envy and resentment, have to be excluded. Harsanyi achieves this by claiming that such preferences partially exclude those people from the moral community:

> Utilitarian ethics makes all of us members of the same moral community. A person displaying ill will toward others does remain a member of this community, but not with his whole personality. That part of his personality that harbours these hostile antisocial feelings must be excluded from membership, and has no claim for a hearing when it comes to defining our concept of social utility.[52]:56

8.5 More varieties of utilitarianism

8.5.1 Negative utilitarianism

Main article: Negative utilitarianism

In *The Open Society and its Enemies* (1945), Karl Popper argued that the principle "maximize pleasure" should be replaced by "minimize pain". He thought "it is not only impossible but very dangerous to attempt to maximize the pleasure or the happiness of the people, since such an attempt must lead to totalitarianism."[55] He claimed that:[56]

> there is, from the ethical point of view, no symmetry between suffering and happiness, or between pain and pleasure... In my opinion human suffering makes a direct moral appeal, namely, the appeal for help, while there is no similar call to increase the happiness of a man who is doing well anyway. A further criticism of the Utilitarian formula "Maximize pleasure" is that it assumes a continuous pleasure-pain scale which allows us to treat degrees of pain as negative degrees of pleasure. But, from the moral point of view, pain cannot be outweighed by pleasure, and especially not one man's pain by another man's pleasure. Instead of the greatest happiness for the greatest number, one should demand, more modestly, the least amount of avoidable suffering for all...

The actual term *negative utilitarianism* was introduced by R.N.Smart as the title to his 1958 reply to Popper[57] in which he argued that the principle would entail seeking the quickest and least painful method of killing the entirety of humanity.

Negative *total* utilitarianism, in contrast, tolerates suffering that can be compensated within the same person.[58][59]

Negative *preference* utilitarianism avoids the problem of moral killing with reference to existing preferences that such killing would violate, while it still demands a justification for the creation of new lives.[60] A possible justification is the reduction of the average level of preference-frustration.[61]

Others see negative utilitarianism as a branch within modern hedonistic utilitarianism, which assigns a higher weight to the avoidance of suffering than to the promotion of happiness.[62] The moral weight of suffering can be increased by using a "compassionate" utilitarian metric, so that the result is the same as in prioritarianism.[63]

Pessimistic representatives of negative utilitarianism can be found in the environment of Buddhism.[64]

8.5.2 Motive utilitarianism

Motive utilitarianism was first proposed by Robert Merrihew Adams in 1976.[65] Whereas act utilitarianism requires us to choose our actions by calculating which action will maximize utility and rule utilitarianism requires us to implement rules which will, on the whole, maximize utility, motive utilitarianism "has the utility calculus being used to select motives and dispositions according to their general felicific effects, and those motives and dispositions then dictate our choices of actions."[66]

The arguments for moving to some form of motive utilitarianism at the personal level can be seen as mirroring the arguments for moving to some form of rule utilitarianism at the social level.[67] Adams refers to Sidgwick's observation that "Happiness (general as well as individual) is likely to be better attained if the extent to which we set ourselves consciously to aim at it be carefully restricted."[68] Trying to apply the utility calculation on each and every occasion is likely to lead to a sub-optimal outcome. Applying carefully selected rules at the social level and encouraging appropriate motives at the personal level is, so it is argued, likely to lead to a better overall outcome even if on some individual occasions it leads to the wrong action when assessed according to act utilitarian standards.[69]

Adams concludes that "right action, by act-utilitarian standards, and right motivation, by motive-utilitarian standards, are incompatible in some cases."[70] The necessity of this conclusion is rejected by Fred Feldman who argues that "the conflict in question results from an inadequate formulation of the utilitarian doctrines; motives play no essential role in it...(and that)... Precisely the same sort of conflict arises even when MU is left out of consideration and AU is applied by itself."[71] Instead, Feldman proposes a variant of act utilitarianism that results in there being no conflict between it and motive utilitarianism.

8.6 Criticisms

Because utilitarianism is not a single theory but a cluster of related theories that have been developed over two hundred years, criticisms can be made for different reasons and have different targets.

8.6.1 Ignores justice

As Rosen[22] has pointed out, claiming that act utilitarians are not concerned about having rules is to set up a "straw man". Similarly, Hare refers to "the crude caricature of act utilitarianism which is the only version of it that many philosophers seem to be acquainted with."[72] Given what Bentham says about second order evils[73] it would be a serious misrepresentation to say that he and similar act utilitarians would be prepared to punish an innocent person for the greater good. Nevertheless, whether they would agree or not, this is what critics of utilitarianism claim is entailed by the theory. A classic version of this criticism was given by H. J. McCloskey:[44]

> Suppose that a sheriff were faced with the choice either of framing a Negro for a rape that had aroused hostility to the Negroes (a particular Negro generally being believed to be guilty but whom the sheriff knows not to be guilty)—and thus preventing serious anti-Negro riots which would probably lead to some loss of life and increased hatred of each other by whites and Negroes—or of hunting for the guilty person and thereby allowing the anti-Negro riots to occur, while doing the best he can to combat them. In such a case the sheriff, if he were an extreme utilitarian, would appear to be committed to framing the Negro.

By "extreme" utilitarian, McCloskey is referring to what later came to be called "act" utilitarianism. He suggests one response might be that the sheriff would not frame the innocent negro because of another rule: "do not punish an innocent person". Another response might be that the riots the sheriff is trying to avoid might have positive utility in the long run by drawing attention to questions of race and resources to help address tensions between the communities.

In a later article, McCloskey says:[74]

> Surely the utilitarian must admit that whatever the facts of the matter may be, it is logically possible that an 'unjust' system of punishment—e.g. a system involving collective punishments, retroactive laws and punishments, or punishments of parents and relations of the offender—may be more useful than a 'just' system of punishment?

8.6.2 Predicting consequences

Some argue that it is impossible to do the calculation that utilitarianism requires because consequences are inherently unknowable. Daniel Dennett describes this as the Three Mile Island effect.[75] Dennett points out that not only is it impossible to assign a precise utility value to the incident, it is impossible to know whether, ultimately, the near-meltdown that occurred was a good or bad thing. He suggests that it would have been a good thing if plant operators learned lessons that prevented future serious incidents.

Russell Hardin rejects such arguments. He argues that it is possible to distinguish the moral impulse of utilitarianism (which is "to define the right as good consequences and to motivate people to achieve these") from our ability to correctly apply rational principles which will among other things "depend on the perceived facts of the case and on the particular moral actor's mental equipment."[76] The fact that the latter is limited and can change doesn't mean that the former has to be rejected. "If we develop a better system for determining relevant causal relations so that we are able to choose actions that better produce our intended ends, it does not follow that we then must change our ethics. The moral impulse of utilitarianism is constant, but our decisions under it are contingent on our knowledge and scientific understanding."[77]

From the beginning, utilitarianism has recognized that certainty in such matters is unobtainable and both Bentham and Mill said that it was necessary to rely on the *tendencies* of actions to bring about consequences. G. E. Moore writing in 1903 said:[78]

> We certainly cannot hope directly to compare their effects except within a limited future; and all the arguments, which have ever been used in Ethics, and upon which we commonly act in common life, directed to shewing that one course is superior to another, are (apart from theological dogmas) confined to pointing out such probable immediate advantages...
>
> An ethical law has the nature not of a scientific law but of a scientific *prediction*: and the latter is always merely probable, although the probability may be very great.

8.6.3 Demandingness objection

Act utilitarianism not only requires everyone to do what they can to maximize utility, but to do so without any favouritism. Mill said, "As between his own happiness and that of others, utilitarianism requires him to be as strictly impartial as a disinterested and benevolent spectator."[79] Critics say that this combination of requirements leads to utilitarianism making unreasonable demands. The well-being of strangers counts just as much as that of friends, family or self. "What makes this requirement so demanding is the gargantuan number of strangers in great need of help and the indefinitely many opportunities to make sacrifices to help them."[80] As Shelly Kagan says, "Given the parameters of the actual world, there is no question that ...(maximally)... promoting the good would require a life of hardship, self-denial, and austerity...a life spent promoting the good would be a severe one indeed."[81]

Hooker describes two aspects to the problem: act utilitarianism requires *huge* sacrifices from those who are relatively better off and also requires sacrifice of your own good even when the aggregate good will be only *slightly* increased.[82] Another way of highlighting the complaint is to say that in utilitarianism, "there is no such thing as morally permissible self-sacrifice that goes above and beyond the call of duty."[82] Mill was quite clear about this, "A sacrifice which does not increase, or tend to increase, the sum total of happiness, it considers as wasted."[79]

One response to the problem is to accept its demands. This is the view taken by Peter Singer, who says: "No doubt we do instinctively prefer to help those who are close to us. Few could stand by and watch a child drown; many can ignore the avoidable deaths of children in Africa or India. The question, however, is not what we usually do, but what we ought to do, and it is difficult to see any sound moral justification for the view that distance, or community membership, makes a crucial difference to our obligations."[83]

Others argue that a moral theory that is so contrary to our deeply held moral convictions must either be rejected or modified.[84] There have been various attempts to modify utilitarianism to escape its seemingly over-demanding requirements.[85] One approach is to drop the demand that utility be maximized. In Satisficing Consequentialism, Michael Slote argues for a form of utilitarianism where "an act might qualify as morally right through having good enough consequences, even though better consequences could have been produced."[86] One advantage of such a system is that it would be able to accommodate the notion of supererogatory actions.

Samuel Scheffler takes a different approach and amends the requirement that everyone be treated the same.[87] In particular, Scheffler suggests that there is an "agent-centered prerogative" such that when the overall utility is being calculated it is permitted to count our own interests more heavily than the interests of others. Kagan suggests that such a procedure might be justified on the grounds that "a general requirement to promote the good would lack the motivational underpinning necessary for genuine moral requirements" and, secondly, that personal independence is necessary for the existence of commitments and close personal relations and that "the value of such commitments yields a positive reason for preserving within moral theory at least some moral independence for the personal point of view."[88]

Robert Goodin takes yet another approach and argues that the demandingness objection can be "blunted" by treating utilitarianism as a guide to public policy rather than one of individual morality. He suggests that many of the problems arise under the traditional formulation because the conscientious utilitarian ends up having to make up for the failings of others and so contributing more than their fair share.[89]

Harsanyi argues that the objection overlooks the fact that "people attach considerable utility to freedom from unduly burdensome moral obligations... most people will prefer a society with a more relaxed moral code, and will feel that such a society will achieve a higher level of average utility—even if adoption of such a moral code should lead to some losses in economic and cultural accomplishments (so long as these losses remain within tolerable limits). This means that utilitarianism, if correctly interpreted, will yield a moral code with a standard of acceptable conduct very much below the level of highest moral perfection, leaving plenty of scope for supererogatory actions exceeding this minimum standard."[90]

8.6.4 Aggregating utility

The objection that "utilitarianism does not take seriously the distinction between persons"[91] came to prominence in 1971 with the publication of John Rawls' *A Theory of Justice*. The concept is also important in animal rights advocate Richard Ryder's rejection of utilitarianism, in which he talks of the "boundary of the individual", through which neither pain nor pleasure may pass.[92] However, a similar objection was noted in 1970 by Thomas Nagel (who claimed that consequentialism "treats the desires, needs, satisfactions, and dissatisfactions of distinct persons as if they were the desires, etc., of a mass person"[93]), and even earlier by David Gauthier, who wrote that utilitarianism supposes "that mankind is a super-person, whose greatest satisfaction is the objective of moral action. . . . But this is absurd. Individuals have wants, not mankind; individuals seek satisfaction, not mankind. A person's satisfaction is not part of any greater satisfaction."[94] Thus, the aggregation of utility becomes futile as both pain and happiness are intrinsic to and inseparable from the consciousness in which they are felt, rendering impossible the task of adding up the various pleasures of multiple individuals.

A response to this criticism is to point out that whilst seeming to resolve some problems it introduces others. Intuitively, there are many cases where people do want to take the numbers involved into account. As Alastair Norcross has said, "suppose that Homer is faced with the painful choice between saving Barney from a burning building or saving both Moe and Apu from the building…it is clearly better for Homer to save the larger number, precisely because it is a larger number… Can anyone who really considers the matter seriously honestly claim to believe that it is worse that one person die than that the entire sentient population of the universe be severely mutilated? Clearly not."[95]

It may be possible to uphold the distinction between persons whilst still aggregating utility, if it accepted that people can be influenced by empathy.[96] This position is advocated by Iain King,[97] who has suggested the evolutionary basis of empathy means humans can take into account the interests of other individuals, but only on a one-to-one basis, "since we can only imagine ourselves in the mind of one other person at a time."[98] King uses this insight to adapt utilitarianism, and it may help reconcile Bentham's philosophy with deontology and virtue ethics.[99]

The philosopher John Taurek also argued that the idea of adding happiness or pleasures across persons is quite unintelligible and that the numbers of persons involved in a situation are morally irrelevant.[100] Taurek's basic concern comes down to this: we cannot explain what it means to say that things would be five times worse if five people die than if one person dies. "I cannot give a satisfactory account of the meaning of judgments of this kind," he wrote (p. 304). He argues that each person can only lose one person's happiness or pleasures. There isn't five times more loss of happiness or pleasure when five die: who would be feeling this happiness or pleasure? "Each person's potential loss has the same significance to me, only as a loss to that person alone. because, by hypothesis, I have an equal concern for each person involved, I am moved to give each of them an equal chance to be spared his loss" (p. 307). Parfit[101] and others[102] have criticized Taurek's line, and it continues to be discussed.[103]

8.6.5 Calculating utility is self-defeating

An early criticism, which was addressed by Mill, is that if time is taken to calculate the best course of action it is likely that the opportunity to take the best course of action will already have passed. Mill responded that there had been ample time to calculate the likely effects:[79]

> …namely, the whole past duration of the human species. During all that time, mankind have been learning by experience the tendencies of actions; on which experience all the prudence, as well as all the morality of life, are dependent…It is a strange notion that the acknowledgment of a first principle is inconsistent with the admission of secondary ones. To inform a traveller respecting the place of his ultimate destination, is not to forbid the use of landmarks and direction-posts on the way. The proposition that happiness is the end and aim of morality, does not mean that no road ought to be laid down to that goal, or that persons going thither should not be advised to take one direction rather than another. Men really ought to leave off talking a kind of nonsense on this subject, which they would neither talk nor listen to on other matters of practical concernment.

More recently , Hardin has made the same point . "
It should
embarrass philosophers that they have ever taken th is ob- jection seriously .
Parallel considerations in other realms are
dismissed with eminently good sense   .
Lord Devlin notes   , 'if the reasonable man   "
worked to rule   "by perusing to the   point of comprehension every form he was handed , the commercial and administrative life of the country would creep to a standstill.'"[77]

It is such considerations that lead even act utilitarians t o rely on"rules of thumb", asSmart[104]
has called them.

8.6.6 Karl Marx's criticism

Karl Marx, in *Das Kapital*, wrote:[105]

> Not even excepting our philosopher, Christian Wolff, in no time and in no country has the most homespun commonplace ever strutted about in so self-satisfied a way. The principle of utility was no discovery of Bentham. He simply reproduced in his dull way what Helvétius and other Frenchmen had said with esprit in the 18th century. To know what is useful for a dog, one must study dog-nature. This nature itself is not to be deduced from the principle of utility. Applying this to man, he who would criticize all human acts, movements, relations, etc., by the principle of utility, must first deal with human nature in general, and then with human nature as modified in each

historical epoch. Bentham makes short work of it. With the driest naivete he takes the modern shopkeeper, especially the English shopkeeper, as the normal man. Whatever is useful to this queer normal man, and to his world, is absolutely useful. This yard-measure, then, he applies to past, present, and future. The Christian religion, e.g., is "useful," "because it forbids in the name of religion the same faults that the penal code condemns in the name of the law." Artistic criticism is "harmful," because it disturbs worthy people in their enjoyment of Martin Tupper, etc. With such rubbish has the brave fellow, with his motto, "nulla dies sine linea [no day without a line]", piled up mountains of books.

8.6.7 John Paul II's personalist criticism

Pope John Paul II, following his personalist philosophy, argued that a danger of utilitarianism is that it tends to make persons, just as much as things, the object of use. "Utilitarianism," he wrote, "is a civilization of production and of use, a civilization of things and not of persons, a civilization in which persons are used in the same way as things are used."[106]

8.7 Additional considerations

8.7.1 Average v. total happiness

Main article: Average and total utilitarianism

In *The Methods of Ethics*, Henry Sidgwick asked, "Is it total or average happiness that we seek to make a maximum?"[107] He noted that aspects of the question had been overlooked and answered the question himself by saying that what had to be maximized was the average multiplied by the number of people living.[108] He also argued that, if the "average happiness enjoyed remains undiminished, Utilitarianism directs us to make the number enjoying it as great as possible."[108] This was also the view taken earlier by Paley. He notes that, although he speaks of the happiness of communities, "the happiness of a people is made up of the happiness of single persons; and the quantity of happiness can only be augmented by increasing the number of the percipients, or the pleasure of their perceptions" and that if extreme cases, such as people held as slaves, are excluded the amount of happiness will usually be in proportion to the number of people. Consequently, "the decay of population is the greatest evil that a state can suffer; and the improvement of it the object which ought, in all countries, to be aimed at in preference to every other political purpose whatsoever."[109] More recently, a similar view has been expressed by Smart, who argued that all other things being equal a universe with two million happy people is better than a universe with only one million happy people.[110]

Since Sidgwick raised the question it has been studied in detail and philosophers have argued that using either total or average happiness can lead to objectionable results.

According to Derek Parfit, using total happiness falls victim to the repugnant conclusion, whereby large numbers of people with very low but non-negative utility values can be seen as a better goal than a population of a less extreme size living in comfort. In other words, according to the theory, it is a moral good to breed more people on the world for as long as total happiness rises.[111]

On the other hand, measuring the utility of a population based on the average utility of that population avoids Parfit's repugnant conclusion but causes other problems. For example, bringing a moderately happy person into a very happy world would be seen as an immoral act; aside from this, the theory implies that it would be a moral good to eliminate all people whose happiness is below average, as this would raise the average happiness.[112]

William Shaw suggests that the problem can be avoided if a distinction is made between potential people, who need not concern us, and actual future people, who should concern us. He says, "utilitarianism values the happiness of people, not the production of units of happiness. Accordingly, one has no positive obligation to have children. However, if you have decided to have a child, then you have an obligation to give birth to the happiest child you can."[113]

8.7.2 Motives, intentions, and actions

Utilitarianism is typically taken to assess the rightness or wrongness of an action by considering just the consequences of that action. Bentham very carefully distinguishes motive from intention and says that motives are not in themselves good or bad but can be referred to as such on account of their tendency to produce pleasure or pain. He adds that, "from every kind of motive, may proceed actions that are good, others that are bad, and others that are indifferent."[114] Mill makes a similar point[115] and explicitly says that "motive has nothing to do with the morality of the action, though much with the worth of the agent. He who saves a fellow creature from drowning does what is morally right, whether his motive be duty, or the hope of being paid for his trouble."[116]

However, with intention the situation is more complex. In a footnote printed in the second edition of *Utilitarianism*, Mill says: "the morality of the action depends entirely upon the intention—that is, upon what the agent wills to do."[116] Elsewhere, he says, "Intention, and motive, are two very different things. But it is the intention, that is, the foresight of consequences, which constitutes the moral rightness or wrongness of the act."[117]

The correct interpretation of Mill's footnote is a matter of some debate. The difficulty in interpretation centres around trying to explain why, since it is consequences that matter, intentions should play a role in the assessment of the morality of an action but motives should not. One possibility "involves supposing that the 'morality' of the act is one thing, probably to do with the praiseworthiness or blameworthiness of the agent, and its rightness or wrongness another."[118] Jonathan Dancy rejects this interpretation on the grounds that Mill is explicitly making intention relevant to an assessment of the act not to an assessment of the agent.

An interpretation given by Roger Crisp draws on a definition given by Mill in *A System of Logic*, where he says that an "intention to produce the effect, is one thing; the effect produced in consequence of the intention, is another thing; the two together constitute the action."[119] Accordingly, whilst two actions may outwardly appear to be the same they will be different actions if there is a different intention. Dancy notes that this does not explain why intentions count but motives do not.

A third interpretation is that an action might be considered a complex action consisting of several stages and it is the intention that determines which of these stages are to be considered part of the action. Although this is the interpretation favoured by Dancy, he recognizes that this might not have been Mill's own view, for Mill "would not even allow that 'p & q' expresses a complex proposition. He wrote in his *System of Logic* I iv. 3, of 'Caesar is dead and Brutus is alive', that 'we might as well call a street a complex house, as these two propositions a complex proposition'."[118]

Finally, whilst motives may not play a role in determining the morality of an action, this does not preclude utilitarians from fostering particular motives if doing so will increase overall happiness.

8.8 Application to specific issues

8.8.1 Nonhuman animals

Further information: Speciesism and Animal welfare
In *An Introduction to the Principles of Morals and Legislation* Bentham wrote "the question is not, Can they reason? nor, Can they talk? but, Can they suffer?"[120] Mill's distinction between higher and lower pleasures might suggest that he gave more status to humans but in *The Methods of*

Peter Singer

Ethics, philosopher Henry Sidgwick says "We have next to consider who the 'all' are, whose happiness is to be taken into account. Are we to extend our concern to all the beings capable of pleasure and pain whose feelings are affected by our conduct? or are we to confine our view to human happiness? The former view is the one adopted by Bentham and Mill, and (I believe) by the Utilitarian school generally: and is obviously most in accordance with the universality that is characteristic of their principle ... it seems arbitrary and unreasonable to exclude from the end, as so conceived, any pleasure of any sentient being."[121]

Moreover, John Stuart Mill himself, in *Whewell on Moral Philosophy*, defends Bentham's advocacy for animal rights, calling it a 'noble anticipation', and writing: "Granted that any practice causes more pain to animals than it gives pleasure to man; is that practice moral or immoral? And if, exactly in proportion as human beings raise their heads out of the slough of selfishness, they do not with one voice answer 'immoral', let the morality of the principle of utility be for ever condemned."[122]

The utilitarian philosopher Peter Singer and many other animal rights activists have continued to argue that the wellbeing of all sentient beings ought to be seriously considered. Singer suggests that rights are conferred according to the level of a creature's self-awareness, regardless of their species. He adds that humans tend to be speciesist (discriminatory against non-humans) in ethical matters, and argues that, on utilitarianism, speciesism cannot be justified as there is no rational distinction that can be made between the suffering of humans and the suffering of nonhuman animals; all suffering ought to be reduced. Singer writes: "The racist violates the principle of equality by giving greater weight to the interests of members of his own race, when there is a clash between their interests and the interests of those of another race. Similarly the speciesist allows the interests of his own species to override the greater interests of members of other species. The pattern is the same in each case ... Most human beings are speciesists."[123]

In his 1990 edition of *Animal Liberation*, Peter Singer said that he no longer ate oysters and mussels, because although the creatures might not suffer, they might, it's not really known, and it's easy enough to avoid eating them in any case[124] (and this aspect of seeking better alternatives is a prominent part of utilitarianism).

This view still might be contrasted with deep ecology, which holds that an intrinsic value is attached to all forms of life and nature, whether currently assumed to be sentient or not. According to utilitarianism, the forms of life that are unable to experience anything akin to either enjoyment or discomfort are denied moral status, because it is impossible to increase the happiness or reduce the suffering of something that cannot feel happiness or suffer. Singer writes:

> The capacity for suffering and enjoying things is a prerequisite for having interests at all, a condition that must be satisfied before we can speak of interests in any meaningful way. It would be nonsense to say that it was not in the interests of a stone to be kicked along the road by a schoolboy. A stone does not have interests because it cannot suffer. Nothing that we can do to it could possibly make any difference to its welfare. A mouse, on the other hand, does have an interest in not being tormented, because it will suffer if it is. If a being suffers, there can be no moral justification for refusing to take that suffering into consideration. No matter what the nature of the being, the principle of equality requires that its suffering be counted equally with the like suffering—in so far as rough comparisons can be made—of any other being. If a being is not capable of suffering, or of experiencing enjoyment or happiness, there is nothing to be taken into account.

Thus, the moral value of one-celled organisms, as well as some multi-cellular organisms, and natural entities like a river, is only in the benefit they provide to sentient beings. Similarly, utilitarianism places no direct intrinsic value on biodiversity, although the benefits that biodiversity bring to sentient beings may mean that, on utilitarianism, biodiversity ought to be maintained in general.

In John Stuart Mill's essay "On Nature"[125] he argues that the welfare of wild animals is to be considered when making utilitarian judgments. Tyler Cowen argues that, if individual animals are carriers of utility, then we should consider limiting the predatory activity of carnivores relative to their victims: "At the very least, we should limit current subsidies to nature's carnivores."[126]

8.8.2 World poverty

An article in the American journal for Economics has addressed the issue of Utilitarian ethics within redistribution of wealth. The journal stated that taxation of the wealthy is the best way to make use of the disposable income they receive. This says that the money creates utility for the most people by funding government services.[127] Many utilitarian philosophers, including Peter Singer and Toby Ord, argue that inhabitants of developed countries in particular have an obligation to help to end extreme poverty across the world, for example by regularly donating some of their income to charity. Peter Singer, for example, argues that donating some of one's income to charity could help to save a life or cure somebody from a poverty-related illness, which is a much better use of the money as it brings someone in extreme poverty far more happiness than it would bring to oneself if one lived in relative comfort. However, Singer not only argues that one ought to donate a significant proportion of one's income to charity, but also that this money should be directed to the most cost-effective charities, in order to bring about the greatest good for the greatest number, consistent with utilitarian thinking.[128] Singer's ideas have formed the basis of the modern effective altruist movement.

8.9 See also

- Altruism (ethical doctrine)
- Applied ethics
- Anti-Utilitarianism
- Appeal to consequences
- Bounded rationality
- Charity International
- Classical liberalism
- Cost–benefit analysis
- Decision analysis
- Decision theory
- Effective altruism
- Gross national happiness

- List of utilitarians
- Pleasure principle (psychology)
- Prioritarianism
- Probabilistic reasoning
- Relative utilitarianism
- State consequentialism
- Uncertainty
- Utility monster
- Utilitarian bioethics
- Utilitarian cake-cutting

8.10 Notes

[1] Habibi, Don (2001). "Chapter 3, Mill's Moral Philosophy". *John Stuart Mill and the Ethic of Human Growth*. Dordrecht: Springer Netherlands. pp. 89–90, 112. ISBN 978-90-481-5668-9.

[2] John Stuart Mill (1861) *Utilitarianism*, footnote 1.

[3] Erica Brindley, The Polarization of the Concepts Si (Private Interest) and Gong (Public Interest) in Early Chinese Thought. p.6, 8, 12-13, 16, 19, 21-22, 24, 27

[4] Fraser, Chris, "Mohism", The Stanford Encyclopedia of Philosophy (Winter 2015 Edition), Edward N. Zalta (ed.), URL = https://plato.stanford.edu/archives/win2015/entries/mohism/

[5] Hansen, Chad. Philosophy East & West. Jul94, Vol. 44 Issue 3, p435. 54p. Fa (standards: laws) and meaning changes in Chinese philosophy.

[6] "SUMMA THEOLOGICA: Man's last end (Prima Secundae Partis, Q. 1)". *newadvent.org*.

[7] "SUMMA THEOLOGICA: Things in which man's happiness consists (Prima Secundae Partis, Q. 2)". *newadvent.org*.

[8] "SUMMA THEOLOGICA: What is happiness (Prima Secundae Partis, Q. 3)". *newadvent.org*.

[9] "SUMMA THEOLOGICA: Things that are required for happiness (Prima Secundae Partis, Q. 4)". *newadvent.org*.

[10] "SUMMA THEOLOGICA: The attainment of happiness (Prima Secundae Partis, Q. 5)". *newadvent.org*.

[11] Fraser, Chris (2011). *The Oxford Handbook of World Philosophy*. Oxford University Press. p. 62. ISBN 978-0-19-532899-8.

[12] Warburton, Nigel (2000). *Reading Political Philosophy: Machiavelli to Mill*. Psychology Press. p. 10. ISBN 978-0-415-21197-0.

[13] Hume, David (2002). "An Enquiry Concerning the Principles of Morals". In Schneewind, J. B. *Moral Philosophy from Montaigne to Kant*. Cambridge University Press. p. 552. ISBN 978-0521003049.

[14] Hutcheson, Francis (2002). "The Original of Our Ideas of Beauty and Virtue". In Schneewind, J. B. *Moral Philosophy from Montaigne to Kant*. Cambridge University Press. p. 515. ISBN 978-0521003049.

[15] Ashcraft, Richard (1991) John Locke: Critical Assessments (Critical assessments of leading political philosophers), Routledge, p. 691

[16] Gay, John (2002). "Concerning the Fundamental Principle of Virtue or Morality". In Schneewind, J. B. *Moral Philosophy from Montaigne to Kant*. Cambridge University Press. p. 408. ISBN 978-0521003049.

[17] Gay, John (2002). "Concerning the Fundamental Principle of Virtue or Morality". In Schneewind, J. B. *Moral Philosophy from Montaigne to Kant*. Cambridge University Press. pp. 404–05. ISBN 978-0521003049.

[18] Schneewind, J. B. (2002). *Moral Philosophy from Montaigne to Kant*. Cambridge University Press. p. 446. ISBN 978-0521003049.

[19] Smith, Wilson (Jul., 1954) William Paley's Theological Utilitarianism in America, The William and Mary Quarterly Third Series, Vol. 11, No. 3, Published by: Omohundro Institute of Early American History and Culture, pp. 402–24

[20] Schneewind, J. B. (1977). *Sidgwick's Ethics and Victorian Moral Philosophy*. Oxford University Press. p. 122. ISBN 978-0198245520.

[21] Paley, William (2002). "The Principles of Moral and Political Philosophy". In Schneewind, J. B. *Moral Philosophy from Montaigne to Kant*. Cambridge University Press. pp. 455–56. ISBN 978-0521003049.

[22] Rosen, Frederick (2003) Classical Utilitarianism from Hume to Mill. Routledge, p. 132

[23] Schneewind, J.B. (1977) Sidgwick's Ethics and Victorian Moral Philosophy, Oxford: Clarendon Press, p. 122

[24] Bentham, Jeremy (January 2009). *An Introduction to the Principles of Morals and Legislation (Dover Philosophical Classics)*. Dover Publications Inc. p. 1. ISBN 978-0486454528.

[25] An Inquiry into the Original of Our Ideas of Beauty and Virtue – Francis Hutcheson, Introduction, 1726

[26] Rosen, Frederick (2003) Classical Utilitarianism from Hume to Mill. Routledge, p. 32

[27] Bentham, Jeremy; Dumont, Etienne; Hildreth, R (November 2005). *Theory of Legislation: Translated from the French of Etienne Dumont*. Adamant Media Corporation. p. 58. ISBN 978-1402170348.

[28] Halevy, Elie (1966). *The Growth of Philosophic Radicalism*. Beacon Press. pp. 282–84. ISBN 0-19-101020-0.

[29] Hinman, Lawrence (2012). *Ethics: A Pluralistic Approach to Moral Theory*. Wadsworth. ISBN 1133050018.

[30] Mill, John Stuart (1998). Crisp, Roger, ed. *Utilitarianism*. Oxford University Press. p. 56. ISBN 0-19-875163-X.

[31] Mill, John Stuart (1998). Crisp, Roger, ed. *Utilitarianism*. Oxford University Press. pp. 56–57. ISBN 0-19-875163-X.

[32] John Stuart Mill,Utilitarianism,Chapter 2

[33] Mill, John Stuart (1998). Crisp, Roger, ed. *Utilitarianism*. Oxford: Oxford University Press. p. 81. ISBN 0-19-875163-X.

[34] Popkin, Richard H. (1950). "A Note on the 'Proof' of Utility in J. S. Mill". *Ethics*. **61**: 66.

[35] Hall, Everett W. (1949). "The 'Proof' of Utility in Bentham and Mill". *Ethics*. **60**: 1–18.

[36] Popkin, Richard H. (1950). "A Note on the 'Proof' of Utility in J. S. Mill". *Ethics*. **61**: 66–68.

[37] Mill, John Stuart (1998). Crisp, Roger, ed. *Utilitarianism*. Oxford: Oxford University Press. p. 82. ISBN 0-19-875163-X.

[38] Mill, *Utilitarianism*, Chapter 2.

[39] Moore, G. E. (1912). *Ethics*, London: Williams and Norgate, Ch. 7

[40] Bayles, M. D., ed. (1968) Contemporary Utilitarianism, Anchor Books, Doubleday

[41] Mill, John Stuart (1998). Crisp, Roger, ed. *Utilitarianism*. Oxford University Press. p. 70. ISBN 0-19-875163-X.

[42] Bayles, M. D., ed. (1968) Contemporary Utilitarianism, Anchor Books, Doubleday, p. 1

[43] Smart, J.J.C. (1956). "Extreme and Restricted Utilitarianism". *The Philosophical Quarterly*. **VI**: 344–54. JSTOR 2216786. doi:10.2307/2216786.

[44] McCloskey, H.J. (October 1957). "An Examination of Restricted Utilitarianism". *The Philosophical Review*. **66** (4): 466–85. JSTOR 2182745. doi:10.2307/2182745.

[45] Urmson, J.O. (1953). "The Interpretation of the Moral Philosophy of J.S.Mill". *The Philosophical Quarterly*. **III**: 33–39. JSTOR 2216697. doi:10.2307/2216697.

[46] Mill, John Stuart, The Collected Works of John Stuart Mill. Gen. Ed. John M. Robson. 33 vols. Toronto: University of Toronto Press, 1963-91. Vol. 17, p. 1881

[47] Oliphant, Jill, OCR Religious Ethics for AS and A2, Routledge, (2007)

[48] David Lyons, *Forms and Limits of Utilitarianism*, 1965

[49] Allen Habib (2008), *Promises*, in the Stanford Encyclopedia of Philosophy.

[50] Hare, R. M. (1972–1973). "The Presidential Address: Principles". *Proceedings of the Aristotelian Society, New Series*. **73**: 1–18. JSTOR 4544830. doi:10.1093/aristotelian/73.1.1.

[51] Hare, R.M. (1981). *Moral thinking: its levels, method, and point*. Oxford New York: Clarendon Press Oxford University Press. ISBN 9780198246602.

[52] Harsanyi, John C. (1982), "Morality and the theory of rational behaviour", in Sen, Amartya; Williams, Bernard, *Utilitarianism and beyond*, Cambridge: Cambridge University Press, pp. 39–62, ISBN 9780511611964.

Originally printed as: Harsanyi, John C. (Winter 1977). "Morality and the theory of rational behavior". *Social Research, special issue: Rationality, Choice, and Morality*. The New School. **44** (4): 623–56. JSTOR 40971169.

[53] Singer, Peter (1979). *Practical ethics* (1st ed.). Cambridge New York: Cambridge University Press. ISBN 9780521297202.

Singer, Peter (1993). *Practical ethics* (2nd ed.). Cambridge New York: Cambridge University Press. ISBN 9780521439718.

[54] Brandt, Richard B. (1979). *A theory of the good and the right*. Oxford New York: Clarendon Press Oxford University Press. ISBN 9780198245506.

[55] Popper, Karl (2002). *The Open Society and Its Enemies: Volume 2*. Routledge. p. 339. ISBN 978-0415278423.

[56] Popper, Karl (2002). *The Open Society and Its Enemies: Volume 1: The Spell of Plato*. Routledge. pp. 284–85. ISBN 978-0415237314.

[57] Smart, R.N. (October 1958). "Negative Utilitarianism". *Mind*. **67** (268): 542–43. JSTOR 2251207. doi:10.1093/mind/lxvii.268.542.

[58] Fricke Fabian (2002), Verschiedene Versionen des negativen Utilitarismus, Kriterion, vol.15, no.1, p. 14

[59] Arrhenius Gustav (2000), Future Generations, A Challenge for Moral Theory, FD-Diss., Uppsala University, Dept. of Philosophy, Uppsala: University Printers, p. 100

[60] Fricke Fabian (2002), Verschiedene Versionen des negativen Utilitarismus, Kriterion, vol.15, no.1, pp. 20–22

[61] {Chao, "Negative Average Preference Utilitarianism", *Journal of Philosophy of Life*, 2012; 2(1): 55–66

[62] Fricke Fabian(2002), Verschiedene Versionen des negativen Utilitarismus, Kriterion, vol.15, no.1, p. 14

[63] Broome John (1991), Weighing Goods, Oxford: Basil Blackwell, p. 222

[64] Bruno Contestabile: *Negative Utilitarianism and Buddhist Intuition*. In: *Contemporary Buddhism* Vol.15, Issue 2, S. 298–311, London 2014.

[65] Robert Merrihew Adams, *Motive Utilitarianism*, The Journal of Philosophy, Vol. 73, No. 14, On Motives and Morals (12 August 1976), pp. 467–81

[66] Goodin, Robert E. "Utilitarianism as a Public Philosophy" (Cambridge Studies in Philosophy and Public Policy), Cambridge University Press, p. 60

[67] Goodin, Robert E. "Utilitarianism as a Public Philosophy" (Cambridge Studies in Philosophy and Public Policy), Cambridge University Press, p. 17

[68] Robert Merrihew Adams, *Motive Utilitarianism*, The Journal of Philosophy, Vol. 73, No. 14, On Motives and Morals (12 August 1976), p. 467

[69] Robert Merrihew Adams, *Motive Utilitarianism*, The Journal of Philosophy, Vol. 73, No. 14, On Motives and Morals (12 August 1976), p. 471

[70] Robert Merrihew Adams, *Motive Utilitarianism*, The Journal of Philosophy, Vol. 73, No. 14, On Motives and Morals (12 August 1976), p. 475

[71] Feldman, Fred, *On the Consistency of Act- and Motive-Utilitarianism: A Reply to Robert Adams*, Philosophical Studies: An International Journal for Philosophy in the Analytic Tradition, Vol. 70, No. 2 (May, 1993), pp. 211–12

[72] Hare, R. M. (1981) Moral Thinking. Oxford Univ. Press, p. 36

[73] Bentham, Jeremy (2009) Theory of Legislation. General Books LLC, p. 58

[74] McCloskey, H.J. (1963) A Note on Utilitarian Punishment, in Mind, 72, 1963, p. 599

[75] Dennett, Daniel (1995), *Darwin's Dangerous Idea*, Simon & Schuster, p. 498 ISBN 0-684-82471-X.

[76] Hardin, Russell (May 1990). *Morality within the Limits of Reason*. University Of Chicago Press. p. 3. ISBN 978-0226316208.

[77] Hardin, Russell (May 1990). *Morality within the Limits of Reason*. University Of Chicago Press. p. 4. ISBN 978-0226316208.

[78] Moore, G.E. (1903). *Principia Ethica*. Prometheus Books UK. pp. 203–04. ISBN 0879754982.

[79] Mill, John Stuart. "Utilitarianism, Chapter 2". Retrieved 24 June 2012.

[80] Hooker, Brad (9 September 2011). "Chapter 8: The Demandingness Objection". In Chappell, Timothy. *The problem of moral demandingness: new philosophical essays*. Palgrave Macmillan. p. 151. ISBN 9780230219403.

[81] Kagan, Shelly (April 1991). *The Limits of Morality (Oxford Ethics Series)*. Clarendon Press. p. 360. ISBN 978-0198239161.

[82] Hooker, Brad (October 2002). *Ideal Code, Real World: A Rule-Consequentialist Theory of Morality*. Clarendon Press. p. 152. ISBN 978-0199256570.

[83] Singer, Peter (February 2011). *Practical Ethics, Third Edition*. Cambridge University Press. pp. 202–03. ISBN 978-0521707688.

[84] Hooker, Brad (9 September 2011). "Chapter 8: The Demandingness Objection". In Chappell, Timothy. *The problem of moral demandingness: new philosophical essays*. Palgrave Macmillan. p. 148. ISBN 9780230219403.

[85] Kagan, Shelly (Summer 1984). "Does Consequentialism Demand too Much? Recent Work on the Limits of Obligation". *Philosophy & Public Affairs*. **13** (3): 239–54. JSTOR 2265413.

[86] Slote, Michael (1984). "Satisficing Consequentialism". *Proceedings of the Aristotelian Society, Supplementary Volumes*. **58**: 140. JSTOR 4106846.

[87] Scheffler, Samuel (August 1994). *The Rejection of Consequentialism: A Philosophical Investigation of the Considerations Underlying Rival Moral Conceptions, Second Edition*. Clarendon Press. ISBN 978-0198235118.

[88] Kagan, Shelly (Summer 1984). "Does Consequentialism Demand too Much? Recent Work on the Limits of Obligation". *Philosophy & Public Affairs*. **13** (3): 254. JSTOR 2265413.

[89] Goodin, Robert E. (May 1995). *Utilitarianism as a Public Philosophy*. Cambridge University Press. p. 66. ISBN 978-0521468060.

[90] Harsanyi, John C. (June 1975). "Can the Maximin Principle Serve as a Basis for Morality? A Critique of John Rawls's Theory A Theory of Justice by John Rawls". *The American Political Science Review*. **69** (2): 601. JSTOR 1959090. doi:10.2307/1959090.

[91] Rawls, John (March 22, 2005). *A Theory of Justice*. Harvard University Press; reissue edition. p. 27. ISBN 978-0674017726.

[92] Ryder, Richard D. *Painism: A Modern Morality*. Centaur Press, 2001. pp. 27–29

[93] Nagel, Thomas (2012). *The Possibility of Altruism*. Princeton University Press, New Ed edition. p. 134. ISBN 978-0691020020.

[94] Gauthier, David (1963). *Practical Reasoning: The Structure and Foundations of Prudential and Moral Arguments and Their Exemplification in Discourse.* Oxford University Press. p. 126. ISBN 978-0198241904.

[95] Norcross, Alastair (2009). "Two Dogmas of Deontology: Aggregation, Rights and the Separateness of Persons" (PDF). *Social Philosophy and Policy.* **26**: 81–82. doi:10.1017/S0265052509090049. Retrieved 2012-06-29.

[96] In Moral Laws of the Jungle (link to Philosophy Now magazine), Iain King argues: "The way I reconcile my interests with those of other people is not for all of us to pour everything we care about into a pot then see which of the combination of satisfied wants would generate the most happiness (benefit). If we did that, I could be completely outnumbered.... No, the way we reconcile interests is through empathy. Empathy is one-to-one, since we only imagine ourselves in the mind of one other person at a time. Even when I empathise with 'the people' here... I am really imagining what it is like to be just one woman. I cannot imagine myself to be more than one person at a time, and neither can you." Link accessed 2014-01-29.

[97] King, Iain (2008). *How to Make Good Decisions and Be Right All the Time.* Continuum. p. 225. ISBN 978-1847063472.

[98] This quote is from Iain King's article in issue 100 of Philosophy Now magazine, Moral Laws of the Jungle (link), accessed 29 January 2014.

[99] Chapter Eight of the book *Ethics Matters* by Charlotte Vardy, ISBN 978-0-334-04391-1 (published by SCM Press, April 2012), entitled "Developments in Utilitarianism", describes Iain King's philosophy as "quasi-utilitarian", and suggests it is an original "development" on the utilitarian theme. Vardy argues King's system is "compatible with consequence-, virtue- and act based ethics." A Google Books link to the reference can be accessed here (link confirmed 2014-01-29.)

[100] John M. Taurek, "Should the Numbers Count?", *Philosophy and Public Affairs*, 6:4 (Summer 1977), pp. 293–316.

[101] Derek Parfit, "Innumerate Ethics", *Philosophy and Public Affairs*, 7:4 (Summer 1978), pp. 285–301.

[102] See for example: (1) Frances Myrna Kamm, "Equal Treatment and Equal Chances", *Philosophy and Public Affairs*, 14:2 (Spring 1985), pp. 177–94; (2) Gregory S. Kavka, "The Numbers Should Count", *Philosophical Studies*, 36:3 (October 1979), pp. 285–94.

[103] See for example: (1) Michael Otsuka, "Skepticism about Saving the Greater Number", *Philosophy and Public Affairs*, 32:4 (Autumn 2004), pp. 413–26; (2) Rob Lawlor, "Taurek, Numbers and Probabilities", *Ethical Theory and Moral Practice*, 9:2 (April 2006), pp. 149–66.

[104] Smart, J.J.C.; Williams, Bernard (January 1973). *Utilitarianism: For and Against.* Cambridge University Press. p. 42. ISBN 978-0521098229.

[105] Das Kapital Volume I Chapter 24 endnote 50

[106] "Archived copy". Archived from the original on 5 April 2011. Retrieved 1 April 2011.

[107] Sidgwick, Henry (January 1981). *Methods of Ethics.* Hackett Publishing Co, Inc; 7th Revised edition. p. xxxvi. ISBN 978-0915145287.

[108] Sidgwick, Henry (January 1981). *Methods of Ethics.* Hackett Publishing Co, Inc; 7th Revised edition. p. 415. ISBN 978-0915145287.

[109] Paley, William (1785). "The Principles of Moral and Political Philosophy". Retrieved 1 July 2012.

[110] Smart, J. J. C.; Williams, Bernard (January 1973). *Utilitarianism: For and Against.* Cambridge University Press. pp. 27–28. ISBN 978-0521098229.

[111] Parfit, Derek (January 1986). *Reasons and Persons.* Oxford Paperbacks. p. 388. ISBN 978-0198249085.

[112] Shaw, William (November 1998). *Contemporary Ethics: Taking Account of Utilitarianism.* Wiley-Blackwell. pp. 31–35. ISBN 978-0631202943.

[113] Shaw, William (November 1998). *Contemporary Ethics: Taking Account of Utilitarianism.* Wiley-Blackwell. p. 34. ISBN 978-0631202943.

[114] Bentham, Jeremy (January 2009). *An Introduction to the Principles of Morals and Legislation (Dover Philosophical Classics).* Dover Publications Inc. p. 102. ISBN 978-0486454528.

[115] Mill, John Stuart (1981). "Autobiography". In Robson, John. *Collected Works, volume XXXI.* University of Toronto Press. p. 51. ISBN 0-7100-0718-3.

[116] Mill, John Stuart (1998). Crisp, Roger, ed. *Utilitarianism.* Oxford University Press. p. 65. ISBN 0-19-875163-X.

[117] Mill, John Stuart (1981). "comments upon James Mill's Analysis of the Phenomena of the Human Mind". In Robson, John. *Collected Works, volume XXXI.* University of Toronto Press. pp. 252–53. ISBN 0-7100-0718-3. and as quoted by Ridge, Michael (2002). "Mill's Intentions and Motives". *Utilitas.* **14**: 54–70. doi:10.1017/S0953820800003393.

[118] Dancy, Jonathan (2000). "Mill's Puzzling Footnote". *Utilitas.* **12**: 219–22. doi:10.1017/S095382080000279X.

[119] Mill, John Stuart (February 2011). *A System of Logic, Ratiocinative and Inductive (Classic Reprint).* Forgotten Books. p. 51. ISBN 978-1440090820.

[120] An Introduction to the Principals of Morals and Legislation, Jeremy Bentham, 1789 ("printed" in 1780, "first published" in 1789, "corrected by the Author" in 1823.) See Chapter I: Of the Principle of Utility. For Bentham on animals, see Ch. XVII Note 122.

[121] Sidgwick, Henry (January 1981). *Methods of Ethics*. Hackett Publishing Co, Inc; 7th Revised edition. p. 414. ISBN 978-0915145287.

[122] Mill, JS. "Whewell on Moral Philosophy" (PDF). *Collected Works*. **X**: 185–87.

[123] Peter Singer, *Animal Liberation*, Chapter I, pp. 7–8, 2nd edition, 1990.

[124] Animal Liberation, Second Edition, Singer, Peter, 1975, 1990, excerpt, pp. 171–74, main passage on oysters, mussels, etc. p. 174 (last paragraph of this excerpt). And in a footnote in the actual book, Singer writes "My change of mind about mollusks stems from conversations with R.I. Sikora."

[125] "Mill's "On Nature"". *www.lancaster.ac.uk*. 1904. Retrieved 2015-08-09.

[126] Cowen, T. (2003). c. Hargrove, Eugene, ed. "Policing Nature". *Environmental Ethics*. **25** (2): 169–. doi:10.5840/enviroethics200325231.

[127] N. Gregory Mankiw; Matthew Weinzierl (2010). "The Optimal Taxation of Height: A Case Study of Utilitarian Income Redistribution". *American Economic Journal: Economic Policy*. American Economic Association. **2**: 155–176.

[128] Peter Singer: The why and how of effective altruism | Talk Video. TED.com.

8.11 References

- Adams, Robert Merrihew (August 1976). "Motive Utilitarianism". *The Journal of Philosophy*. **73** (14): 467. JSTOR 2025783. doi:10.2307/2025783.

- Alican, Necip Fikri (1994). *Mill's Principle of Utility: A Defense of John Stuart Mill's Notorious Proof*. Amsterdam and Atlanta: Editions Rodopi B.V. ISBN 978-90-518-3748-3.

- Anscombe, G. E. M. (January 1958). "Modern Moral Philosophy". *Philosophy*. **33** (124): 1. JSTOR 3749051. doi:10.1017/s0031819100037943.

- Ashcraft, Richard (1991). *John Locke: Critical Assessments*. Routledge.

- Bayles, M. D. (1968). *Contemporary Utilitarianism*. Anchor Books, Doubleday.

- Bentham, Jeremy (January 2009). *An Introduction to the Principles of Morals and Legislation (Dover Philosophical Classics)*. Dover Publications Inc. ISBN 978-0486454528.

- Bentham, Jeremy (2001). *The Works of Jeremy Bentham: Published under the Superintendence of His Executor, John Bowring. Volume 1*. Adamant Media Corporation. ISBN 978-1402163937.

- Bentham, Jeremy; Dumont, Etienne; Hildreth, R (November 2005). *Theory of Legislation: Translated from the French of Etienne Dumont*. Adamant Media Corporation. ISBN 978-1402170348.

- Brandt, Richard B. (1979). *A Theory of the Good and the Right*. Clarendon Press. ISBN 0-19-824550-5.

- Bredeson, Dean (2011). "Utilitarianism vgs. Dentological Ethics". *Applied Business Ethics: A Skills-Based Approach*. Cengage Learning. ISBN 978-0-538-45398-1.

- Broome, John (1991). *Weighing Goods*. Oxford: Basil Blackwell.

- Dancy, Jonathan (2000). "Mill's Puzzling Footnote". *Utilitas*. **12**: 219. doi:10.1017/S095382080000279X.

- Dennett, Daniel (1995). *Darwin's Dangerous Idea*. Simon & Schuster. ISBN 0-684-82471-X.

- Feldman, Fred (May 1993). "On the Consistency of Act- and Motive-Utilitarianism: A Reply to Robert Adams". *Philosophical Studies: An International Journal for Philosophy in the Analytic Tradition*. **70** (2): 201–12. doi:10.1007/BF00989590.

- Gauthier, David (1963). *Practical Reasoning: The Structure and Foundations of Prudential and Moral Arguments and Their Exemplification in Discourse*. Oxford University Press. ISBN 978-0198241904.

- Gay, John (2002). "Concerning the Fundamental Principle of Virtue or Morality". In Schneewind, J. B. *Moral Philosophy from Montaigne to Kant*. Cambridge University Press. ISBN 978-0521003049.

- Goodin, Robert E. (May 1995). *Utilitarianism as a Public Philosophy*. Cambridge University Press. ISBN 978-0521468060.

- Goodstein, Eban (2011). "Chapter 2: Ethics and Economics". *Economics and the Environment*. Wiley. ISBN 978-0-470-56109-6.

- Habib, Allen (2008), *Promises*, in the Stanford Encyclopedia of Philosophy.

- Halevy, Elie (1966). *The Growth of Philosophic Radicalism*. Beacon Press. ISBN 0-19-101020-0.

- Hall, Everett W. (1949). "The 'Proof' of Utility in Bentham and Mill". *Ethics*. **60**: 1–18. JSTOR 2378436. doi:10.1086/290691.

- Hardin, Russell (May 1990). *Morality within the Limits of Reason*. University Of Chicago Press. ISBN 978-0226316208.

- Hare, R. M. (1972–1973). "The Presidential Address: Principles". *Proceedings of the Aristotelian Society, New Series*. **73**: 1–18. JSTOR 4544830. doi:10.1093/aristotelian/73.1.1.

- Hare, R. M. (1981). *Moral thinking: its levels, method, and point*. Oxford New York: Clarendon Press Oxford University Press. ISBN 9780198246602.

- Harsanyi, John C. (Winter 1977). "Morality and the theory of rational behavior". *Social Research, special issue: Rationality, Choice, and Morality*. The New School. **44** (4): 623–56. JSTOR 40971169.

    > *Reprinted as*: Harsanyi, John C. (1982), "Morality and the theory of rational behaviour", in Sen, Amartya; Williams, Bernard, *Utilitarianism and beyond*, Cambridge: Cambridge University Press, pp. 39–62, ISBN 9780511611964.

- Harsanyi, John C. (June 1975). "Can the Maximin Principle Serve as a Basis for Morality? A Critique of John Rawls's Theory of Justice". *The American Political Science Review*. **69** (2): 594. JSTOR 1959090. doi:10.2307/1959090.

- Hooker, Brad (October 2002). *Ideal Code, Real World: A Rule-Consequentialist Theory of Morality*. Clarendon Press. ISBN 978-0199256570.

- Hooker, Brad (9 September 2011). "Chapter 8: The Demandingness Objection". In Chappell, Timothy. *The problem of moral demandingness: new philosophical essays*. Palgrave Macmillan. ISBN 9780230219403.

- Hume, David (2002). "An Enquiry Concerning the Principles of Morals". In Schneewind, J. B. *Moral Philosophy from Montaigne to Kant*. Cambridge University Press. ISBN 978-0521003049.

- Hutcheson, Francis (2002). "The Original of Our Ideas of Beauty and Virtue". In Schneewind, J. B. *Moral Philosophy from Montaigne to Kant*. Cambridge University Press. ISBN 978-0521003049.

- Kagan, Shelly (April 1991). *The Limits of Morality (Oxford Ethics Series)*. Clarendon Press. ISBN 978-0198239161.

- Kagan, Shelly (Summer 1984). "Does Consequentialism Demand too Much? Recent Work on the Limits of Obligation". *Philosophy & Public Affairs*. **13** (3). JSTOR 2265413.

- Lyons, David (November 1965). *Forms and Limits of Utilitarianism*. Oxford University Press(UK). ISBN 978-0198241973.

- McCloskey, H.J. (1963). "A Note on Utilitarian Punishment". *Mind*. **72**: 599. JSTOR 2251880. doi:10.1093/mind/LXXII.288.599.

- McCloskey, H.J. (October 1957). "An Examination of Restricted Utilitarianism". *The Philosophical Review*. **66** (4): 466–85. JSTOR 2182745. doi:10.2307/2182745.

- Mill, John Stuart (1998). Crisp, Roger, ed. *Utilitarianism*. Oxford University Press. ISBN 0-19-875163-X.

- Mill, John Stuart (February 2011). *A System of Logic, Ratiocinative and Inductive (Classic Reprint)*. Forgotten Books. ISBN 978-1440090820.

- Mill, John Stuart (1981). "Autobiography". In Robson, John. *Collected Works, volume XXXI*. University of Toronto Press. ISBN 0-7100-0718-3.

- Moore, G.E. (1903). *Principia Ethica*. Prometheus Books UK. ISBN 0879754982.

- Nagel, Thomas (2012). *The Possibility of Altruism*. Princeton University Press, New Ed edition. ISBN 978-0691020020.

- Norcross, Alastair (2009). "Two Dogmas of Deontology: Aggregation, Rights and the Separateness of Persons" (PDF). *Social Philosophy and Policy*. **26**: 76. doi:10.1017/S0265052509090049. Retrieved 2012-06-29.

- Oliphant,, Jill (2007). *OCR Religious Ethics for AS and A2*. Routledge.

- Paley, William (2002). "The Principles of Moral and Political Philosophy". In Schneewind, J. B. *Moral Philosophy from Montaigne to Kant*. Cambridge University Press. ISBN 978-0521003049.

- Parfit, Derek (January 1986). *Reasons and Persons*. Oxford Paperbacks. ISBN 978-0198249085.

- Popkin, Richard H. (October 1950). "A Note on the 'Proof' of Utility in J. S. Mill". *Ethics*. **61** (1): 66–68. JSTOR 2379052. doi:10.1086/290751.

- Popper, Karl (2002). *The Open Society and Its Enemies*. Routledge. ISBN 0-415-29063-5.

- Rawls, John (22 March 2005). *A Theory of Justice*. Harvard University Press; reissue edition. ISBN 978-0674017726.

- Rosen, Frederick (2003). *Classical Utilitarianism from Hume to Mill*. Routledge.

- Ryder, Richard D (2002). *Painism: A Modern Morality*. Centaur Press.

- Scheffler, Samuel (August 1994). *The Rejection of Consequentialism: A Philosophical Investigation of the Considerations Underlying Rival Moral Conceptions, Second Edition*. Clarendon Press. ISBN 978-0198235118.

- Schneewind, J. B. (1977). *Sidgwick's Ethics and Victorian Moral Philosophy*. Oxford University Press. ISBN 978-0198245520.

- Shaw, William (November 1998). *Contemporary Ethics: Taking Account of Utilitarianism*. Wiley-Blackwell. ISBN 978-0631202943.

- Sidgwick, Henry (January 1981). *Methods of Ethics*. Hackett Publishing Co, Inc; 7th Revised edition. ISBN 978-0915145287.

- Singer, Peter (2001). *Animal Liberation*. Ecco Press. ISBN 978-0060011574.

- Singer, Peter (February 2011). *Practical Ethics, Third Edition*. Cambridge University Press. ISBN 978-0521707688.

- Slote, Michael (1995). "The Main Issue between Unitarianism and Virtue Ethics". *From Morality to Virtue*. Oxford University Press. ISBN 978-0-19-509392-6.

- Slote, Michael (1984). "Satisficing Consequentialism". *Proceedings of the Aristotelian Society, Supplementary Volumes*. **58**: 139–76. JSTOR 4106846. doi:10.1093/aristoteliansupp/58.1.139.

- Smart, J. J. C.; Williams, Bernard (January 1973). *Utilitarianism: For and Against*. Cambridge University Press. ISBN 978-0521098229.

- Smart, J.J.C. (1956). "Extreme and Restricted Utilitarianism". *The Philosophical Quarterly*. **VI**: 344–54. JSTOR 2216786. doi:10.2307/2216786.

- Smart, R.N. (October 1958). "Negative Utilitarianism". *Mind*. **67** (268): 542–43. JSTOR 2251207. doi:10.1093/mind/LXVII.268.542.

- Smith, Wilson (July 1954). "William Paley's Theological Utilitarianism in America". *William and Mary Quarterly*. Third Series. **11** (3): 402. doi:10.2307/1943313.

- Soifer, Eldon (2009). *Ethical Issues: Perspectives for Canadians*. Broadview Press. ISBN 978-1-55111-874-1.

- Urmson, J.O. (1953). "The Interpretation of the Moral Philosophy of J.S.Mill". *The Philosophical Quarterly*. **III**: 33–39. JSTOR 2216697. doi:10.2307/2216697.

8.12 Further reading

- Cornman, James, et al. (1992). *Philosophical Problems and Arguments – An Introduction*, 4th edition Indianapolis, IN: Hackett Publishing Co.

- Glover, Jonathan (1977). *Causing Death and Saving Lives*, Penguin Books. ISBN 9780140220032. OCLC 4468071

- Hansas, John (2008). "Utilitarianism". In Hamowy, Ronald. *The Encyclopedia of Libertarianism*. Thousand Oaks, CA: SAGE; Cato Institute. pp. 518–19. ISBN 978-1-4129-6580-4. LCCN 2008009151. OCLC 750831024. doi:10.4135/9781412965811.n317.

- Harwood, Sterling (2009). "Ch. 11. Eleven Objections to Utilitarianism". In Pojman, Louis P.; Tramel, Peter. *Moral Philosophy: a reader* (4th ed.). Indianapolis, IN: Hackett. ISBN 978-0872209626. OCLC 488531841.

- Mackie, J. L. (1991). "esp. Chapter 6, Utilitarianism". *Ethics: Inventing Right and Wrong*. Penguin Books. ISBN 978-0140135589.

- Martin, Michael (1970). "A Utilitarian Kantian Principle," Philosophical Studies, (with H. Ruf), 21. pp. 90–91.

- Rachels, James; Rachels, Stuart (2012). "esp. Chapters 7&8, The Utilitarian Approach & The Debate of Utilitarianism". *The Elements of Moral Philosophy*. McGraw-Hill Higher Education. ISBN 978-0078038242.

- Scheffler, Samuel (1988). *Consequentialism and its Critics*. Oxford University Press. ISBN 978-0198750734.

- Silverstein, Harry S. (1972). *A Defence of Cornman's Utilitarian Kantian Principle*, Philosophical Studies (Dordrecht u.a.) 23, pp. 212–15.

- Singer, Peter (1993). "esp. Chapter 19 & 20, Consequentialism & The Utility and the Good". *A Companion to Ethics (Blackwell Companions to Philosophy)*. Wiley-Blackwell. ISBN 978-0631187851.

- Singer, Peter (1981). *The Expanding Circle: Ethics and Sociobiology*, New York: Farrar, Straus & Giroux. ISBN 0-374-15112-1

- Stokes, Eric (1959, plus reprints). *The English Utilitarians and India*, Clarendon Press. OCLC 930495493

- Sumner, L. Wayne. *Abortion: A Third Way*, Princeton, NJ: Princeton University Press.

- Vergara, Francisco « Bentham and Mill on the "Quality" of Pleasures », *Revue d'études benthamiennes*, Paris, 2011.

- Vergara, Francisco « A Critique of Elie Halévy; refutation of an important distortion of British moral philosophy », *Philosophy*, Journal of The Royal Institute of Philosophy, London, 1998.

- Williams, Bernard (1993). "esp. Chapter 10, Utilitarianism". *Morality: An Introduction to Ethics.* Cambridge University Press. ISBN 978-0521457293.

8.13 External links

- Nathanson, Stephen. "Act and Rule Utilitarianism". *Internet Encyclopedia of Philosophy*.

- Sinnott-Armstrong, Walter. "Consequentialism". *Stanford Encyclopedia of Philosophy*.

- Driver, Julia. "The History of Utilitarianism". *Stanford Encyclopedia of Philosophy*.

- *Wiki Felicifia*, the collaboratively edited encyclopaedia for utilitarians

- utilitarian.org FAQ A FAQ by Nigel Phillips on utilitarianism by a Web Site affiliated to David Pearce

- *A Utilitiarian FAQ*, by Ian Montgomerie

- *The English Utilitarians*, Volume I by Sir Leslie Stephen

- *The English Utilitarians*, Volume II by Sir Leslie Stephen

- Utilitarian Philosophers Large compendium of writings by and about the major utilitarian philosophers, both classic and contemporary.

- Utilitarianism A summary of classical utilitarianism, and modern alternatives, with application to ethical issues and criticisms

- Utilitarian Resources Collection of definitions, articles and links.

- Primer on the Elements and Forms of Utilitarianism A convenient summary of the major points of utilitarianism.

- International Website for Utilitarianism and Utilitarian Scholar's Conferences and Research

- Utilitarianism as Secondary Ethic A concise review of Utilitarianism, its proponents and critics.

- Essays on Reducing Suffering

Chapter 9

Ethical movement

Brooklyn Society for Ethical Culture building on Prospect Park West, originally designed by architect William Tubby as a home for William H. Childs (inventor of Bon Ami Cleaning Powder)

The **Ethical movement**, also referred to as the **Ethical Culture movement**, **Ethical Humanism** or simply **Ethical Culture**, is an ethical, educational, and religious movement that is usually traced back to Felix Adler (1851–1933).[1] Individual chapter organizations are generically referred to as "Ethical Societies", though their names may include "Ethical Society", "Ethical Culture Society", "Society for Ethical Culture", "Ethical Humanist Society", or other variations on the theme of "Ethical".

The Ethical movement is an outgrowth of secular moral traditions in the 19th century, principally in Europe and the United States. While some in this movement went on to organise for a non-congregational secular humanist movement, others attempted to build a secular moral movement that was emphatically "religious" in its approach to developing humanist ethical codes, in the sense of encouraging congregational structures and religious rites and practices. While in the United States, these movements formed as separate education organisations (the American Humanist Association and the American Ethical Union), the American Ethical Union's British equivalents, the South Place Ethical Society and the British Ethical Union consciously moved away from a congregational model to become Conway Hall and the British Humanist Association respectively. Subsequent "godless" congregational movements include the Sunday Assembly. At the international level, Ethical Culture and secular humanist groups have always organised jointly; the American Ethical Union was a founding organisation of the International Humanist and Ethical Union.

Ethical Culture is premised on the idea that honoring and living in accordance with *ethical principles* is central to what it takes to live meaningful and fulfilling lives, and to creating a world that is good for all. Practitioners of Ethical Culture focus on supporting one another in becoming better people, and on doing good in the world.[2][3]

9.1 History

9.1.1 Background

The Ethical movement was an outgrowth of the general loss of faith among the intellectuals of the Victorian era. A precursor to the doctrines of the ethical movement can be found in the South Place Ethical Society, founded in 1793 as the South Place Chapel on Finsbury Square, on the edge of the City of London.[4]

In the early nineteenth century, the chapel became known as "a radical gathering-place".[5] At that point it was a Unitarian chapel, and that movement, like Quakers, supported female equality.[6] Under the leadership of Reverend William Johnson Fox,[7] it lent its pulpit to activists

ciety's goals was drawn up by Maurice Adams:

> We, recognizing the evils and wrongs that must beset men so long as our social life is based upon selfishness, rivalry, and ignorance, and desiring above all things to supplant it by a life based upon unselfishness, love, and wisdom, unite, for the purpose of realizing the higher life among ourselves, and of inducing and enabling others to do the same.
>
> And we now form ourselves into a Society, to be called the Guild [Fellowship] of the New Life, to carry out this purpose.[12]

Although the Fellowship was a short-lived organization, it spawned the Fabian Society, which split in 1884 from the Fellowship of the New Life.[13][14]

9.1.2 Ethical movement

The Fabian Society was an outgrowth from the Fellowship of the New Life.

such as Anna Wheeler, one of the first women to campaign for feminism at public meetings in England, who spoke in 1829 on "Rights of Women." In later decades, the chapel moved away from Unitarianism, changing its name first to the South Place Religious Society, then the South Place Ethical Society (a name it held formally, though it was better known as Conway Hall from 1929) and is now Conway Hall Ethical Society.

The Fellowship of the New Life was established in 1883 by the Scottish intellectual Thomas Davidson.[8] Fellowship members included poets Edward Carpenter and John Davidson, animal rights activist Henry Stephens Salt,[9] sexologist Havelock Ellis, feminist Edith Lees (who later married Ellis), novelist Olive Schreiner[10] and Edward R. Pease.

Its objective was "The cultivation of a perfect character in each and all." They wanted to transform society by setting an example of clean simplified living for others to follow. Davidson was a major proponent of a structured philosophy about religion, ethics, and social reform.[11]

At a meeting on 16 November 1883, a summary of the so-

Felix Adler, founder of the ethical movement.

In his youth, Felix Adler was being trained to be a rabbi like his father, Samuel Adler, the rabbi of the Reform Jewish Temple Emanu-El in New York. As part of his education, he enrolled at the University of Heidelberg, where he was influenced by neo-Kantian philosophy. He was especially drawn to the Kantian ideas that one could not prove the existence or non-existence of deities or immortality and that morality could be established independently of theology.[15]

During this time he was also exposed to the moral problems caused by the exploitation of women and labor. These experiences laid the intellectual groundwork for the ethical movement. Upon his return from Germany, in 1873, he shared his ethical vision with his father's congregation in the form of a sermon. Due to the negative reaction he elicited it became his first and last sermon as a rabbi in training.[16] Instead he took up a professorship at Cornell University and in 1876 gave a follow up sermon that led to the 1877 founding of the **New York Society for Ethical Culture**, which was the first of its kind.[15] By 1886, similar societies had sprouted up in Philadelphia, Chicago and St. Louis.[16]

These societies all adopted the same statement of principles:

- The belief that morality is independent of theology;
- The affirmation that new moral problems have arisen in modern industrial society which have not been adequately dealt with by the world's religions;
- The duty to engage in philanthropy in the advancement of morality;
- The belief that self-reform should go in lock step with social reform;
- The establishment of republican rather than monarchical governance of Ethical societies
- The agreement that educating the young is the most important aim.

In effect, the movement responded to the religious crisis of the time by replacing theology with unadulterated morality. It aimed to "disentangle moral ideas from religious doctrines, metaphysical systems, and ethical theories, and to make them an independent force in personal life and social relations."[16] Adler was also particularly critical of the religious emphasis on creed, believing it to be the source of sectarian bigotry. He therefore attempted to provide a universal fellowship devoid of ritual and ceremony, for those who would otherwise be divided by creeds. For the same reasons the movement also adopted a neutral position on religious beliefs, advocating neither atheism nor theism, agnosticism nor deism.[16]

The Adlerian emphasis on "deed not creed" translated into several public service projects. The year after it was founded, the New York society started a kindergarten, a district nursing service and a tenement-house building company. Later they opened the Ethical Culture School, then

Ethical Culture School (red) and Ethical Culture Society (white) buildings.

called the "Workingman's School," a Sunday school and a summer home for children, and other Ethical societies soon followed suit with similar projects. Unlike the philanthropic efforts of the established religious institutions of the time, the Ethical societies did not attempt to proselytize those they helped. In fact, they rarely attempted to convert anyone. New members had to be sponsored by existing members, and women were not allowed to join at all until 1893. They also resisted formalization, though nevertheless slowly adopted certain traditional practices, like Sunday meetings and life cycle ceremonies, yet did so in a modern humanistic context. In 1893, the four existing societies unified under the umbrella organization, the **American Ethical Union**.[16]

After some initial success the movement stagnated until after World War II. In 1946 efforts were made to revitalize and societies were created in New Jersey and Washington D.C., along with the inauguration of the Encampment for Citizenship. By 1968 there were thirty societies with a total national membership of over 5,500. However, the resuscitated movement differed from its predecessor in a few ways. The newer groups were being created in suburban locales and often to provide alternative Sunday schools for children, with adult activities as an afterthought.

There was also a greater focus on organization and bureaucracy, along with an inward turn emphasizing the needs of the group members over the more general social issues that had originally concerned Adler. The result was a transformation of American ethical societies into something much more akin to small Christian congregations in which the minister's most pressing concern is to tend to his or her flock.[16]

9.1.3 In Britain

Stanton Coit led the ethical movement in Britain.

In 1885 the ten-year-old American Ethical Culture movement helped to stimulate similar social activity in Great Britain, when American sociologist John Graham Brooks distributed pamphlets by Chicago ethical society leader William Salter to a group of British philosophers, including Bernard Bosanquet, John Henry Muirhead, and John Stuart MacKenzie.

One of Felix Adler's colleagues, Stanton Coit, visited them in London to discuss the "aims and principles" of their American counterparts. In 1886 the first British ethical society was founded. Coit took over the leadership of South Place for a few years. Ethical societies flourished in Britain. By 1896 the four London societies formed the Union of Ethical Societies, and between 1905 and 1910 there were over fifty societies in Great Britain, seventeen of which were affiliated with the Union. Part of this rapid growth was due to Coit, who left his role as leader of South Place in 1892 after being denied the power and authority he was vying for.

Because he was firmly entrenched in British ethicism, Coit remained in London and formed the West London Ethical Society, which was almost completely under his control. Coit worked quickly to shape the West London society not only around Ethical Culture but also the trappings of religious practice, renaming the society in 1914 to the Ethical Church. He transformed his meetings into services, and their space into something akin to a church. In a series of books Coit also began to argue for the transformation of the Anglican Church into an Ethical Church, while holding up the virtue of ethical ritual. He felt that the Anglican Church was in the unique position to harness the natural moral impulse that stemmed from society itself, as long as the Church replaced theology with science, abandoned supernatural beliefs, expanded its bible to include a cross-cultural selection of ethical literature and reinterpreted its creeds and liturgy in light of modern ethics and psychology. His attempt to reform the Anglican church failed, and ten years after his death in 1944, the Ethical Church building was sold to the Roman Catholic Church.[16]

During Stanton Coit's lifetime, the Ethical Church never officially affiliated with the Union of Ethical Societies, nor did South Place. In 1920 the Union of Ethical Societies changed its name to the Ethical Union.[17] Harold Blackham, who had taken over leadership of the London Ethical Church, then promoted its merger with the Rationalist Press Association and the South Place Ethical Society, and, in 1957, a Humanist Council was set up to explore amalgamation. Although issues over charitable status prevented a full amalgamation, the Ethical Union under Blackham changed its name in 1967 to become the British Humanist Association. The BHA is thus the legal successor body to the Union of Ethical Societies.[18]

Between 1886 and 1927 seventy-four ethical societies were started in Great Britain, although this rapid growth did not last long. The numbers declined steadily throughout the 1920s and early 30s, until there were only ten societies left in 1934. By 1954 there were only four. The situation became such that in 1971, sociologist Colin Campbell even suggested that one could say, "that when the South Place Ethical Society discussed changing its name to the South Place Humanist society in 1969, the English ethical movement ceased to exist."[16]

9.2 Ethical perspective

While Ethical Culturists generally share common beliefs about what constitutes ethical behavior and the good, individuals are encouraged to develop their own personal understanding of these ideas. This does not mean that Ethical Culturists condone moral relativism, which would relegate ethics to mere preferences or social conventions. Ethical principles are viewed as being related to deep truths about the way the world works, and hence not arbitrary. However, it is recognized that complexities render the understanding of ethical nuances subject to continued dialogue, exploration, and learning.

While the founder of Ethical Culture, Felix Adler, was a transcendentalist, Ethical Culturists may have a variety of understandings as to the theoretical origins of ethics. Key to the founding of Ethical Culture was the observation that too often disputes over religious or philosophical doctrines have distracted people from actually living ethically and doing good. Consequently, *"Deed before creed"* has long been a motto of the movement.[3][19]

9.3 Religious aspect

Pews and stained glass

Functionally, Ethical Societies are similar to churches or synagogues and are headed by "leaders" as clergy. Ethical Societies typically have Sunday morning meetings, offer moral instruction for children and teens, and do charitable work and social action. They may offer a variety of educational and other programs. They conduct weddings, commitment ceremonies, baby namings, and memorial services.

Individual Ethical Society members may or may not believe in a deity or regard Ethical Culture as their religion. Felix Adler said "Ethical Culture is religious to those who are religiously minded, and merely ethical to those who are not so minded." The movement does consider itself a religion in the sense that

> Religion is that set of beliefs and/or institutions, behaviors and emotions which bind human beings to something beyond their individual selves and foster in its adherents a sense of humility and gratitude that, in turn, sets the tone of one's world-view and requires certain behavioral dispositions relative to that which transcends personal interests.[20]

The Ethical Culture 2003 ethical identity statement states:

> It is a chief belief of Ethical religion that if we relate to others in a way that brings out their best, we will at the same time elicit the best in ourselves. By the "best" in each person, we refer to his or her unique talents and abilities that affirm and nurture life. We use the term "spirit" to refer to a person's unique personality and to the love, hope, and empathy that exists in human beings. When we act to elicit the best in others, we encourage the growing edge of their ethical development, their perhaps as-yet untapped but inexhaustible worth.

Since around 1950 the Ethical Culture movement has been increasingly identified as part of the modern Humanist movement. Specifically, in 1952, the American Ethical Union, the national umbrella organization for Ethical Culture societies in the United States, became one of the founding member organizations of the International Humanist and Ethical Union.

9.4 Key ideas

While Ethical Culture does not regard its founder's views as necessarily the final word, Adler identified focal ideas that remain important within Ethical Culture. These ideas include:

- *Human Worth and Uniqueness* – All people are taken to have inherent worth, not dependent on the value of what they do. They are deserving of respect and dignity, and their unique gifts are to be encouraged and celebrated.[2]

- *Eliciting the Best* – "Always act so as to Elicit the best in others, and thereby yourself" is as close as Ethical Culture comes to having a Golden Rule.[2]

- *Interrelatedness* – Adler used the term *The Ethical Manifold* to refer to his conception of the universe as made up of myriad unique and indispensable moral agents (individual human beings), each of whom has an inestimable influence on all the others. In other words, we are all interrelated, with each person playing a role in the whole and the whole affecting each person. Our interrelatedness is at the heart of ethics.

Many Ethical Societies prominently display a sign that says "The Place Where People Meet to Seek the Highest is Holy Ground".[21]

9.5 Locations

The largest concentration of Ethical Societies is in the New York metropolitan area, including Societies in New York, Manhattan, the Bronx,[22] Brooklyn, Queens, Westchester and Nassau County; and New Jersey, such as Bergen and Essex Counties, New Jersey.[23][24]

Ethical Societies exist in several U.S. cities and counties, including Austin, Texas; Baltimore; Boston; Chapel Hill; Asheville, North Carolina; Chicago; San Jose, California; Philadelphia; St. Louis; St. Peters, Missouri; Washington, D.C.; Lewisburg, Pennsylvania, and Vienna, Virginia.

Ethical Societies also exist outside the U.S. Conway Hall in London is home to the South Place Ethical Society, which was founded in 1787.[25]

9.6 Structure and events

Ethical societies are typically led by "Leaders" elected from the body of society members by the same members. A board of executives handles day-to-day affairs, and committees of members focus on specific activities and involvements of the society.

Ethical societies usually hold weekly meetings on Sundays, with the main event of each meeting being the "Platform", which involves a half-hour speech by the Leader of the Ethical Society, a member of the society or by guests. Sunday school for minors is also held at most ethical societies concurrent with the Platform.

The American Ethical Union holds an annual AEU Assembly bringing together Ethical societies from across the US.

9.7 Legal challenges

The tax status of Ethical Societies as religious organizations has been upheld in court cases in Washington, D.C. (1957), and in Austin, Texas (2003). The Texas State Appeals Court said of the challenge by the state comptroller Carole Keeton Strayhorn, "the Comptroller's test [requiring a group to demonstrate its belief in a Supreme Being] fails to include the whole range of belief systems that may, in our diverse and pluralistic society, merit the First Amendment's protection."[26]

9.8 Advocates

Albert Einstein was a supporter of Ethical Culture. On the seventy-fifth anniversary of the New York Society for Ethical Culture he noted that the idea of Ethical Culture embodied his personal conception of what is most valuable and enduring in religious idealism. Humanity requires such a belief to survive, Einstein argued. He observed, "Without 'ethical culture' there is no salvation for humanity."[27]

9.9 See also

- Arthur E. Briggs, Los Angeles City Council member, 1939–41, Ethical Society leader
- British Humanist Association, which inherited many British ethical societies
- Religious humanism
- Unitarian Universalism
- Washington Ethical Society v. District of Columbia

9.10 References

[1] From Reform Judaism to ethical culture: the religious evolution of Felix Adler Benny Kraut, Hebrew Union College Press, 1979

[2] Brown, Stuart C; Collinson, Diané, "Adler", *Biographical dictionary of twentieth-century philosophers*, Books, Google, p. 7

[3] The conservator, Volumes 3-4, Horace Traubel, Volume 3, page 31

[4] City of London page on Finsbury Circus Conservation Area Character Summary Archived 8 October 2006 at the Wayback Machine.

[5] *The Sexual Contract*, by Carole Patema. P160

[6] "Women's Politics in Britain 1780-1870: Claiming Citizenship" by Jane Rendall, esp. "72. The religious backgrounds of feminist activists"

[7] "Ethical Society history page". Ethicalsoc.org.uk. Retrieved 2013-09-29.

[8] Good, James A. "The Development of Thomas Davidson's Religious and Social Thought".

[9] George Hendrick, *Henry Salt: Humanitarian Reformer and Man of Letters*, University of Illinois Press, pg. 47 (1977).

[10] Jeffrey Weeks, *Making Sexual History*, Wiley-Blackwell, pg. 20, (2000).

- [11] Knight, William. *Memorials of Thomas Davidson.*(Boston: Ginn & Company, 1907), 18

- [12] Knight, William. *Memorials of Thomas Davidson.*(Boston: Ginn & Company, 1907), 19

- [13] William A. Knight, *Memorials of Thomas Davidson: The Wandering Scholar* (Boston and London: Ginn and Co, 1907). p. 16, 19, 46.

- [14] Pease, Edward R. (1916). *The History of the Fabian Society.* New York: E.P. Dutton and Co.

- [15] Howard B. Radest. 1969. *Toward Common Ground: The Story of the Ethical Societies in the United States.* New York: Fredrick Unger Publishing Co.

- [16] Colin Campbell. 1971. *Towards a Sociology of Irreligion.* London: McMillan Press.

- [17] I.D. MacKillop. 1986. *The British Ethical Societies.* Cambridge: Cambridge University Press.

- [18] British Humanist Association: Our History since 1896 Archived 9 August 2013 at WebCite

- [19] Ethics as a Religion, David Saville Muzzey, 273 pages, 1951, 1967, 1986

- [20] Arthur Dobrin, quoted in "Ethical Culture as Religion", Jone Johnson Lewis, 2003, American Ethical Union Library

- [21] Goldberger, Paul (August 12, 2010), *Architecture, Sacred Space, and the Challenge of the Modern*, Chautauqua Institution

- [22] "Riverdale Yonkers Society for Ethical Culture". Rysec.org. 2012-08-24. Retrieved 2013-09-29.

- [23] Ethical Societies.

- [24] Bergen, NJ Society

- [25] South Place Ethical Society, About the Society.

- [26] Report on Texas Court of Appeals decision, rutgers.edu, 2003

- [27] Ericson, Edward L. *The Humanist Way: An Introduction to Ethical Humanist Religion.* The American Ethical Union. Retrieved 2008-07-23.

This article incorporates text from a publication now in the public domain: Edward William Bennett (1901–1906). "Ethical Culture, Society for". In Singer, Isidore; et al. *Jewish Encyclopedia*. New York: Funk & Wagnalls Company.

9.11 Further reading

- Ericson, Edward L. *The Humanist Way: An Introduction to Ethical Humanist Religion.* A Frederick Ungar book, The Continuum Publishing Company. 205 pages, 1988.

- Radest, Howard. *Toward Common Ground: The Story of the Ethical Societies in the United States.* Ungar, 1969

- Muzzey, David Saville. *Ethics as a Religion*, 273 pages, 1951, 1967, 1986.

9.12 External links

- Official website

- *Comptroller of Public Accounts v. Ethical Society of Austin*

Chapter 10

Ethical naturalism

Ethical naturalism (also called **moral naturalism** or **naturalistic cognitivistic definism**)[1] is the meta-ethical view which claims that:

Reductive Naturalism

1. Ethical sentences express propositions.
2. Some such propositions are true.
3. Those propositions are made true by objective features of the world, independent of human opinion.
4. These moral features of the world *are* reducible to some set of non-moral features

or

Non-Reductive Naturalism

1. Ethical sentences express propositions.
2. Some such propositions are true.
3. Those propositions are made true by objective features of the world, independent of human opinion.
4. These moral features of the world are not reducible to some set of non-moral features, but are supervened by some set of non-moral features

This makes ethical naturalism a definist form of moral realism, which is in turn a form of cognitivism. Ethical naturalism stands in opposition to ethical non-naturalism, which denies that moral terms refer to anything other than irreducible moral properties, as well as to all forms of moral anti-realism, including ethical subjectivism (which denies that moral propositions refer to objective facts), error theory (which denies that any moral propositions are true), and non-cognitivism (which denies that moral sentences express propositions at all).

It is important to distinguish the versions of ethical naturalism which have received the most sustained philosophical interest, for example, Cornell Realism, from the position that "the way things are is always the way they ought to be", which few ethical naturalists hold. Ethical naturalism does, however, reject the fact-value distinction: it suggests that inquiry into the natural world can increase our moral knowledge in just the same way it increases our scientific knowledge. Indeed, proponents of ethical naturalism have argued that humanity needs to invest in their science of morality – although the existence of such a science is debated.

Ethical naturalism encompasses any reduction of ethical properties, such as 'goodness', to non-ethical properties; there are many different examples of such reductions, and thus many different varieties of ethical naturalism. Hedonism, for example, is the view that goodness is ultimately just pleasure.

10.1 Ethical theories that can be naturalistic

- Altruism
- Consequentialism
- Consequentialist libertarianism
- Cornell realism
- Ethical egoism
- Evolutionary ethics
- Hedonism
- Humanistic ethics
- Natural-rights libertarianism
- Objectivism
- Utilitarianism
- Virtue ethics

10.2 Criticisms

Ethical naturalism has been criticized most prominently by ethical non-naturalist G. E. Moore, who formulated the open-question argument. Garner and Rosen say that a common definition of "natural property" is one "which can be discovered by sense observation or experience, experiment, or through any of the available means of science." They also say that a good definition of "natural property" is problematic but that "it is only in criticism of naturalism, or in an attempt to distinguish between naturalistic and nonnaturalistic definist theories, that such a concept is needed."[2] R. M. Hare also criticised ethical naturalism because of its fallacious definition of the terms 'good' or 'right' explaining how value-terms being part of our prescriptive moral language are not reducible to descriptive terms: "Value-terms have a special function in language, that of commending; and so they plainly cannot be defined in terms of other words which themselves do not perform this function"[3]

10.2.1 Moral relativism

When it comes to the moral questions that we might ask, it can be difficult to argue that there is not necessarily some level of meta-ethical relativism – and failure to address this matter is criticized as ethnocentrism.

Could torture under certain conditions be "wrong" for a species?

As a broad example of relativism, we would no doubt see very different moral systems in an alien race that can only survive by occasionally ingesting one another. As a narrow example, there would be further specific moral opinions for each individual of that species.

Some forms of moral realism are compatible with some degree of meta-ethical relativism. This argument rests on the assumption that one can have a "moral" discussion on various scales; that is, what is "good" for: a certain part of your being (leaving open the possibility of conflicting motives), you as a single individual, your family, your society, your species, your type of species. For example, a moral universalist (and certainly an absolutist) might argue that, just as one can discuss what is 'good and evil' at an individual's level, so too can one make certain "moral" propositions with truth values relative at the level of the species. In other words, the moral relativist need not deem *all* moral propositions as necessarily subjective. The answer to "*is free speech normally good for human societies?*" is relative in a sense, but the moral realist would argue that an individual can be incorrect in this matter. This may be the philosophical equivalent of the more pragmatic arguments made by some scientists.

10.2.2 Moral nihilism

Moral nihilists maintain that any talk of an objective morality is incoherent and better off using other terms. Proponents of moral science like Ronald A. Lindsay have counter-argued that their way of understanding "morality" as a practical enterprise is the way we ought to have understood it in the first place. He holds the position that the alternative seems to be the elaborate philosophical reduction of the word "moral" into a vacuous, useless term.[4] Lindsay adds that it is important to reclaim the specific word "Morality" because of the connotations it holds with many individuals.

10.3 Morality as a science

Main article: Science of morality

Author Sam Harris has argued that we overestimate the relevance of many arguments against the science of morality, arguments he believes scientists happily and rightly disregard in other domains of science like physics. For example, a scientist may find herself attempting to argue against philosophical skeptics, when Harris says she should be practically asking – as scientists would in any other domain – "why would we listen to a solipsist in the first place?" This, Harris contends, is part of what it means to practice a science of morality.

Physicist Sean Carroll believes that conceiving of morality as a science could be a case of scientific imperialism and insists that what is "good for conscious creatures" is not an adequate working definition of "moral".[5] In opposition, Vice President at the Center for Inquiry, John Shook, claims that this working definition is more than adequate for science at present, and that disagreement should not immobilize the scientific study of ethics.[6]

Richard Carrier's chapter *"Moral Facts Naturally Exist (and*

Science Could Find Them)" sets out to prove a Moral realism centered around human satisfaction. It was peer reviewed by four philosophers.

Sam Harris argues that there are societally optimal "moral peaks" to discover.

In modern times, many thinkers discussing the fact-value distinction and the Is-ought problem have settled on the idea that one cannot derive *ought* from *is*. Conversely, Harris maintains that the fact-value distinction is a confusion, proposing that values are really a certain kind of fact. Specifically, Harris suggests that values amount to empirical statements about "the flourishing of conscious creatures in a society". He argues that there are objective answers to moral questions, even if some are difficult or impossible to possess in practice. In this way, he says, science can tell us what to value. Harris adds that we do not demand absolute certainty from predictions in physics so we should not demand that of a science studying morality.[7]

10.4 References

[1] Garner & Rosen 1967, p. 228

[2] Garner & Rosen 1967, p. 239

[3] Hare & M.R. 1964, p. 91

[4] Center Stage | Episode 24 – Bioethics and Public Policy, Part 1. Center for Inquiry (2010-04-12). Retrieved on 2011-04-30.

[5] Sam Harris Responds | Cosmic Variance | Discover Magazine. Blogs.discovermagazine.com. Retrieved on 2011-04-30.

[6] Sam Harris vs. The Philosophers on Morality. Center for Inquiry (2010-05-14). Retrieved on 2011-04-30.

[7] Sam Harris: Science can answer moral questions | Video on. Ted.com. Retrieved on 2011-04-30.

10.5 Other sources

- Garner, Richard T.; Rosen, Bernard (1967). *Moral Philosophy: A Systematic Introduction to Normative Ethics and Meta-ethics*. New York: Macmillan. OCLC 362952.

- Hare, H. R. (1964). *The Language of Morals*. Oxford: Oxford University Press.

10.6 External links

- Lenman, James (August 7, 2006). "Moral Naturalism". *Stanford Encyclopedia of Philosophy*.

- Philosophy 302: Naturalistic Ethics

Chapter 11

Evolutionary ethics

Evolutionary ethics is a field of inquiry that explores how evolutionary theory might bear on our understanding of ethics or morality.[1] The range of issues investigated by evolutionary ethics is quite broad. Supporters of evolutionary ethics have claimed that it has important implications in the fields of descriptive ethics, normative ethics, and metaethics.

Descriptive evolutionary ethics consists of biological approaches to morality based on the alleged role of evolution in shaping human psychology and behavior. Such approaches may be based in scientific fields such as evolutionary psychology, sociobiology, or ethology, and seek to explain certain human moral behaviors, capacities, and tendencies in evolutionary terms. For example, the nearly universal belief that incest is morally wrong might be explained as an evolutionary adaptation that furthered human survival.

Normative (or prescriptive) evolutionary ethics, by contrast, seeks not to explain moral behavior, but to justify or debunk certain normative ethical theories or claims. For instance, some proponents of normative evolutionary ethics have argued that evolutionary theory undermines certain widely held views of humans' moral superiority over other animals.

Evolutionary metaethics asks how evolutionary theory bears on theories of ethical discourse, the question of whether objective moral values exist, and the possibility of objective moral knowledge. For example, some evolutionary ethicists have appealed to evolutionary theory to defend various forms of moral anti-realism (the claim, roughly, that objective moral facts do not exist) and moral skepticism.

11.1 History

The first notable attempt to explore links between evolution and ethics was made by Charles Darwin in *The Descent of Man* (1871). In Chapters IV and V of that work Darwin set out to explain the origin of human morality in order to show that there was no absolute gap between man and animals. Darwin sought to show how a refined moral sense, or conscience, could have developed through a natural evolutionary process that began with social instincts rooted in our nature as social animals.

Not long after the publication of Darwin's *The Descent of Man*, evolutionary ethics took a very different—and far more dubious—turn in the form of Social Darwinism. Leading Social Darwinists such as Herbert Spencer and William Graham Sumner sought to apply the lessons of biological evolution to social and political life. Just as in nature, they claimed, progress occurs through a ruthless process of competitive struggle and "survival of the fittest," so human progress will occur only if government allows unrestricted business competition and makes no effort to protect the "weak" or "unfit" by means of social welfare laws.[2] Critics such as Thomas Henry Huxley, G. E. Moore, William James, and John Dewey roundly criticized such attempts to draw ethical and political lessons from Darwinism, and by the early decades of the twentieth century Social Darwinism was widely viewed as discredited.[3]

The modern revival of evolutionary ethics owes much to E. O. Wilson's 1975 book, *Sociobiology: The New Synthesis*. In that work, Wilson argues that there is a genetic basis for a wide variety of human and nonhuman social behaviors. In recent decades, evolutionary ethics has become a lively topic of debate in both scientific and philosophical circles.

11.2 Descriptive evolutionary ethics

See also: Evolution of morality

The most widely accepted form of evolutionary ethics is descriptive evolutionary ethics. Descriptive evolutionary ethics seeks to explain various kinds of moral phenomena wholly or partly in genetic terms. Ethical topics addressed include altruistic behaviors, an innate sense of fairness, a capacity for normative guidance, feelings of kindness or love, self-sacrifice, incest-avoidance, parental care, in-group loyalty, monogamy, feelings related to competitiveness and retribution, moral "cheating," and hypocrisy.

A key issue in evolutionary psychology has been how altruistic feelings and behaviors could have evolved, in both humans and nonhumans, when the process of natural selection is based on the multiplication over time only of those genes that adapt better to changes in the environment of the species. Theories addressing this have included kin selection, group selection, and reciprocal altruism (both direct and indirect, and on a society-wide scale). Descriptive evolutionary ethicists have also debated whether various types of moral phenomena should be seen as adaptations which have evolved because of their direct adaptive benefits, or spin-offs that evolved as side-effects of adaptive behaviors.

11.3 Normative evolutionary ethics

Normative evolutionary ethics is the most controversial branch of evolutionary ethics. Normative evolutionary ethics aims at defining which acts are right or wrong, and which things are good or bad, in evolutionary terms. It is not merely *describing*, but it is *prescribing* goals, values and obligations. Social Darwinism, discussed above, is the most historically influential version of normative evolutionary ethics. As philosopher G. E. Moore famously argued, many early versions of normative evolutionary ethics seemed to commit a logical mistake that Moore dubbed the *naturalistic fallacy*. This was the mistake of defining a normative property, such as goodness, in terms of some non-normative, naturalistic property, such as pleasure or survival. Many early critics of normative evolutionary ethics also argued that such ethics commits the "is-ought fallacy" of drawing an ethical conclusion (e.g., "Social cooperation is good") directly from a non-ethical premise (e.g., "Social cooperation contributes to human survival").

More sophisticated forms of normative evolutionary ethics need not commit either the naturalistic fallacy or the is-ought fallacy. But all varieties of normative evolutionary ethics face the difficult challenge of explaining how evolutionary facts can have normative authority for rational agents. "Regardless of why one has a given trait, the question for a rational agent is always: is it right for me to exercise it, or should I instead renounce and resist it as far as I am able?"[4]

11.4 Evolutionary Metaethics

Evolutionary theory may not be able to tell us what is morally right or wrong, but it might be able to illuminate our use of moral language, or to cast doubt on the existence of objective moral facts or the possibility of moral knowledge. Evolutionary ethicists such as Michael Ruse, E. O. Wilson, Richard Joyce, and Sharon Street have defended such claims.

Some philosophers who support evolutionary meta-ethics use it to undermine views of human well-being that rely upon Aristotelian teleology, or other goal-directed accounts of human flourishing. A number of thinkers have appealed to evolutionary theory is to attempt to debunk moral realism or support moral skepticism. Sharon Street is one prominent ethicist who argues that evolutionary psychology undercuts moral realism. According to Street, human moral decision-making is "thoroughly saturated" with evolutionary influences. Natural selection, she argues, would have rewarded moral dispositions that increased fitness, not ones that track moral truths, should they exist. It would be a remarkable and unlikely coincidence if "morally blind" ethical traits aimed solely at survival and reproduction aligned closely with independent moral truths. So we cannot be confident that our moral beliefs accurately track objective moral truth. Consequently, realism forces us to embrace moral skepticism. Such skepticism, Street claims, is implausible. So we should reject realism and instead embrace some antirealist view that allows for rationally justified moral beliefs.[5]

Defenders of moral realism have offered two sorts of replies. One is to deny that evolved moral responses would likely diverge sharply from moral truth. According to David Copp, for example, evolution would favor moral responses that promote social peace, harmony, and cooperation. But such qualities are precisely those that lie at the core of any plausible theory of objective moral truth. So Street's alleged "dilemma"—deny evolution or embrace moral skepticism—is a false choice.[6]

A second response to Street is to deny that morality is as "saturated" with evolutionary influences as Street claims. William Fitzpatrick, for instance, argues that "[e]ven if there is significant evolutionary influence on the content of many of our moral beliefs, it remains possible that many of our moral beliefs are arrived at partly (or in some cases wholly) through autonomous moral reflection and reasoning, just as with our mathematical, scientific and philosophical beliefs."[7] The wide variability of moral codes, both across cultures and historical time periods, is difficult to explain if morality is as pervasively shaped by genetic factors as Street claims.

Another common argument evolutionary ethicists use to debunk moral realism is to claim that the success of evolutionary psychology in explaining human ethical responses makes the notion of moral truth "explanatorily superfluous." If we can fully explain, for example, why parents naturally love and care for their children in purely evolutionary terms, there is no need to invoke any "spooky" realist moral truths to do any explanatory work. Thus, for reasons of theoretical simplicity we should not posit the existence of such truths and, instead, should explain the widely held belief in objective moral truth as "an illusion fobbed off on us by our genes in order to get us to cooperate with one another (so that our genes survive)."[8]

Here again the central question is whether the influence of evolution on morality is as pervasive as the critics of moral realism claim. If, as seems likely, there are important aspects of morality that cannot be explained in genetic terms, appeals to moral truth may not be explanatory fifth-wheels.

11.5 See also

- Animal Faith
- Appeal to nature
- Bioethics
- Eugenics
- Evolution of morality
- Game theory
- Social Darwinism
- Universal Darwinism

11.6 Notes

[1] William Fitzpatrick, "Morality and Evolutionary Biology." *Stanford Encyclopedia of Philosophy* Available online at: https://plato.stanford.edu/entries/morality-biology/.

[2] Gregory Bassham, *The Philosophy Book: From the Vedas to the New Atheists, 250 Milestones in the History of Philosophy*. New York: Sterling, 2015, p. 318.

[3] Richard Hofstadter, *Social Darwinism in American Thought*, rev. ed. Boston: Beacon Press, 1955, p. 203.

[4] Fitzpatrick, "Morality and Evolutionary Biology," Section 3.2.

[5] Sharon Street, "A Darwinian Dilemma for Realist Theories of Value." *Philosophical Studies*, 127: 109-66.

[6] David Copp, "Darwinian Skepticism about Moral Realism." *Philosophical Issues*, 18: 186-206.

[7] Fitzpatrick, "Morality and Evolutionary Biology," Section 4.1.

[8] Michael Ruse and E. O. Wilson, "The Evolution of Ethics." *New Scientist*, 102: 1478 (17 October 1985): 51-52.

11.7 References

- Huxley, Thomas Henry (1893). "Evolution and Ethics". In Nitecki, Matthew H.; Nitecki, Doris V. *Evolutionary Ethics*. Albany: State University of New York (published 1993). ISBN 0-7914-1499-X.

- Ruse, Michael (1995). "Evolutionary Ethics: A Phoenix Arisen". In Thompson, Paul. *Issues in Evolutionary Ethics*. Albany: State University of New York. ISBN 0-7914-2027-2.

11.8 Further reading

- Curry, O. (2006). Who's afraid of the naturalistic fallacy? *Evolutionary Psychology, 4*, 234-247. Full text

- Dawkins, Richard (1976). *The Selfish Gene*. ISBN 1-155-16265-X.

- Duntley, J.D., & Buss, D.M. (2004). The evolution of evil. In A. Miller (Ed.), *The social psychology of good and evil*. New York: Guilford. 102-123. Full text

- Hauser, Marc (2006). *Moral Minds*. ISBN 0-06-078070-3.

- Huxley, Julian. *Evolutionary Ethics 1893-1943*. Pilot, London. In USA as *Touchstone for ethics* Harper, N.Y. (1947) [includes text from both T.H. Huxley and Julian Huxley]

- Katz, L. (Ed.) Evolutionary Origins of Morality: Cross-Disciplinary Perspectives Imprint Academic, 2000 ISBN 0-907845-07-X

- Kitcher, Philip (1995) "Four Ways of "Biologicizing" Ethics" in Elliott Sober (ed.) Conceptual Issues in Evolutionary Biology, The MIT Press

- Kitcher, Philip (2005) "Biology and Ethics" in David Copp (ed.) The Oxford Handbook of Ethical Theory, Oxford University Press

- Krebs, D. L. & Denton, K. (2005). Toward a more pragmatic approach to morality: A critical evaluation of Kohlberg's model. *Psychological Review, 112*, 629-649. Full text

- Krebs, D. L. (2005). An evolutionary reconceptualization of Kohlberg's model of moral development. In R. Burgess & K. MacDonald (Eds.) *Evolutionary Perspectives on Human Development*, (pp. 243–274). CA: Sage Publications. Full text

- Mascaro, S., Korb, K.B., Nicholson, A.E., Woodberry, O. (2010). Evolving Ethics: The New Science of Good and Evil. Exeter, UK: Imprint Academic.

- Richerson, P.J. & Boyd, R. (2004). Darwinian Evolutionary Ethics: Between Patriotism and Sympathy. In Philip Clayton and Jeffrey Schloss, (Eds.), *Evolution and Ethics: Human Morality in Biological and Religious Perspective*, pp. 50–77. Full text ISBN 0-8028-2695-4

- Ridley, Matt (1996). *The Origins of Virtue*. Viking. ISBN 0-14-026445-0.

- Ruse, Michael (1993). "The New Evolutionary Ethics". In Nitecki, Matthew H.; Nitecki, Doris V. *Evolutionary Ethics*. Albany: State University of New York. ISBN 0-7914-1499-X.

- Shermer, Michael (2004). *The Science of Good and Evil: Why People Cheat, Gossip, Care, Share, and Follow the Golden Rule*. New York: Henry Holt and Company. ISBN 0-8050-7520-8.

- Teehan, J. & diCarlo, C. (2004). On the Naturalistic Fallacy: A conceptual basis for evolutionary ethics. *Evolutionary Psychology, 2*, 32-46. Full text

- de Waal, Frans (1996). *Good Natured: The Origins of Right and Wrong in Humans and Other Animals*. London: Harvard University Press. ISBN 0-674-35660-8.

- Walter, A. (2006). The anti-naturalistic fallacy: Evolutionary moral psychology and the insistence of brute facts. *Evolutionary Psychology, 4*, 33-48. Full text

- Wilson, D. S., E. Dietrich, et al. (2003). On the inappropriate use of the naturalistic fallacy in evolutionary psychology. *Biology and Philosophy 18:* 669-682. Full text

- Wilson, D. S. (2002). Evolution, morality and human potential. *Evolutionary Psychology: Alternative Approaches*. S. J. Scher and F. Rauscher, Kluwer Press: 55-70 Full text

- Wilson, E. O. (1979). *On Human Nature*. ISBN 0-671-54130-7.

- Wright, Robert (1995). *The Moral Animal*. ISBN 0-679-40773-1.

11.9 External links

- The Evolution of Ethics: An Introduction to Cybernetic Ethics by S. E. Bromberg

- Evolutionary Ethics at the Internet Encyclopedia of Philosophy

- FitzPatrick, William. "Morality and Evolutionary Biology". *Stanford Encyclopedia of Philosophy*.

- Okasha, Samir. "Biological Altruism". *Stanford Encyclopedia of Philosophy*.

Chapter 12

Secular ethics

Secular ethics is a branch of moral philosophy in which ethics is based solely on human faculties such as logic, empathy, reason or moral intuition, and not derived from supernatural revelation or guidance—the source of ethics in many religions. Secular ethics refers to any ethical system that does not draw on the supernatural, such as humanism, secularism and freethinking. A classical example of literature on secular ethics is the Kural text, authored by the ancient Indian philosopher Valluvar who lived around the 4th and 1st centuries BCE.

Secular ethical systems comprise a wide variety of ideas to include the normativity of social contracts, some form of attribution of intrinsic moral value, intuition-based deontology, cultural moral relativism, and the idea that scientific reasoning can reveal objective moral truth (known as science of morality).

Secular ethics frameworks are not always mutually exclusive from theological values. For example, the Golden Rule or a commitment to non-violence, could be supported by both religious and secular frameworks. Secular ethics systems can also vary within the societal and cultural norms of a specific time period.

12.1 Tenets of secular ethics

Despite the width and diversity of their philosophical views, secular ethicists generally share one or more principles:

- Human beings, through their ability to empathize, are capable of determining ethical grounds.

- The well-being of others is central to ethical decision-making

- Human beings, through logic and reason, are capable of deriving normative principles of behavior.

- This may lead to a behavior preferable to that propagated or condoned based on religious texts. Alternatively, this may lead to the advocacy of a system of moral principles that a broad group of people, both religious and non-religious, can agree upon.

- Human beings have the moral responsibility to ensure that societies and individuals act based on these ethical principles.

- Societies should, if at all possible, advance from a less ethical and just form to a more ethical and just form.

Many of these tenets are applied in the science of morality, the use of the scientific method to answer moral questions. Various thinkers have framed morality as questions of empirical truth to be explored in a scientific context. The science is related to ethical naturalism, a type of ethical realism.

In *How Good People Make Tough Choices: Resolving the Dilemmas of Ethical Living*, Rushworth Kidder identifies four general characteristics of an ethical code:

1. It is brief
2. It is usually not explanatory
3. Can be expressed in a number of forms (e.g. positive or negative, single words or a list of sentences)
4. Centers on moral values[1]

12.1.1 Humanist ethics

Humanists endorse universal morality based on the commonality of human nature, and that knowledge of right and wrong is based on our best understanding of our individual and joint interests, rather than stemming from a transcendental or arbitrarily local source, therefore rejecting faith completely as a basis for action. The humanist ethics goal is a search for viable individual, social and political principles of conduct, judging them on their ability to enhance human well-being and individual responsibility, ultimately eliminating human suffering.

The International Humanist and Ethical Union (IHEU) is the world-wide umbrella organization for those adhering to the Humanist life stance.

> *Humanism is a democratic and ethical life stance, which affirms that human beings have the right and responsibility to give meaning and shape to their own lives. It stands for the building of a more humane society through an ethic based on human and other natural values in the spirit of reason and free inquiry through human capabilities. It is not theistic, and it does not accept supernatural views of reality.*[2]

Humanism is known to adopt principles of the Golden Rule.

12.1.2 Secular ethics and religion

See also: Morality § Morality and religion

There are those who state that religion is not necessary for moral behavior at all.[3] The Dalai Lama has said that compassion and affection are human values independent of religion: "We need these human values. I call these secular ethics, secular beliefs. There's no relationship with any particular religion. Even without religion, even as nonbelievers, we have the capacity to promote these things."[4]

Those who are unhappy with the negative orientation of traditional religious ethics believe that prohibitions can only set the absolute limits of what a society is willing to tolerate from people at their worst, not guide them towards achieving their best. In other words, someone who follows all these prohibitions has just barely avoided being a criminal, not acted as a positive influence on the world. They conclude that rational ethics can lead to a fully expressed ethical life, while religious prohibitions are insufficient.

That does not mean secular ethics and religion are mutually exclusive. In fact, many principles, such as the Golden Rule, are present in both systems, and some religious people, as well as some Deists, prefer to adopt a rational approach to ethics.

12.2 Examples of secular ethical codes

12.2.1 Humanist Manifestos

The Humanist Manifestos are three manifestos, the first published in 1933, that outline the philosophical views and stances of humanists. Integral to the manifestos is a lack of supernatural guidance.

12.2.2 Alternatives to the Ten Commandments

There are numerous versions of Alternatives to the Ten Commandments

12.2.3 Girl Scout law

The Girl Scout law is as follows:

> I will do my best to be
>
>> honest and fair,
>> friendly and helpful,
>> considerate and caring,
>> courageous and strong, and
>> responsible for what I say and do,
>
> and to
>
>> respect myself and others,
>> respect authority,
>> use resources wisely,
>> make the world a better place, and
>> be a sister to every Girl Scout.[5]

12.2.4 United States Naval Academy honor concept

"Midshipmen are persons of integrity: They stand for that which is right.

> They tell the truth and ensure that the full truth is known. **They do not lie**.
> They embrace fairness in all actions. They ensure that work submitted as their own is their own, and that assistance received from any source is authorized and properly documented. **They do not cheat**.
> They respect the property of others and ensure that others are able to benefit from

the use of their own property. **They do not steal.**[6]

12.2.5 Minnesota Principles

The Minnesota Principles were proposed "by the Minnesota Center for Corporate Responsibility in 1992 as a guide to international business activities":

1. Business activities must be characterized by fairness. We understand fairness to include equitable treatment and equality of opportunity for all participants in the marketplace.

2. Business activities must be characterized by honesty. We understand honesty to include candor, truthfulness and promise-keeping.

3. Business activities must be characterized by respect for human dignity. We understand this to mean that business activities should show a special concern for the less powerful and the disadvantaged.

4. Business activities must be characterized by respect for the environment. We understand this to mean that business activities should promote sustainable development and prevent environmental degradation and waste of resources.[7]

12.2.6 Rotary Four-Way Test

The Four-Way Test test is the "linchpin of Rotary International's ethical practice." It acts as a test of thoughts as well as actions. It asks, "Of the things we think, say, or do":

1. Is it the truth?
2. Is it fair to all concerned?
3. Will it build goodwill and better friendships?
4. Will it be beneficial to all concerned?[1]

12.2.7 Military codes

See also: Ranger Creed

As the United States Constitution prohibits the establishment of a government religion, US military codes of conduct typically contain no religious overtones.

West Point Honor Code

The West Point honor code states that "A cadet will not lie, cheat, steal, or tolerate those who do." The non-toleration clause is key in differentiating it from numerous other codes.[8]

12.3 Nature and ethics

See also: Social effect of evolutionary theory and evolutionary ethics

Whether or not the relationships between animals found in nature and between people in early human evolution can provide a basis for human morality is a persistently unresolved question. Thomas Henry Huxley wrote in *Evolution and Ethics* in 1893 that people make a grave error in trying to create moral ideas from the behavior of animals in nature. He remarked:

> The practice of that which is ethically best — what we call goodness or virtue — involves a course of conduct which, in all respects, is opposed to that which leads to success in the cosmic struggle for existence. In place of ruthless self-assertion it demands self-restraint; in place of thrusting aside, or treading down, all competitors, it requires that the individual shall not merely respect, but shall help his fellows... It repudiates the gladiatorial theory of existence... Laws and moral precepts are directed to the end of curbing the cosmic process.[9]

Famous biologist and writer Stephen Jay Gould has stated that "answers will not be read passively from nature" and "[t]he factual state of the world does not teach us how we, with our powers for good and evil, should alter or preserve it in the most ethical manner". Thus, he concluded that ideas of morality should come from a form of higher mental reason, with nature viewed as an independent phenomenon.[9]

Evolutionary ethics is not the only way to involve nature with ethics. For example, there are ethically realist theories like ethical naturalism. Related to ethical naturalism is also the idea that ethics are best explored, not just using the lens of philosophy, but science as well (a science of morality).

12.4 Key philosophers and philosophical texts

12.4.1 Valluvar

Valluvar (a theist who wrote a secular text) statue in SOAS, University of London.

Thiruvalluvar, an Indian poet-philosopher of the pre-Christian era and the author of the Kural, a non-denominational work on secular ethics and morality, is believed to have lived around 5th and 2nd centuries BCE. While others of his time chiefly focused on the praise of God, culture and the ruler of the land, Valluvar focused on the moral behaviors of the common individual.[10] Valluvar limits his theistic teachings to the introductory chapter of the Kural text, the "Praise of God."[11] Throughout the text thereafter, he focuses on the everyday moral behaviors of an individual, thus making the text a secular one. Even in the introductory chapter, he refrains from mentioning the name of any particular god but only addresses God in generic terms as "the Creator," "the truly Wise One," "the One of eight-fold excellence," and so forth.[12] Translated into about 80 world languages, the Kural text remains one of the most widely translated non-religious works in the world.[13] Praised as "the Universal Veda,"[14] it emphasizes on the ethical edifices of non-violence, vegetarianism, casteless human brotherhood, absence of desires, path of righteousness and truth, and so forth, besides covering a wide range of subjects such as moral codes of rulers, friendship, agriculture, knowledge and wisdom, sobriety, love, and domestic life.[12]

12.4.2 Holyoake

George Jacob Holyoake's 1896 publication *English Secularism* defines secularism thus:

"Secularism is a code of duty pertaining to this life, founded on considerations purely human, and intended mainly for those who find theology indefinite or inadequate, unreliable or unbelievable. Its essential principles are three: (1) The improvement of this life by material means. (2) That science is the available Providence of man. (3) That it is

Holyoake, agnostic

good to do good. Whether there be other good or not, the good of the present life is good, and it is good to seek that good."[15]

Holyoake held that secularism should take no interest at all in religious questions (as they were irrelevant), and was thus to be distinguished from strong freethought and atheism. In this he disagreed with Charles Bradlaugh, and the disagreement split the secularist movement between those who argued that anti-religious movements and activism was not necessary or desirable and those who argued that it was.

12.4.3 Nietzsche

Friedrich Nietzsche based his work on ethics on the rejection of Christianity and authority in general, or on moral nihilism. Nietzsche's many works spoke of a Master-Slave Morality, The Will to Power, or something stronger that overcomes the weaker and Darwinistic adaptation and will to live. Nietzsche expressed his moral philosophy throughout his collection of works; the most important of these to secular ethics being *The Gay Science* (in which the famous God is dead phrase was first used), *Thus Spoke Zarathustra*, *Beyond Good and Evil* and *On The Genealogy of Morals*.

Nietzsche, atheist

Kant, theist (disputably Christian)

12.4.4 Kant

Main article: Kantian ethics

On ethics, Kant wrote works that both described the nature of universal principles and also sought to demonstrate the procedure of their application. Kant maintained that only a "good will" is morally praiseworthy, so that doing what appears to be ethical for the wrong reasons is not a morally good act. Kant's emphasis on one's intent or reasons for acting is usually contrasted with the utilitarian tenet that the goodness of an action is to be judged solely by its results. Utilitarianism is a hypothetical imperative, if one wants _____, they must do _____. Contrast this with the Kantian ethic of the categorical imperative, where the moral act is done for its own sake, and is framed: One must do _____ or alternatively, one must not do _____.

For instance, under Kantian ethics, if a person were to give money to charity because failure to do so would result in some sort of punishment from a god or Supreme Being, then the charitable donation would not be a morally good act. A dutiful action must be performed solely out of a sense of duty; any other motivation profanes the act and strips it of its moral quality.

12.4.5 Utilitarianism

Main article: Utilitarianism

Utilitarianism (from the Latin utilis, useful) is a theory of ethics that prescribes the quantitative maximization of good consequences for a population. It is a form of consequentialism. This good to be maximized is usually happiness, pleasure, or preference satisfaction. Though some utilitarian theories might seek to maximize other consequences, these consequences generally have something to do with the welfare of people (or of people and nonhuman animals). For this reason, utilitarianism is often associated with the term welfarist consequentialism.

In utilitarianism it is the "end result" which is fundamental (as opposed to Kantian ethics discussed above). Thus using the same scenario as above, it would be irrelevant whether the person giving money to charity was doing so out of personal or religious conviction, the mere fact that the charitable donation is being made is sufficient for it to be classified as *morally good*.

John Stuart Mill, developer of Jeremy Bentham's utility-based theory

12.5 See also

- Anthropocentrism
- Anarchism
- Brights movement
- Cognitivism with subcategories Ethical naturalism & Ethical non-naturalism, and opponent Non-cognitivism
- Environmentalism
- Ethical subjectivism
- Hedonism
- Liberalism
- Marxism
- Moral realism
- Moral relativism
- Moral skepticism
- Moral Zeitgeist
- Normative ethics
- Objectivism (Ayn Rand)
- Peter Singer
- Secular humanism
- Secular morality
- Secular religion
- Socialism
- Utilitarian bioethics

12.6 References

[1] Kidder 2003. 82

[2] Humanism's Unfinished Agenda

[3] Is Atheism Consistent With Morality?, paper (2001) by Mark I. Vuletic

[4] Interview with the Dalai Lama, *The Progressive* (January 2006), scroll to *Question*: Apart from Buddhism, what are your sources of inspiration? *The Dalai Lama*: Human values.

[5] Girl Scouts of the USA (2010). "The Girl Scout Promise and Law". Retrieved 16 March 2010.

[6] "Officer Development". Retrieved 15 September 2010.

[7] Kidder 2003. 83–84

[8] "Information Paper on "Honor" – A Bedrock of Military Leadership". Retrieved 16 March 2010.

[9] Stephen Jay Gould. "Nonmoral Nature". stephenjaygould.org. Retrieved January 5, 2009. External link in |publisher= (help)

[10] Vettriazhagan (2015). *Pathinen Keelkanakku Noolgal* (in Tamil). **1** (5 ed.). Chennai: Saratha Pathippagam. pp. iv–vi.

[11] Pope, GU (1886). *Thirukkural English Translation and Commentary* (PDF). W.H. Allen, & Co. p. 160.

[12] Lal, Mohan (1992). *Encyclopaedia of Indian Literature*. **V**. New Delhi: Sahitya Akademi. pp. 4333–4334. ISBN 81-260-1221-8.

[13] "Thirukkural translations in different languages of the world". Retrieved 13 August 2016.

[14] Kamil Zvelebil (1973). *The smile of Murugan on Tamil literature of South India*. BRILL. pp. 156–. ISBN 978-90-04-03591-1. Retrieved 11 December 2010.

[15] Holyoake, George J. (1896). *English Secularism*. Chicago: The Open Court Publishing Company.

12.7 Bibliography

- Kidder, Rushworth M. Kidder (2003). *How Good People Make Tough Choices: Resolving the Dilemmas of Ethical Living.* New York: Harper. ISBN 0-688-17590-2.

Chapter 13

Law of three stages

For Kierkegaard's theory of the three stages, see Three stages of life of Søren Kierkegaard.

The **Law of Three Stages** is an idea developed by Auguste Comte in his work *The Course in Positive Philosophy*. It states that society as a whole, and each particular science, develops through three mentally conceived stages: (1) the theological stage, (2) the metaphysical stage, and (3) the positive stage.

13.1 The progression of the three stages of Sociology

(1) The Theological stage refers to explanation by personified deities. During the earlier stages, people believe that all the phenomena of nature are the creation of the divine or supernatural. Men and children failed to discover the natural causes of various phenomena and hence attributed them to a supernatural or divine power.[1] Comte broke this stage into 3 sub-stages:

> 1A. Fetishism - Fetishism was the primary stage of the theological stage of thinking. Throughout this stage, primitive people believe that inanimate objects have living spirit in them, also known as animism. People worship inanimate objects like trees, stones, a piece of wood, volcanic eruptions, etc.[1]
>
> 1B. Polytheism - The explanation of things through the use of many Gods. Primitive people believe that all natural forces are controlled by different Gods; a few examples would be God of water, God of rain, God of fire, God of air, God of earth, etc.[1]
>
> 1C. Monotheism - Monotheism means believing in one God or God in one; attributing all to a single, supreme deity. Primitive people believe a single theistic entity is responsible for the existence of the universe.[1]

(2) The Metaphysical stage is the extension of the theological stage. Metaphysical stage refers to explanation by impersonal abstract concepts. People often tried to believe that God is an abstract being.[1] They believe that an abstract power or force guides and determines events in the world. Metaphysical thinking discards belief in a concrete God. The nature of inquiry was legal and rational in nature. For example: In Classical Hindu Indian society the principle of the transmigration of the soul, the conception of rebirth, notions of pursuant were largely governed by metaphysical uphill.[1]

(3) The Positivity stage, also known as the scientific stage, refers to scientific explanation based on observation, experiment, and comparison. Positive explanations rely upon a distinct method, the scientific method, for their justification. Today people attempt to establish cause and effect relationships. Positivism is a purely intellectual way of looking at the world; as well, it emphasizes observation and classification of data and facts. This is the highest, most evolved behavior according to Comte. [1]

Comte, however, was conscious of the fact that the three stages of thinking may or do coexist in the same society or in the same mind and may not always be successive.

Comte proposed a hierarchy of the sciences based on historical sequence, with areas of knowledge passing through these stages in order of complexity. The simplest and most remote areas of knowledge — mechanical or physical — are the first to become scientific. These are followed by the more complex sciences, those considered closest to us.

The sciences, then, according to Comte's "law", developed in this order: Mathematics; Astronomy; Physics; Chemistry; Biology; Sociology. A science *of society* is thus the "Queen science" in Comte's hierarchy as it would be the most fundamentally complex. Through social science, Comte believed all human social ills could be remedied.

13.2 Critiques of the law

Historian William Whewell wrote "Mr. Comte's arrangement of the progress of science as successively metaphysical and positive, is contrary to history in fact, and contrary to sound philosophy in principle." [2] The historian of science H. Floris Cohen has made a significant effort to draw the modern eye towards this first debate on the foundations of positivism.[3]

In contrast, within an entry dated early October 1838 Charles Darwin wrote in one of his then private notebooks that "M. Comte's idea of a theological state of science [is a] grand idea." [4]

13.3 See also

- Antipositivism
- Religion of Humanity
- Sociological positivism

13.4 References

[1] ""What Are the Major Contributions of Auguste Comte to Sociology?"". PreserveArticles.com: Preserving Your Articles for Eternity. Retrieved 2012-02-24.

[2] p.233 of *On the Philosophy of Discovery: Chapters Historical and Critical (Including completion of the third edition of the philosophy of the inductive sciences)*, William Whewell, New York: Burt Franklin, 1860

[3] H. Floris Cohen, *The Scientific Revolution: A Historiographical Inquiry*, University of Chicago Press 1994, p.35-39

[4] *Notebook N (Metaphysics and Expression)*. Charles Darwin. Journal's timespan: fall 1838—summer 1839. page[leaf] 12.

13.5 External links

- History Guide

Chapter 14

Science of morality

For the comparative study of moral systems across cultures or species, see moral psychology and evolution of morality.

The **science of morality** may refer to various forms of ethical naturalism grounding morality in rational, empirical consideration of the natural world.[1]

14.1 Overview

Moral science may refer to the consideration of what is best for, and how to maximize the flourishing of, either particular individuals[2] or all conscious creatures.[3][4] It has been proposed that "morality" can be appropriately defined on the basis of fundamental premises necessary for *any* empirical, secular, and philosophical discussion and that societies can use the methods of science to provide answers to moral questions.[5]

In sum, from the perspective of neuroscience and brain evolution, the routine rejection of scientific approaches to moral behavior based on Hume's warning against deriving *ought* from *is* seems unfortunate, especially as the warning is limited to deductive inferences. The dictum can be set aside for a deeper, albeit programmatic, neurobiological perspective on what reasoning and problem-solving are, how social navigation works, how evaluation is accomplished by nervous systems, and how mammalian brains make decisions.
"
"

-Patricia Churchland in her book *Braintrust* (emphasis added)

The norms advocated by moral scientists (e.g. rights to abortion, euthanasia, and drug liberalization under certain circumstances) would be founded upon the shifting and growing collection of human understanding.[6] Even with science's admitted degree of ignorance, and the various semantic issues, moral scientists can meaningfully discuss things as being *almost certainly* "better" or "worse" for promoting flourishing.[7]

14.2 History

14.2.1 In philosophy

Utilitarian Jeremy Bentham discussed some of the ways moral investigations are a science.[8] He criticizes deontological ethics for failing to recognize that it needed to make the same presumptions as his science of morality to really work – whilst pursuing rules that were to be obeyed in every situation (something that worried Bentham).

W.V.O. Quine advocated naturalizing epistemology by looking to natural sciences like psychology for a full explanation of knowledge. His work contributed to a resurgence of moral naturalism in the last half of the 20th century. Paul Kurtz, who believes that the careful, secular pursuit of normative rules is vital to society, coined the term *eupraxophy* to refer to his approach to normative ethics. Steven Pinker, Sam Harris, and Peter Singer believe that we learn what is right and wrong through reason and empirical methodology.[9][10]

Maria Ossowska thought that sociology was inextricably related to philosophical reflections on morality, including normative ethics. She proposed that science analyze: (a) existing social norms and their history, (b) the psychology of morality, and the way that individuals interact with moral matters and prescriptions, and (c) the sociology of morality.[11]

14.2.2 Popular literature

The theory and methods of a normative science of morality are explicitly discussed in Joseph Daleiden's *The Science of Morality: The Individual, Community, and Future*

Maria Ossowska used the methods of science to understand the origins of moral norms.

Generations (1998). Daleiden's book, in contrast to Harris, extensively discusses the relevant philosophical literature. In *The Moral Landscape: How Science Can Determine Human Values*, Sam Harris's goal is to show how moral truth can be backed by "science", or more specifically, empirical knowledge, critical thinking, philosophy, but most controversially, the scientific method.

Patricia Churchland offers that, accepting Hume's is-ought problem, the use of induction from premises and definitions remains a valid way of reasoning in life and science.[12]

> Our moral behavior, while more complex than the social behavior of other animals, is similar in that it represents our attempt to manage well in the existing social ecology....from the perspective of neuroscience and brain evolution, the routine rejection of scientific approaches to moral behavior based on Hume's warning against deriving *ought* from *is* seems unfortunate, especially as the warning is limited to deductive inferences....The truth seems to be that values rooted in the circuitry for caring—for well-being of self, offspring, mates, kin, and others—shape social reasoning about many issues: conflict resolutions, keeping the peace, defense, trade, resource distribution, and many other aspects of social life in all its vast richness.[13]
> — Patricia Churchland, Braintrust: What Neuroscience Tells Us About Morality

Daleiden and Leonard Carmichael warn that science is probabilistic, and that certainty is not possible. One should therefore expect that moral prescriptions will change as humans gain understanding.[14][note 1]

14.3 Views in Scientific Morality

See also: Moral psychology

14.3.1 Training to promote good behaviour

The science of morality may aim to discover the best ways to motivate and shape individuals. Methods to accomplish this include instilling explicit virtues, building character strengths, and forming mental associations. These generally require some level of practical reason. James Rest suggested that abstract reasoning is also a factor in making moral judgements[16] and emphasized that moral judgements alone do not predict moral behaviour: "Moral judgement may be closely related to advocacy behaviour, which in turn influences social institutions, which in turn creates a system of norms and sanctions that influences people's behaviour."[16] Daleiden suggested that religions instill a practical sense of virtue and justice, right and wrong. They also effectively use art and myths to educate people about moral situations.[17]

The role of government

Harris argues that moral science does not imply an "Orwellian future" with "scientists at every door". Instead, Harris imagines data about normative moral issues being shared in the same way as other sciences (e.g. peer-reviewed journals on medicine).[18]

Daleiden specifies that government, like any organization, should have limited power. He says "centralization of power irrevocably in the hands of one person or an elite has always ultimately led to great evil for the human race. It was the novel experiment of democracy—a clear break with tradition—that ended the long tradition of tyranny."[19] He is also explicit that government should only use law to enforce the most basic, reasonable, evidence and widely supported moral norms. In other words, there are a great many moral norms that should never be the task of the government to enforce.[20]

The role of punishment

Main articles: Differential reinforcement and Prison

One author has argued that to attain a society where people are motivated by conditioned self-interest, punishment must go hand-in-hand with reward.[21] For instance, in this line of reasoning, prison remains necessary for many perpetrators of crimes. This is so, even if libertarian free will is false. This is because punishment can still serve its purposes: it deters others from committing their own crimes, educates and reminds everyone about what the society stands for, incapacitates the criminal from doing more harm, goes some way to relieving or repaying the victim, and corrects the criminal (also see recidivism). This author argues that, at least, any prison system *should* be pursuing those goals, and that it is an empirical question as to what sorts of punishment realize these goals most effectively, and how well various prison systems actually serve these purposes.[22]

14.3.2 Research

See also: Positive psychology and Moral development

The brain areas that are consistently involved when humans reason about moral issues have been investigated.[23] The neural network underlying moral decisions overlaps with the network pertaining to representing others' intentions (i.e., theory of mind) and the network pertaining to representing others' (vicariously experienced) emotional states (i.e., empathy). This supports the notion that moral reasoning is related to both seeing things from other persons' points of view and to grasping others' feelings. These results provide evidence that the neural network underlying moral decisions is probably domain-global (i.e., there might be no such things as a "moral module" in the human brain) and might be dissociable into cognitive and affective subsystems.

14.3.3 Other implications

Daleiden provides examples of how science can use empirical evidence to assess the effect that specific behaviors can have on the well-being of individuals and society with regard to various moral issues. He argues that science supports decriminalization and regulation of drugs, euthanasia under some circumstances, and the permission of sexual behaviors that are not tolerated in some cultures (he cites homosexuality as an example). Daleiden further argues that in seeking to reduce human suffering, abortion should not only be permissible, but at times a moral obligation (as in the case of a mother of a potential child who would face the probability of much suffering). Like all moral claims in his book, however, Daleiden is adamant that these decisions remain grounded in, and contingent on empirical evidence.[6][note 2]

The ideas of cultural relativity, to Daleiden, do offer some lessons: investigators must be careful not to judge a person's behaviour without understanding the environmental context. An action may be necessary and more moral once we are aware of circumstances.[24] However, Daleiden emphasizes that this does not mean all ethical norms or systems are equally effective at promoting flourishing[24] and he often offers the equal treatment of women as a reliably superior norm, wherever it is practiced.

14.4 Criticisms

The idea of a normative science of morality has met with some criticisms. Critics include physicist Sean M. Carroll, who argues that morality cannot be part of science.[25] He and other critics cite the widely held "fact-value distinction", that the scientific method cannot answer "moral" questions, although it can describe the norms of different cultures. In contrast, moral scientists defend the position that such a division between values and scientific facts ("moral relativism") is not only arbitrary and illusory, but impeding progress towards taking action against documented cases of human rights violations in different cultures.[26]

Stephen Jay Gould argued that science and religion occupy "non-overlapping magisteria". To Gould, science is concerned with questions of fact and theory, but not with meaning and morality – the magisteria of religion. In the same vein, Edward Teller proposed that politics decides what is right, whereas science decides what is true.[27]

During a discussion on the role that naturalism might play in professions like nursing, Philosopher Trevor Hussey calls the popular view that science is unconcerned with morality "too simplistic". Although his main focus in the paper is naturalism in nursing, he goes on to explain that science can, at very least, be interested in morality at a descriptive level. He even briefly entertains the idea that morality could itself be a scientific subject, writing that one *might* argue "..that moral judgements are subject to the same kinds of rational, empirical examination as the rest of the world: they are a subject for science – although a difficult one. If this could be shown to be so, morality would be contained within naturalism. However, I will not assume the truth of moral realism here." [note 3]

14.5 See also

- Ethical calculus
- Evolutionary ethics
- Felicific calculus
- Meta-ethics
- Moral skepticism
- Pareto efficiency
- Scientism
- Sociobiology
- Social Darwinism
- Value (personal and cultural)
- Welfare economics

14.6 Notes

[1] To quote Carmichael: "We do not turn aside from what we know about astronomy at any time because there is still a great deal we do not know, or because so much of what we once thought we knew is no longer recognized as true. May not the same argument be accepted in our thinking about ethical and esthetic judgements?"[15]

[2] Joseph Daleiden's final word regarding his book, *The Science of Morality,* is that "[The study of ethics] should be included with the social sciences and be subject to as rigorous a scientific program of research as any other area of human behaviour. Lacking this scientific rigour, the moral conclusions drawn in this volume must be considered as working hypotheses, some with greater degree of evidentiary support than others. It is the process by which to assess and transmit moral norms that was the primary focus of this work, and I hope it will serve as a new way of deciding moral issues."

[3] Hussey writes "The relationship between naturalism and morality and politics is complicated, and is difficult to state in a few sentences because it involves deep philosophical issues. Only the briefest discussion is possible here. The most popular view is that science, and hence naturalism, is concerned with objective facts and not with values: with what is the case rather than what ought to be. But this is too simplistic." He gives a reason immediately: "First, at the very least, science can study morality and politics at a descriptive level and try to understand their workings within societies and in the lives of individuals, and investigate their evolutionary origins, their social propagation, and so on." Hussey then describes how scientists must adhere to certain values, but also how values guide what it is that science may investigate. His real interest in the paper is to justify naturalism as a nursing practice, yet he does eventually write: "Finally, the idea that science and morality are separate realms, one dealing with facts the other with values, is not as certain and clear-cut as it seems. Various versions of moral realism are now widely discussed among philosophers (e.g. Railton, 1986, 1996, 2003; Sayre-McCord, 1988; Dancy, 1993; Casebeer, 2003; Shafer-Landau, 2003; Baghramian, 2004; Smith, 1994, 2004). Despite their differences, moral realists generally agree on two principles. First, that our moral utterances, such as 'Murder is morally wrong' or 'We ought to be honest' are genuine statements and hence they are capable of being either true or false. Second, what makes them either true or false are aspects of the real world, open to objective examination. It can be argued that it is an implication of this thesis that moral judgements are subject to the same kinds of rational, empirical examination as the rest of the world: they are a subject for science – although a difficult one." He continues "If this could be shown to be so, morality would be contained within naturalism. However, I will not assume the truth of moral realism here. It is sufficient to say that it has at least as much credibility as any theory claiming a supernatural or divine foundation for morality: views which, while popular among the general public, do not have widespread support among moral philosophers – for what that is worth." Hussey thus directs discussion back towards Naturalism in nursing because his main point in all this was, in the end, to prove that naturalistic moralities are not necessarily less credible than supernatural ones, and may even be more credible.[28]

14.7 References

[1] Lenman, James (2008). Edward N. Zalta, ed. "Moral Naturalism". *The Stanford Encyclopedia of Philosophy* (Winter 2008 ed.).

[2] Carrier

[3] Ted.com, "Sam Harris: Science Can Answer Moral Questions."

[4] Harris, *The Moral Landscape*, pp. 39ff

[5] Joseph Daleiden *The Science of Morality: The Individual, Community, and Future Generations*; Sam Harris *The Moral Landscape*, 2010; and Richard Carrier *Moral Facts Naturally Exist (and Science Could Find Them)*

[6] Daleiden, Joseph (1998). Chapter 20: Summary and conclusions. Pages 485–500

[7] Sam Harris (2010), page 183: "Much of the skepticism I encounter when speaking about these issues comes from people who think "happiness" is a superficial state of mind and that there are far more important things in life than "being happy." Some reasers may think that concepts like "well-being" and "flourishing" are similarly effete. However, I don't know of any better terms with which to signify the most positive states of being to which we can aspire. One of the virtues of thinking about a moral landscape, the heights of which remain to be discovered, is that it frees us from these semantic difficulties. Generally speaking, we need only worry about what it will mean to move "up" as opposed to "down".

[8] *Deontology, or The Science of Morality*

[9] http://thesciencenetwork.org/programs/the-great-debate/steven-pinker-3

[10] At 11:25 in the video debate at http://thesciencenetwork.org/programs/the-great-debate/the-great-debate-panel-1

[11] Marcin T. Zdrenka. (2006). "Moral philosopher or sociologist of morals?". Journal of Classical Sociology.

[12] http://thesciencenetwork.org/programs/the-great-debate/the-great-debate-panel-1

[13] Churchland, Patricia Smith (2011). *Braintrust: What Neuroscience Tells Us About Morality*. Princeton University Press. pp. 7–9. ISBN 978-0-691-13703-2. LCCN 2010043584.

[14] p502, Daleiden (1998)

[15] Leaonard Carmichael, the chapter "Absolutes, Relativism and the Scientific Psychology of Human Nature", H. Schoeck and J. Wiggins (eds), in the book "Relativism and the Study of Man, Princeton, NJ: D. Van Nostrand, 1961, page 16

[16] James R. Rest, Development in Judging Moral Issues. (1979). Minneapolis: University of Minnesota Press.

[17] 323, 326, Daleiden (1998)

[18] www.salon.com Asked "Let's say scientists do end up discovering moral truths. How are they supposed to enforce their findings? Would they become something like policemen or priests?" Harris writes *"They wouldn't necessarily enforce them any more than they enforce their knowledge about human health. What are scientists doing with the knowledge that smoking causes cancer or obesity is bad for your health, or that the common cold is spread by not washing your hands? We're not living in some Orwellian world where we have scientists in lab coats at every door. Imagine we discovered that there is a best way to teach your children to be compassionate, or to defer short-term gratification in the service of a long-term goal. What if it turns out to be true that calcium intake in the first two years of life has a significant effect on a child's emotional life? If we learn that, what parent wouldn't want that knowledge? The fear of a "Brave New World" component to this argument is unfounded."*

[19] 219, Daleiden (1998)

[20] 273–274, Daleiden (1998)

[21] 77, Daleiden (1998), quote "We use rewards and punishments, praise and blame, in training any animal. The human species is only different in degree in this regard, not in kind."

[22] 289, Daleiden (1998)

[23] "Bzdok, D. et al. Parsing the neural correlates of moral cognition: ALE meta-analysis on morality, theory of mind, and empathy. Brain Struct Funct, 2011."

[24] 100, Daleiden

[25] Sean Carroll (2010-05-04). "Science And Morality: You Can't Derive 'Ought' From 'Is'". *NPR*. Retrieved 2010-06-14. Casting morality as a maximization problem might seem overly restrictive at first glance, but the procedure can potentially account for a wide variety of approaches. A libertarian might want to maximize a feeling of personal freedom, while a traditional utilitarian might want to maximize some version of happiness. The point is simply that the goal of morality should be to create certain conditions that are, in principle, directly measurable by empirical means. ...Nevertheless, I want to argue that this program is simply not possible. ... Morality is not part of science, however much we would like it to be. There are a large number of arguments one could advance for in support of this claim, but I'll stick to three.

[26] Sam Harris (2010-03-29). "Moral confusion in the name of "science"". *PROJECT REASON*. Retrieved 2014-12-06. There are also very practical, moral concerns that follow from the glib idea that anyone is free to value anything— the most consequential being that it is precisely what allows highly educated, secular, and otherwise well-intentioned people to pause thoughtfully, and often interminably, before condemning practices like compulsory veiling, genital excision, bride-burning, forced marriage, and the other cheerful products of alternative "morality" found elsewhere in the world. Fanciers of Hume's is/ought distinction never seem to realize what the stakes are, and they do not see what an abject failure of compassion their intellectual "tolerance" of moral difference amounts to. While much of this debate must be had in academic terms, this is not merely an academic debate. There are women and girls getting their faces burned off with acid at this moment for daring to learn to read, or for not consenting to marry men they have never met, or even for the crime of getting raped.

[27] Essays on Science and Society. "Science and Morality".

[28] Naturalistic nursing, Trevor Hussey (2011), Nursing Philosophy, Vol 12, Pg.45–52.

14.8 Further reading

- Kohlberg's stages of moral development

14.8. FURTHER READING

- *The Science of Good and Evil*, a book by Michael Shermer

Chapter 15

Secular morality

Secular morality is the aspect of philosophy that deals with morality outside of religious traditions. Modern examples include humanism, freethinking, and most versions of consequentialism. Additional philosophies with ancient roots include those such as skepticism and virtue ethics. Greg M. Epstein also states that, "much of ancient Far Eastern thought is deeply concerned with human goodness without placing much if any stock in the importance of gods or spirits."[1]:45 An example is the non-denominational Kural text of Valluvar, an ancient Indian theistic poet-philosopher whose work remains secular.[2][3][4] Other philosophers have proposed various ideas about how to determine right and wrong actions. An example is Immanuel Kant's categorical imperative.

A variety of positions are apparent regarding the relationship between religion and morality. Some believe that religion is necessary as a guide to a moral life. According to some, this idea has been with us for nearly 2,000 years.[1]:5 According to others, the idea goes back as far as 4,000 years, with the ancient Egyptians' 42 Principles of Ma'at.[5] There are various thoughts regarding how this idea has arisen. For example, Greg Epstein suggests that this idea is connected to a concerted effort by theists to question nonreligious ideas: "conservative authorities have, since ancient days, had a clever counterstrategy against religious skepticism—convincing people that atheism is evil, and then accusing their enemies of being atheists."[1]:7

Others eschew the idea that religion is required to provide a guide to right and wrong behavior, such as the *Westminster Dictionary of Christian Ethics* which states that religion and morality "are to be defined differently and have no definitional connections with each other".[6]:401 Some believe that religions provide poor guides to moral behavior. Various commentators, such as Richard Dawkins (*The God Delusion*) and Christopher Hitchens are among those who have asserted this view.

15.1 Secular moral frameworks

15.1.1 Consequentialism

Main article: Consequentialism
See also: Utilitarianism

"Consequentialists", as described by Peter Singer, "start not with moral rules, but with goals. They assess actions by the extent to which they further those goals."[7]:3 Singer also notes that utilitarianism is "the best-known, though not the only, consequentialist theory."[7]:3 Consequentialism is the class of normative ethical theories holding that the consequences of one's conduct are the ultimate basis for any judgment about the rightness of that conduct. Thus, from a consequentialist standpoint, a morally right act (or omission) is one that will produce a good outcome, or consequence. In his 2010 book, *The Moral Landscape*, Sam Harris describes a utilitarian science of morality.

15.1.2 Freethinking

Main article: Freethinking

Freethought is a philosophical viewpoint that holds that opinions should be formed on the basis of science, logic, and reason, and should not be influenced by authority, tradition, or other dogmas. Freethinkers strive to build their opinions on the basis of facts, scientific inquiry, and logical principles, independent of any logical fallacies or intellectually limiting effects of authority, confirmation bias, cognitive bias, conventional wisdom, popular culture, prejudice, sectarianism, tradition, urban legend, and all other dogmas.

15.1.3 Secular humanism

Main article: Secular humanism

Secular humanism focuses on the way human beings can

lead happy and functional lives. It posits that human beings are capable of being ethical and moral without religion or God, it neither assumes humans to be inherently evil or innately good, nor presents humans as "above nature" or superior to it. Rather, the humanist life stance emphasizes the unique responsibility facing humanity and the ethical consequences of human decisions. Fundamental to the concept of secular humanism is the strongly held viewpoint that ideology—be it religious or political—must be thoroughly examined by each individual and not simply accepted or rejected on faith. Along with this, an essential part of secular humanism is a continually adapting search for truth, primarily through science and philosophy.

15.2 Positions on religion and morality

See also: Ethics in religion and Secular ethics

The subject of secular morality has been discussed by prominent secular scholars as well as popular culture-based atheist and anti-religious writers. These include Paul Chamberlain's *Can We Be Good Without God?* (1996), Richard Holloway's *Godless Morality: Keeping Religion Out of Ethics* (1999), Robert Buckman's *Can We Be Good Without God?* (2002), Michael Shermer's *The Science of Good and Evil* (2004), Richard Dawkins's *The God Delusion* (2006), Christopher Hitchens's *God Is Not Great* (2007), Greg Epstein's *Good Without God: What A Billion Nonreligious People Do Believe* (2010), and Sam Harris's *The Moral Landscape: How Science Can Determine Human Values* (2011).

15.2.1 Morality requires religious tenets

According to Greg Epstein, "the idea that we can't be 'good without God' " has been with us for nearly 2,000 years.[1]:5 This idea is seen in various holy books, for example in Psalms 14 of the Christian Bible: "The fool says in his heart, 'there is no God.' They are corrupt, they do abominable deeds, there is none who does good ... not even one."[8] And this idea is still present today. "Many today ... argue that religious beliefs are necessary to provide moral guidance and standards of virtuous conduct in an otherwise corrupt, materialistic, and degenerate world."[9]:115 For example, Christian writer and medievalist C. S. Lewis made the argument in his popular book *Mere Christianity* that if a supernatural, objective standard of right and wrong does not exist outside of the natural world, then right and wrong becomes mired in the is-ought problem. Thus, he wrote, preferences for one moral standard over another become as inherently indefensible and arbitrary as preferring a certain flavor of food over another or choosing to drive on a certain side of a road.[10]:3–28 In the same vein, Christian theologian Ron Rhodes has remarked that "it is impossible to distinguish evil from good unless one has an infinite reference point which is absolutely good."[11] Peter Singer states that, "Traditionally, the more important link between religion and ethics was that religion was thought to provide a reason for doing what is right, the reason being that those who are virtuous will be rewarded by an eternity of bliss while the rest roast in hell."[7]:4

Proponents of theism argue that without a God or gods it is impossible to *justify* moral behavior on metaphysical grounds and thus to make a coherent case for abiding by moral standards. C. S. Lewis makes such an argument in *Mere Christianity*. Peter Robinson, a political author and commentator with Stanford's Hoover Institution, has commented that, if an inner moral conscience is just another adaptive or evolved feeling in the human mind like simple emotional urges, then no inherent reason exists to consider morality as over and above other urges.[12] According to Thomas Dixon, "Religions certainly do provide a framework within which people can learn the difference between right and wrong."[9]

15.2.2 Morality does not rely on religion

"A man's ethical behavior should be based effectually on sympathy, education, and social ties and needs; no religious basis is necessary. Man would indeed be in a poor way if he had to be restrained by fear of punishment and hopes of reward after death."
— Albert Einstein, "Religion and Science," New York Times Magazine, 1930

Various commentators have stated that morality does not require religion as a guide. *The Westminster Dictionary of Christian Ethics* states that, "it is not hard to imagine a society of people that has no religion but has a morality, as well as a legal system, just because it says that people cannot live together without rules against killing, etc., and that it is not desirable for these all to be legally enforced. There have also certainly been people who have had a morality but no religious beliefs."[6]:400 Bernard Williams, an English philosopher, stated that the secular "utilitarian outlook"—a popular ethical position wherein the morally right action is defined as that action which effects the greatest amount of happiness or pleasure for the greatest number of people— is "non-transcendental, and makes no appeal outside human life, in particular not to religious considerations."[13]:83 Williams also argued that, "Either one's motives for following the moral word of God are moral motives, or they are not. If they are, then one is already equipped with moral motivations, and the introduction of God adds nothing extra. But if they are not moral motives, then they will be motives of such a kind that they cannot appropriately motivate *morality* at all ... we reach the conclusion that any appeal to God in this connection either adds to nothing at all, or it adds the wrong sort of thing."[13]:64–65

Socrates' "Euthyphro dilemma" is often considered one of the earliest refutations of the idea that morality requires religion. This line of reasoning is described by Peter Singer:

> "Some theists say that ethics cannot do without religion because the very meaning of 'good' is nothing other than 'what God approves'. Plato refuted a similar claim more than two thousand years ago by arguing that if the gods approve of some actions it must be because those actions are good, in which case it cannot be the gods' approval that makes them good. The alternative view makes divine approval entirely arbitrary: if the gods had happened to approve of torture and disapprove of helping our neighbors, torture would have been good and helping our neighbors bad. Some modern theists have attempted to extricate themselves from this type of dilemma by maintaining that God is good and so could not possibly approve of torture; but these theists are caught in a trap of their own making, for what can they possibly mean by the assertion that God is good? That God is approved of by God?"[7]:3–4

Greg Epstein, a Humanist chaplain at Harvard University, dismisses the question of whether God is needed to be good "because that question does not need to be answered—it needs to be rejected outright," adding, "To suggest that one *can't* be good without belief in God is not just an opinion ... it is a prejudice. It may even be discrimination."[1]:ix This is in line with the *Westminster Dictionary of Christian Ethics* which states that religion and morality "are to be defined differently and have no definitional connections with each other. Conceptually and in principle, morality and a religious value system are two distinct kinds of value systems or action guides."[6]:401 Others share this view. Singer states that morality "is not something intelligible only in the context of religion".[7][a] Atheistic philosopher Julian Baggini stated that "there is nothing to stop atheists believing in morality, a meaning for life, or human goodness. Atheism is only intrinsically negative when it comes to belief about God. It is as capable of a positive view of other aspects of life as any other belief."[14]:3 He also states that "Morality is more than possible without God, it is entirely independent of him. That means atheists are not only more than capable of leading moral lives, they may even be able to lead more moral lives than religious believers who confuse divine law and punishment with right and wrong.[14]:37

Popular atheist author and *Vanity Fair* writer Christopher Hitchens remarked on the program *Uncommon Knowledge*:

> "I think our knowledge of right and wrong is innate in us. Religion gets its morality from humans. We know that we can't get along if we permit perjury, theft, murder, rape, all societies at all times, well before the advent of monarchies and certainly, have forbidden it... Socrates called his daemon, it was an inner voice that stopped him when he was trying to take advantage of someone... Why don't we just assume that we do have some internal compass?"[12]

Daniel Dennett says it is a "pernicious" myth that religion or God are needed for people to fulfill their desires to be good. However, he offers that secular and humanist groups are still learning how to organize effectively.[15]

Philosopher Daniel Dennett says that secular organizations need to learn more 'marketing' lessons from religion—and from effective secular organizations like the TED conferences. This is partly because Dennett says that the idea that people need God to be morally good is an extremely harmful, yet popular myth. He believes it is a falsehood that persists because churches are currently much better at organizing people to do morally good work.[15] In Dennett's words:

> "What is particularly pernicious about it [the myth] is that it exploits a wonderful human trait; people want to be good. They want to lead good lives... So then along come religions that say *'Well you can't be good without God'* to convince people that they have to do this. That may be the main motivation for people to take religions seriously—to try to take religions seriously, to try and establish an allegiance to the church—because they want to lead good lives."[15]

15.2.3 Religion is a poor moral guide

Popular atheist author and biologist Richard Dawkins, writing in *The God Delusion*, has stated that religious people have committed a wide variety of acts and held certain beliefs through history that are considered today to be morally repugnant. He has stated that Adolf Hitler and the Nazis held broadly Christian religious beliefs that inspired the Holocaust on account of antisemitic Christian doctrine, that Christians have traditionally imposed unfair restrictions on the legal and civil rights of women, and that Christians have condoned slavery of some form or description throughout most of Christianity's history. Dawkins insists that, since Jewish and Christian interpretations of the Bible have changed over the span of history so that what was formerly seen as permissible is now seen as impermissible, it is intellectually dishonest for them to believe theism provides an absolute moral foundation apart from secular intuition. In addition, he argued that since Christians and other religious groups do not acknowledge the binding authority of all parts of their holy texts (e.g., The books of Exodus and Leviticus state that those who work on the Sabbath[16] and those caught performing acts of homosexuality,[17] respectively, were to be put to death.), they are already capable of distinguishing "right" from "wrong."[18]:281

The well-known passage from Dostoyevsky's *The Brothers Karamazov*, "If God is dead, all is permitted,"[1]:63 suggests that non-believers would not hold moral lives without the possibility of punishment by a God. Greg M. Epstein notes a similar theme in reverse. Famous apologies by Christians who have "sinned" (such as Bill Clinton and Jimmy Swaggart) "must embolden some who take enormous risks for the thrill of a little immoral behavior: their Lord will forgive them, if they only ask nicely enough when—or if—they are eventually caught. If you're going to do something naughty, you're going to do it, and all the theology in the world isn't going to stop you."[1]:115–116 Some survey and sociological literature suggests that theists do no better than their secular counterparts in the percentage adhering to widely held moral standards (e.g., lying, theft and sexual infidelity).[e]

15.2.4 Evidential findings

Cases can also be seen in nature of animals exhibiting behavior we might classify as "moral" without religious directives to guide them. These include "detailed studies of the complex systems of altruism and cooperation that operate among social insects" and "the posting of altruistic sentinels by some species of bird and mammal, who risk their own lives to warn the rest of the group of imminent danger."[9]:117

Greg Epstein states that "sociologists have recently begun to pay more attention to the fact that some of the world's most secular countries, such as those in Scandinavia, are among the least violent, best educated, and most likely to care for the poor".[19] He adds that, "scientists are beginning to document, though religion may have benefits for the brain, so may secularism and Humanism."[19]

On April 26, 2012, the results of a study which tested their subjects' pro-social sentiments were published in the Social Psychological and Personality Science journal in which non-religious people had higher scores showing that they were more inclined to show generosity in random acts of kindness, such as lending their possessions and offering a seat on a crowded bus or train. Religious people also had lower scores when it came to seeing how much compassion motivated participants to be charitable in other ways, such as in giving money or food to a homeless person and to non-believers.[20][21]

A number of studies have been conducted on the empirics of morality in various countries, and the overall relationship between faith and crime is unclear.[b] A 2001 review of studies on this topic found "The existing evidence surrounding the effect of religion on crime is varied, contested, and inconclusive, and currently no persuasive answer exists as to the empirical relationship between religion and crime."[22] Phil Zuckerman's 2008 book, *Society without God*, notes that Denmark and Sweden, "which are probably the least religious countries in the world, and possibly in the history of the world", enjoy "among the lowest violent crime rates in the world [and] the lowest rates of corruption in the world".[23][c] Dozens of studies have been conducted on this topic since the twentieth century. A 2005 study by Gregory S. Paul published in the *Journal of Religion and Society* stated that, "In general, higher rates of belief in and worship of a creator correlate with higher rates of homicide, juvenile and early adult mortality, STD infection rates, teen pregnancy, and abortion in the prosperous democracies," and "In all secular developing democracies a centuries long-term trend has seen homicide rates drop to historical lows" with the exceptions being the United States (with a high religiosity level) and "theistic" Portugal.[24][d] In a response, Gary Jensen builds on and refines Paul's study.[25] His conclusion is that a "complex relationship" exists between religiosity and homicide "with some dimensions of religiosity encouraging homicide and other dimensions discouraging it".

15.2.5 Other views

Some non-religious nihilistic and existentialist thinkers have affirmed the prominent theistic position that the existence of the personal God of theism is linked to the existence of an objective moral standard, asserting that questions of right and wrong inherently have no meaning and, thus, any notions of morality are nothing but an anthropogenic fantasy. Agnostic author and Absurdist philosopher Albert Camus discussed the issue of what he saw as the universe's indifference towards humankind and the meaninglessness of life in his prominent novel *The Stranger*, in which the protagonist accepts death via execution without sadness or feelings of injustice. In his philosophical work, *The Myth of Sisyphus*, Camus argues that human beings must choose to live defiantly in spite of their longing for purpose or direction and the apparent lack of evidence for God or moral imperatives. The atheistic existentialist philosopher Jean-Paul Sartre proposed that the individual must create his own essence and therefore must freely and independently create his own subjective moral standards by which to live.

15.3 See also

- Morality and religion
- Secular ethics
- Science of morality

15.4 Notes

a.^ Singer uses the word "ethics", but states in the same work that he uses the words ethics and morals "interchangeably" (p. 1).

b.^ Some studies appear to show positive links in the relationship between religiosity and moral behavior[26][27]—for example, surveys suggesting a positive connection between faith and altruism.[28] Modern research in criminology also suggests an inverse relationship between religion and crime,[29] with some studies establishing this connection.[30] A meta-analysis of 60 studies on religion and crime concluded, "religious behaviors and beliefs exert a moderate deterrent effect on individuals' criminal behavior".[31]

c.^ Zuckerman identifies that Scandinavians have "relatively high rates of petty crime and burglary", but "their overall rates of violent crime—such as murder, aggravated assault, and rape—are among the lowest on earth" (Zuckerman 2008, pp. 5–6).

d.^ The authors also state that "A few hundred years ago rates of homicide were astronomical in Christian Europe and the American colonies,"[32] and "[t]he least theistic secular developing democracies such as Japan, France, and Scandinavia have been most successful in these regards."[33] They argue for a positive correlation between the degree of public religiosity in a society and certain measures of dysfunction,[34] an analysis published later in the same journal argues that a number of methodological problems undermine any findings or conclusions in the research.[35]

e.^ See, for instance, Ronald J. Sider, *The Scandal of the Evangelical Conscience: Why Are Christians Living Just Like the Rest of the World?* (Grand Rapids, Mich.: Baker, 2005). Sider quotes extensively from polling research by The Barna Group showing that the moral behavior of evangelical Christians is anything but exemplary.

15.5 References

[1] Epstein, Greg M. (2010). *Good Without God: What a Billion Nonreligious People Do Believe*. New York: HarperCollins. ISBN 978-0-06-167011-4.

[2] Pope, GU (1886). *Thirukkural English Translation and Commentary* (PDF). W.H. Allen, & Co. p. 160.

[3] Lal, Mohan (1992). *Encyclopaedia of Indian Literature*. **V**. New Delhi: Sahitya Akademi. pp. 4333–4334. ISBN 81-260-1221-8.

[4] Ramasamy, V. (2001). *On Translating Tirukkural* (First ed.). Chennai: International Institute of Tamil Studies.

[5] Richard, Lottie. "42 Principles Of God Maat 2000 Years Before Ten Commandments". *Liberal America*. Retrieved 13 October 2015.

[6] Childress, James F.; Macquarrie, John, eds. (1986). *The Westminster Dictionary of Christian Ethics*. Philadelphia: The Westminster Press. ISBN 0-664-20940-8.

[7] Singer, Peter (2010). *Practical Ethics* (Second ed.). New York: Cambridge University Press. ISBN 978-0-521-43971-8.

[8] ESVBible.org. "Psalm 14 - ESVBible.org". Crossway. Retrieved 4 September 2012.

[9] Dixon, Thomas (2008). *Science and Religion: A Very Short Introduction*. Oxford: Oxford University Press. ISBN 978-0-19-929551-7.

[10] Lewis, C.S. (2001). *Mere Christianity*. HarperCollins.

[11] Ron Rhodes. "Strategies for Dialoguing with Atheists". Reasoning from the Scriptures Ministries. Retrieved January 4, 2010. External link in |publisher= (help)

[12] "Hitchens—The Morals of an Atheist". *Uncommon Knowledge*. August 23, 2007. Retrieved January 4, 2010.

15.5. REFERENCES

[13] Williams, Bernard (1972). *Morality*. Cambridge: Cambridge University Press. ISBN 0-521-45729-7.

[14] Baggini, Julian (2003). *Atheism: A Very Short Introduction*. Oxford: Oxford University Press. ISBN 978-0-19-280424-2.

[15] Dennett, Daniel (December 12, 2011). "The Scientific Study of Religion". Point of Inquiry. Discussion of morality starts especially at 39min

[16] Biblos.com (2004–2011). "Exodus 31:15". Biblos.com. Retrieved 6 September 2012. Exodus 35:2 is similarly worded.

[17] Biblos.com (2004–2011). "Leviticus 20:13". Biblos.com. Retrieved 6 September 2012.

[18] Dawkins, Richard (2006). *The God Delusion*. Bantam Books. ISBN 978-0-618-68000-9.

[19] Zuckerman, Phil (2008). *Society Without God: What the Least Religious Nations Can Tell Us About Contentment*. New York: New York University Press.

[20] Highly Religious People Are Less Motivated by Compassion Than Are Non-Believers by Science Daily

[21] Laura R. Saslow, Robb Willer, Matthew Feinberg, Paul K. Piff, Katharine Clark, Dacher Keltner and Sarina R. Saturn My Brother's Keeper? Compassion Predicts Generosity More Among Less Religious Individuals

[22] Baier, Colin J.; Wright, Bradley R. E. (February 2001). ""If You Love Me, Keep My Commandments": A Meta-analysis of the Effect of Religion on Crime" (PDF). 38. No. 1. Journal of Research in Crime and Delinquency: 3. Retrieved 20 November 2011. Original in italics.

[23] Zuckerman, Phil. *Society Without God: What the Least Religious Nations Can Tell us about Contentment*. New York: New York University Press. p. 2. ISBN 978-0-8147-9714-3. Zuckerman's work is based on his studies conducted during a 14-month period in Scandinavia in 2005–2006.

[24] Paul, Gregory S. (2005). "Cross-National Correlations of Quantifiable Societal Health with Popular Religiosity and Secularism in the Prosperous Democracies: A First Look". *Journal of Religion and Society*. Baltimore, Maryland. **7**: 4, 5, 8, and 10.

[25] Gary F. Jensen (2006) Department of Sociology, Vanderbilt University *Religious Cosmologies and Homicide Rates among Nations: A Closer Look* http://moses.creighton.edu/JRS/2006/2006-7.html http://moses.creighton.edu/JRS/pdf/2006-7.pdf Journal of Religion and Society, Volume 8, ISSN 1522-5658 http://purl.org/JRS

[26] KERLEY, KENT R.; MATTHEWS; BLANCHARD, TROY C. (2005). "Religiosity, Religious Participation, and Negative Prison Behaviors". *Journal for the Scientific Study of Religion*. **44** (4): 443–457. doi:10.1111/j.1468-5906.2005.00296.x.

[27] SAROGLOU, VASSILIS; PICHON; DERNELLE, REBECCA (2005). "Prosocial Behavior and Religion: New Evidence Based on Projective Measures and Peer Ratings". *Journal for the Scientific Study of Religion*. **44** (3): 323–348. doi:10.1111/j.1468-5906.2005.00289.x.

[28] e.g. a survey by Robert Putnam showing that membership of religious groups was positively correlated with membership of voluntary organisations

[29] As is stated in: Chu, Doris C. (2007). "Religiosity and Desistance From Drug Use". *Criminal Justice and Behavior*. **34**: 661. doi:10.1177/0093854806293485.

[30] For example:

- Albrecht, S. I.; Chadwick, B. A.; Alcorn, D. S. (1977). "Religiosity and deviance: Application of an attitude-behavior contingent consistency model". *Journal for the Scientific Study of Religion*. **16**: 263–274. doi:10.2307/1385697.

- Burkett, S.; White, M. (1974). "Hellfire and delinquency:Another look". *Journal for the Scientific Study of Religion*. **13**: 455–462. doi:10.2307/1384608.

- Chard-Wierschem, D. (1998). In pursuit of the "true" relationship: A longitudinal study of the effects of religiosity on delinquency and substance abuse. Ann Arbor, MI: UMI Dissertation.

- Cochran, J. K.; Akers, R. L. (1989). "Beyond Hellfire:An explanation of the variable effects of religiosity on adolescent marijuana and alcohol use". *Journal of Research in Crime and Delinquency*. **26**: 198–225. doi:10.1177/0022427889026003002.

- Evans, T. D.; Cullen, F. T.; Burton, V. S.; Jr; Dunaway, R. G.; Payne, G. L.; Kethineni, S. R. (1996). "Religion, social bonds, and delinquency". *Deviant Behavior*. **17**: 43–70. doi:10.1080/01639625.1996.9968014.

- Grasmick, H. G.; Bursik, R. J.; Cochran, J. K. (1991). "Render unto Caesar what is Caesar's": Religiosity and taxpayer's inclinations to cheat". *The Sociological Quarterly*. **32**: 251–266. doi:10.1111/j.1533-8525.1991.tb00356.x.

- Higgins, P. C.; Albrecht, G. L. (1977). "Hellfire and delinquency revisited". *Social Forces*. **55**: 952–958. doi:10.1093/sf/55.4.952.

- Johnson, B. R.; Larson, D. B.; DeLi, S.; Jang, S. J. (2000). "Escaping from the crime of inner cities: Church attendance and religious salience among disadvantaged youth". *Justice Quarterly*. **17**: 377–391. doi:10.1080/07418820000096371.

- Johnson, R. E.; Marcos, A. C.; Bahr, S. J. (1987). "The role of peers in the complex etiology of adolescent drug use". *Criminology*. **25**: 323–340. doi:10.1111/j.1745-9125.1987.tb00800.x.

- Powell, K. (1997). "Correlates of violent and nonviolent behavior among vulnerable inner-city youths". *Family and Community Health*. **20**: 38–47. doi:10.1097/00003727-199707000-00006.

[31] Baier, C. J.; Wright, B. R. (2001). "If you love me, keep my commandments":A meta-analysis of the effect of religion on crime". *Journal of Research in Crime and Delinquency*. **38**: 3–21. doi:10.1177/0022427801038001001.

[32] Paul, Gregory S. (2005). "Cross-National Correlations of Quantifiable Societal Health with Popular Religiosity and Secularism in the Prosperous Democracies: A First Look". *Journal of Religion and Society*. Baltimore, Maryland. **7**: 4, 5, 8.

[33] Paul, Gregory S. (2005). "Cross-National Correlations of Quantifiable Societal Health with Popular Religiosity and Secularism in the Prosperous Democracies: A First Look". *Journal of Religion and Society*. Baltimore, Maryland. **7**: 11.

[34] Paul, Gregory S. (2005). "Cross-National Correlations of Quantifiable Societal Health with Popular Religiosity and Secularism in the Prosperous Democracies: A First Look". *Journal of Religion and Society*. Baltimore, Maryland. **7**.

[35] Gerson Moreno-Riaño; Mark Caleb Smith; Thomas Mach (2006). "Religiosity, Secularism, and Social Health". *Journal of Religion and Society*. Cedarville University. **8**.

15.6 External links

- Morality without religion, by Marc Hauser
- *Can we be good without God* 1996, Paul Chamberlain ISBN 0-8308-1686-0
- Video: Marc Hauser, Pt 3 Is there morality without religion?

Chapter 16

Metaphysical naturalism

This article is about the worldview. For the working assumption without suggesting ultimate truth, see Methodological naturalism.

Metaphysical naturalism, also called **ontological naturalism**, **philosophical naturalism**, and **scientific materialism** is a philosophical worldview, which holds that there is nothing but natural elements, principles, and relations of the kind studied by the natural sciences. Methodological naturalism is a philosophical basis for science, for which metaphysical naturalism provides only one possible ontological foundation. Broadly, the corresponding theological perspective is religious naturalism or spiritual naturalism. More specifically, metaphysical naturalism rejects the supernatural concepts and explanations that are part of many religions.

16.1 Definition

According to Steven Schafersman, geologist and president of Texas Citizens for Science metaphysical naturalism is a philosophy that maintains that; 1. Nature encompasses all that exists throughout space and time; and 2. Nature (the universe or cosmos) consists only of natural elements, that is, of spatiotemporal physical substance—mass–energy. Non-physical or quasi-physical substance, such as information, ideas, values, logic, mathematics, intellect, and other emergent phenomena, either supervene upon the physical or can be reduced to a physical account; and 3. Nature operates by the laws of physics and in principle, can be explained and understood by science and philosophy; and 4. the supernatural does not exist, i.e., only nature is real. Naturalism is therefore a metaphysical philosophy opposed primarily by Biblical creationism."[1]

> Naturalism, in recent usage, is a species of philosophical monism according to which whatever exists or happens is *natural* in the sense of being susceptible to explanation through methods which, although paradigmatically exemplified in the natural sciences, are continuous from domain to domain of objects and events. Hence, naturalism is polemically defined as repudiating the view that there exists or could exist any entities which lie, in principle, beyond the scope of scientific explanation.
> — Arthur C. Danto, The Encyclopedia of Philosophy, Naturalism[2]

Regarding the vagueness of the general term "naturalism", David Papineau traces the current usage to philosophers in early 20th century America such as John Dewey, Ernest Nagel, Sidney Hook and Roy Wood Sellars: "So understood, 'naturalism' is not a particularly informative term as applied to contemporary philosophers. The great majority of contemporary philosophers would happily accept naturalism as just characterized—that is, they would both reject 'supernatural' entities, and allow that science is a possible route (if not necessarily the only one) to important truths about the 'human spirit'."[3] Papineau remarks that philosophers widely regard naturalism as a "positive" term, and "few active philosophers nowadays are happy to announce themselves as 'non-naturalists'", while noting that "philosophers concerned with religion tend to be less enthusiastic about 'naturalism'" and that despite an "inevitable" divergence due to its popularity, if more narrowly construed, (to the chagrin of John McDowell, David Chalmers and Jennifer Hornsby, for example), those not so disqualified remain nonetheless content "to set the bar for 'naturalism' higher".[3]

Philosopher and theologian Alvin Plantinga, a well-known critic of naturalism in general, comments: "Naturalism is presumably not a religion. In one very important respect, however, it resembles religion: it can be said to perform the cognitive function of a religion. There is that range of deep human questions to which a religion typically provides an answer ... Like a typical religion, naturalism gives a set of answers to these and similar questions."[4]

16.1.1 Methodological Naturalism

Metaphysical naturalism is an approach to metaphysics or ontology, which deals with existence *per se*. It should not be confused with methodological naturalism, which sees empiricism as the basis for the scientific method.

Regarding science and evolution, Eugenie C. Scott, a notable opponent of teaching creationism or intelligent design in US public schools, stresses the importance of separating metaphysical from methodological naturalism:

> If it is important for Americans to learn about science and evolution, decoupling the two forms of naturalism is essential strategy. ... I suggest that scientists can defuse some of the opposition to evolution by first recognizing that the vast majority of Americans are believers, and that most Americans want to retain their faith. It is demonstrable that individuals can retain religious beliefs and still accept evolution as science. Scientists should avoid confusing the methodological naturalism of science with metaphysical naturalism.[5]
>
> — Eugenie C. Scott, Creationism, Ideology, and Science

16.1.2 Lack of necessity for worship

The historian Richard Carrier, in his book *Sense and Goodness without God: A Defense of Metaphysical Naturalism*, describes metaphysical naturalism thus: as a philosophy "wherein worship is replaced with curiosity, devotion with diligence, holiness with sincerity, ritual with study, and scripture with the whole world and the whole of human learning". Carrier wrote that it is the naturalist's duty "to question all things and have a well-grounded faith in what is well-investigated and well-proved, rather than what is merely well-asserted or well-liked."[6]

16.2 Science and naturalism

Main article: Uniformitarianism

While not metaphysical naturalism *per se*, in the more general sense of naturalism and philosophy expressed by Kate and Vitaly (2000) "there are certain philosophical assumptions made at the base of the scientific method - namely, that reality is objective and consistent, that humans have the capacity to perceive reality accurately, and that rational explanations exist for elements of the real world. These assumptions are the basis of naturalism, the philosophy on which science is grounded."[7] As noted by Steven Schafersman, methodological naturalism is "the adoption or assumption of philosophical naturalism within scientific method with or without fully accepting or believing it ... science is not metaphysical and does not depend on the ultimate truth of any metaphysics for its success (although science does have metaphysical implications), but methodological naturalism must be adopted as a strategy or working hypothesis for science to succeed. We may therefore be agnostic about the ultimate truth of naturalism, but must nevertheless adopt it and investigate nature as if nature is all that there is."[1] Contrary to other notable opponents of teaching Creationism or Intelligent Design in US public schools such as Eugenie Scott, Schafersman asserts that "while science as a process only requires methodological naturalism, I think that the assumption of methodological naturalism by scientists and others logically and morally entails ontological naturalism."[1] as well as the similarly controversial assertion: "I maintain that the practice or adoption of methodological naturalism entails a logical and moral belief in ontological naturalism, so they are not logically decoupled."[1] On the other hand, Scott argues:

> that a clear distinction must be drawn between science as a way of knowing about the natural world and science as a foundation for philosophical views. One should be taught to our children in school, and the other can optionally be taught to our children at home. Once this view is explained, I have found far more support than disagreement among my university colleagues. Even someone who may disagree with my logic or understanding of philosophy of science often understands the strategic reasons for separating methodological from philosophical materialism — if we want more Americans to understand evolution.[5][8]
>
> — Eugenie C. Scott, Science and Religion, Methodology and Humanism

However, there are other controversies, Arthur Newell Strahler embeds peculiar anthropic distinctions in the name of naturalism: "The naturalistic view is that the particular universe we observe came into existence and has operated through all time and in all its parts without the impetus or guidance of any supernatural agency. The naturalistic view is espoused by science as its fundamental assumption."[9] Variously known as background independence, the cosmological principle, the principle of universality, the principle of uniformity, or uniformitarianism, there *are* important philosophical assumptions that cannot be derived from nature. As noted by Stephen Jay Gould:

"You cannot go to a rocky outcrop and observe either the constancy of nature's laws or the working of unknown processes. It works the other way around." You first assume these propositions and "then you go to the out crop of rock."[10][11] "The assumption of spatial and temporal invariance of natural laws is by no means unique to geology since it amounts to a warrant for inductive inference which, as Bacon showed nearly four hundred years ago, is the basic mode of reasoning in empirical science. Without assuming this spatial and temporal invariance, we have no basis for extrapolating from the known to the unknown and, therefore, no way of reaching general conclusions from a finite number of observations. (Since the assumption is itself vindicated by induction, it can in no way "prove" the validity of induction - an endeavor virtually abandoned after Hume demonstrated its futility two centuries ago)."[12] Gould also notes that natural processes such as Lyell's "uniformity of process" are an assumption: "As such, it is another *a priori* assumption shared by all scientists and not a statement about the empirical world."[13] Such assumptions across time and space are needed for scientists to extrapolate into the unobservable past, according to G.G. Simpson: "Uniformity is an unprovable postulate justified, or indeed required, on two grounds. First, nothing in our incomplete but extensive knowledge of history disagrees with it. Second, only with this postulate is a rational interpretation of history possible, and we are justified in seeking—as scientists we must seek—such a rational interpretation."[14] and according to R. Hooykaas: "The principle of uniformity is not a law, not a rule established after comparison of facts, but a principle, preceding the observation of facts . . . It is the logical principle of parsimony of causes and of economy of scientific notions. By explaining past changes by analogy with present phenomena, a limit is set to conjecture, for there is only one way in which two things are equal, but there are an infinity of ways in which they could be supposed different."[15]

16.3 Various associated beliefs

Contemporary naturalists possess a wide diversity of beliefs within metaphysical naturalism. Most metaphysical naturalists have adopted some form of materialism or physicalism.[16]

16.3.1 Undesigned universe

Metaphysical naturalists argue that the scientific facts and theories that we have to explain the origins of the universe provide no evidence for supernatural beings or deities.[17] As Richard Carrier explains:

> ...no other worldview is directly and substantially supported by any scientific evidence, whereas all scientific evidence so far does support Metaphysical Naturalism, often directly, sometimes substantially. Though naturalism has not yet been proved, it is the best bet going.[17]

One might say that either it has always existed or it had a purely natural origin, being neither created nor designed.

16.3.2 Abiogenesis and evolution

Since nature is all there is, and there was once no life, abiogenesis is implied: that life arose spontaneously from natural causes.[18][19] Naturalists reason about *how*, not *if* evolution happened. They maintain that humanity's existence is not by intelligent design but rather a natural process of emergence.

16.3.3 Ethics and meta-ethics

Some embrace virtue ethics and many see no compelling argument against ethical naturalism.[20] Some may advocate for a Science of morality. One example of an attempt to ground a naturalist Meta-Ethical system is Richard Carrier's chapter *"Moral Facts Naturally Exist (and Science Could Find Them)"* which was peer reviewed by four philosophers. It sets out to prove a Moral realism centered around human satisfaction. Alexander Rosenberg has expressed a contrary position that naturalists, in general, have to accept moral nihilism.[21]

16.3.4 The mind is a natural phenomenon

If any variety of metaphysical naturalism is true, any mental properties that exist are caused by and ontologically dependent upon nature. However, some metaphysical naturalists consider the mental to be out-of-bounds, just like the supernatural.[22]

Metaphysical naturalists do not believe in a soul or spirit, nor in ghosts, and when explaining what constitutes the mind they rarely appeal to substance dualism. If one's mind, or rather one's identity and existence as a person, is entirely the product of natural processes, three conclusions follow according to W.T. Stace. First, all mental contents (such as ideas, theories, emotions, moral and personal values, or aesthetic response) exist solely as computational constructions of one's brain and genetics, not as things that exist independently of these. Second, damage to the brain (regardless of how) should be of great concern. Third, death or destruction of one's brain cannot be survived, which is to say, all humans are mortal. Stace, however, believes that ecstatic mysticism calls into question the assumption that awareness is impossible without data processing.[23]

16.3.5 Utility of reason

Metaphysical naturalists hold that reason is the refinement and improvement of naturally evolved faculties. The certitude of deductive logic remains unexplained by this essentially probabilistic view. Nevertheless, naturalists believe anyone who wishes to have more beliefs that are true than are false should seek to perfect and consistently employ their reason in testing and forming beliefs. Empirical methods (especially those of proven use in the sciences) are unsurpassed for discovering the facts of reality, while methods of pure reason alone can securely discover logical errors.[24]

16.3.6 Value of society

Humans are social animals, which is why humanity developed culture and civilization. In terms of evolution, this means that differential reproductive success somehow depended on traits that permit the development and maintenance of a healthy and productive culture and civilization.

16.4 History

16.4.1 Ancient period

Metaphysical naturalism appears to have originated in early Greek philosophy. The earliest presocratic philosophers, such as Thales, Anaxagoras or especially the atomist Democritus, were labeled by their peers and successors "the *physikoi*" (from the Greek φυσικός or *physikos*, meaning "natural philosopher," borrowing on the word φύσις or *physis*, meaning "nature") because they investigated natural causes, often excluding any role for gods in the creation or operation of the world. This eventually led to fully developed systems such as Epicureanism, which sought to explain everything that exists as the product of atoms falling and swerving in a void.

> Plato's world of eternal and unchanging Forms, imperfectly represented in matter by a divine Artisan, contrasts sharply with the various mechanistic Weltanschauungen, of which atomism was, by the fourth century at least, the most prominent... This debate was to persist throughout the ancient world. Atomistic mechanism got a shot in the arm from Epicurus... while the Stoics adopted a divine teleology... The choice seems simple: either show how a structured, regular world could arise out of undirected processes, or inject intelligence into the system. This was how Aristotle (384–322 bc), when still a young acolyte of Plato, saw matters. Cicero (On the Nature of the Gods 2. 95 = Fr. 12) preserves Aristotle's own cave-image: if troglodytes were brought on a sudden into the upper world, they would immediately suppose it to have been intelligently arranged. But Aristotle grew to abandon this view; although he believes in a divine being, the Prime Mover is not the efficient cause of action in the Universe, and plays no part in constructing or arranging it... But, although he rejects the divine Artificer, Aristotle does not resort to a pure mechanism of random forces. Instead he seeks to find a middle way between the two positions, one which relies heavily on the notion of Nature, or phusis.[25]
>
> — R. J. Hankinson, Cause and Explanation in Ancient Greek Thought

Metaphysical naturalism is most notably a Western phenomenon, but an equivalent idea has long existed in the East. Though unnamed and never articulated into a coherent system, one tradition within Confucian philosophy embraced a form of metaphysical naturalism dating to the Wang Chong in the 1st century, if not earlier, but it arose independently and had little influence on the development of modern naturalist philosophy or on Eastern or Western culture.

16.4.2 Middle ages to modernity

With the rise and dominance of Christianity in the West and the later spread of Islam, metaphysical naturalism was generally abandoned by intellectuals. Thus, there is little evidence for it in the Middle Ages. The reintroduction of Aristotle's empirical epistemology as well as previously lost treatises by Greco-Roman natural philosophers during the Renaissance contributed to Scientific Revolution which was begun by the medieval Scholastics without resulting in any noticeable increase in commitment to naturalism. It was not until the early modern era and Age of Enlightenment that naturalism, like that of Benedict Spinoza, David Hume, Denis Diderot, Julien La Mettrie, and Baron d'Holbach, among others, started to emerge again in the 17th and 18th centuries.

In this period, some metaphysical naturalists adhered to a distinct doctrine, materialism, which became the only category of metaphysical naturalism widely defended until the 20th century, when advances in physics resulted in widespread abandonment of prior formulations of materialism. 19th century physics added electromagnetic force fields, and in the 20th century matter was found to be a form of energy and therefore not fundamental as materialists had assumed. (See History of physics.) In philosophy, renewed attention to the problem of universals, philosophy of mathematics, the development of mathematical logic, and the post-positivist revival of metaphysics and the philosophy of religion, initially by way of Wittgensteinian linguistic philosophy, further called the naturalistic paradigm into question. Developments such as these, along with those within science and the philosophy of science brought new advancements and revisions of naturalistic doctrines by naturalistic philosophers into metaphysics, ethics, the philosophy of language, the philosophy of mind, epistemology, etc., the products of which include physicalism and eliminative materialism, supervenience, causal theories of reference, anomalous monism, naturalized epistemology (e.g. reliabilism), internalism and externalism, ethical naturalism, and property dualism, for example.

Currently, metaphysical naturalism is more widely embraced than in previous centuries, especially but not exclusively in the natural sciences and the Anglo-American, analytic philosophical communities. While the vast majority of the population of the world remains firmly committed to non-naturalistic worldviews, prominent contemporary defenders of naturalism and/or naturalistic theses and doctrines today include J. J. C. Smart, David Malet Armstrong, David Papineau, Paul Kurtz, Brian Leiter, Daniel Dennett, Michael Devitt, Fred Dretske, Paul and Patricia Churchland, Mario Bunge, Jonathan Schaffer, Hilary Kornblith, Quentin Smith, Paul Draper and Michael Martin, among many other academic philosophers.

According to David Papineau, contemporary naturalism is a consequence of the build-up of scientific evidence during the twentieth century for the "causal closure of the physical", the doctrine that all physical effects can be accounted for by physical causes.[26]

According to Steven Schafersman, president of Texas Citizens for Science, an advocacy group opposing creationism in public schools,[27] the progressive adoption of methodological naturalism—and later of metaphysical naturalism—followed the advances of science and the increase of its explanatory power.[28] These advances also caused the diffusion of positions associated with metaphysical naturalism, such as existentialism.[29]

> By the middle of the twentieth century, the acceptance of the causal closure of the physical realm led to even stronger naturalist views. The causal closure thesis implies that any mental and biological causes must themselves be physically constituted, if they are to produce physical effects. It thus gives rise to a particularly strong form of ontological naturalism, namely the physicalist doctrine that any state that has physical effects must itself be physical.
> From the 1950s onwards, philosophers began to formulate arguments for ontological physicalism. Some of these arguments appealed explicitly to the causal closure of the physical realm (Feigl 1958, Oppenheim and Putnam 1958). In other cases, the reliance on causal closure lay below the surface. However, it is not hard to see that even in these latter cases the causal closure thesis played a crucial role.[30]

— David Papineau, "Naturalism" in the *Stanford Encyclopedia of Philosophy*

16.4.3 Marxism, Objectivism, and secular humanism

A number of politicized versions of naturalism have arisen in the Western world, most notably Marxism in the 19th century and Objectivism in the 20th century. Marxism is an expression of communist or socialist materialism within a naturalistic framework. Objectivism is an expression of capitalist idealism within a naturalistic framework. Most proponents of metaphysical naturalism in First World countries, however, are neither Marxists nor Objectivists, and instead embrace the more moderate political ideals of secular humanism or cultural moral relativism.

16.5 Arguments for metaphysical naturalism

In the context of creation and evolution debates, Internet Infidels co-founder Jeffery Jay Lowder argues against what he calls "the argument from bias", that *a priori*, the supernatural is merely ruled out due to an unexamined stipulation. Lowder believes "there are good empirical reasons for believing that metaphysical naturalism is true, and therefore a denial of the supernatural need not be based upon an *a priori* assumption".[31]

Richard Carrier argues in *Sense and Goodness Without God: A Defense of Metaphysical Naturalism* that Metaphysical Naturalism is true. Topics covered include metaphilosophy, semantics, epistemology, the nature and origin of the universe (including a proposal that spacetime may be the ground of all being and a rejection of the logical possibility for any ultimate answer), free will compatibilism, the nature of mind, abstract objects, ontological reductionism, the nature of emotions, the meaning of life, the nature of reason, atheism, aesthetics, morality (including ethical naturalism and a recommendation for a science of morality), and politics.

16.5.1 Argument from physical minds

Several Metaphysical Naturalists have used the trends in scientific discoveries about minds to argue that no supernatural minds exist. For instance, Lowder says, "Since all known mental activity has a physical basis, there are probably no disembodied minds. But God is conceived of as a disembodied mind. Therefore, God probably does not exist."[32] Lowder argues the correlation between mind and brain implies that supernatural souls do not exist because the theist position, according to Lowder, is that the mind depends upon this soul instead of the brain.[31]

16.5.2 Cosmological argument for naturalism

> [Elegance] goes directly to the question of how the laws of nature are constructed. Nobody knows the answer to that. Nobody! It's a perfectly legitimate hypothesis, in my view, to say that some extremely elegant creator made those laws. But I think if you go down that road, you must have the courage to ask the next question, which is: Where did that creator come from? And where did his, her, or its elegance come from? And if you say it was always there, then why not say that the laws of nature were always there and save a step?[33]
> — Carl Sagan, *Conversations with Carl Sagan*

> There is no plausible reason why an Almighty would need billions of years and trillions of galaxies to accomplish his ends through long, deterministic causal processes. But that is exactly what we should expect if there is no god, but only nature.[34]
> — Richard Carrier, *Sense And Goodness Without God*

16.6 Arguments against metaphysical naturalism

Arguments against metaphysical naturalism include the following examples.

16.6.1 Evolutionary argument against naturalism

Main article: Evolutionary argument against naturalism

Alvin Plantinga is the John A. O'Brien Professor of Philosophy Emeritus at the University of Notre Dame, and the inaugural holder of the Jellema Chair in Philosophy at Calvin College. He is a Christian, and a well-known critic of naturalism. He argues, in his evolutionary argument against naturalism, that the probability that evolution has produced humans with reliable true beliefs, is low or inscrutable, unless their evolution was guided, for example, by God. According to David Kahan of the University of Glasgow, in order to understand how beliefs are warranted, a justification must be found in the context of supernatural theism, as in Plantinga's epistemology.[35][36][37] *(See also supernormal stimuli).*

Plantinga argues that together, naturalism and evolution provide an insurmountable "*defeater* for the belief that our cognitive faculties are reliable", i.e., a skeptical argument along the lines of Descartes' Evil demon or Brain in a vat.[38]

> Take *philosophical naturalism* to be the belief that there aren't any supernatural entities--no such person as God, for example, but also no other supernatural entities, and nothing at all like God. My claim was that naturalism and contemporary evolutionary theory are at serious odds with one another--and this despite the fact that the latter is ordinarily thought to be one of the main pillars supporting the edifice of the former. (Of course I am *not* attacking the theory of evolution, or anything in that neighborhood; I am instead attacking the conjunction of *naturalism* with the view that human beings have evolved in that way. I see no similar problems with the conjunction of *theism* and the idea that human beings have evolved in the way contemporary evolutionary science suggests.) More particularly, I argued that the conjunction of naturalism with the belief that we human beings have evolved in conformity with current evolutionary doctrine... is in a certain interesting way self-defeating or self-referentially incoherent.[38]
> — Alvin Plantinga, "Introduction" in *Naturalism Defeated?: Essays on Plantinga's Evolutionary Argument Against Naturalism*

Branden Fitelson of the University of California, Berkeley and Elliott Sober of the University of Wisconsin–Madison argue that Plantinga must show that the combination of evolution and naturalism also defeats the more modest claim that "at least a non-negligible minority of our beliefs are true", and that defects such as cognitive bias are nonetheless consistent with being made in the image of a rational God. Whereas evolutionary science already acknowledges that cognitive processes are unreliable, including the fallibility of the scientific enterprise itself, Plantinga's hyperbolic doubt is no more a defeater for naturalism than it is for theistic metaphysics founded upon a non-deceiving God who designed the human mind: "[neither] can construct a non-question-begging argument that refutes global skepticism."[39] Plantinga's argument has also been criticized by philosopher Daniel Dennett and historian Richard Carrier who argue that a cognitive apparatus for truth-finding can result from natural selection.[40]

16.6.2 Pre-Modern philosophy

Main articles: Theory of Forms, Hylomorphism, and Scholasticism

Edward Feser, in his book **The Last Superstition: A Refutation of the New Atheism**, lays a plenary case against naturalism by re-examining pre-Modern philosophy. Beginning in the second chapter, Feser cites the Platonic[41] and Aristotelian[42] answers to the problem of universals - that is, realism. Feser also offers arguments against nominalism.[43] And by defending realism and rejecting nominalism, he rejects eliminative materialism - and thus naturalism.

In the third chapter, Feser summarizes three of Thomas Aquinas's arguments for the existence of God.[44] These include arguments for an unmoved mover,[45] first, uncaused cause [46] and (supernatural) supreme intelligence,[47] concluding that these must exist not as a matter of probability - as in the intelligent design view, particularly of irreducible complexity[48] - but as a necessary consequence of "obvious, though empirical, starting points".[49]

16.7 See also

- Atheism
- Daoism
- Dysteleology
- Ethical naturalism
- Liberal naturalism
- Religious naturalism
- Spiritual naturalism

16.8 Further reading

16.8.1 Historical overview

- Edward B. Davis and Robin Collins, "Scientific Naturalism." In *Science and Religion: A Historical Introduction,* ed. Gary B. Ferngren, Johns Hopkins University Press, 2002, pp. 322–34.

16.8.2 Pro

- Gary Drescher, *Good and Real*, The MIT Press, 2006. ISBN 0-262-04233-9

- David Malet Armstrong, *A World of States of Affairs*, Cambridge: Cambridge University Press, 1997. ISBN 0-521-58064-1

- Mario Bunge, 2006, *Chasing Reality: Strife over Realism*, University of Toronto Press. ISBN 0-8020-9075-3 and 2001, *Scientific Realism: Selected Essays of Mario Bunge*, Prometheus Books. ISBN 1-57392-892-5

- Richard Carrier, 2005, Sense and Goodness without God: A Defense of Metaphysical Naturalism, AuthorHouse. ISBN 1-4208-0293-3

- Mario De Caro & David Macarthur (eds), 2004. *Naturalism in Question*. Cambridge, Mass: Harvard University Press. ISBN 0-674-01295-X

- Daniel Dennett, 2003, *Freedom Evolves*, Penguin. ISBN 0-14-200384-0 and 2006

- Andrew Melnyk, 2003, *A Physicalist Manifesto: Thoroughly Modern Materialism*, Cambridge University Press. ISBN 0-521-82711-6

- David Mills, 2004, *Atheist Universe: Why God Didn't Have A Thing To Do With It*, Xlibris. ISBN 1-4134-3481-9

- Jeffrey Poland, 1994, *Physicalism: The Philosophical Foundations*, Oxford University Press. ISBN 0-19-824980-2

16.8.3 Con

- James Beilby, ed., 2002, *Naturalism Defeated? Essays on Plantinga's Evolutionary Argument Against Naturalism*, Cornell University Press. ISBN 0-8014-8763-3

- William Lane Craig and J.P. Moreland, eds., 2000, *Naturalism: A Critical Analysis*, Routledge. ISBN 0-415-23524-3

- Stewart Goetz and Charles Taliaferro, 2008, *Naturalism*, Eerdmans Publishing. ISBN 978-0-8028-0768-7

- Phillip E. Johnson, 1998, *Reason in the Balance: The Case Against Naturalism in Science, Law & Education*, InterVarsity Press. ISBN 0-8308-1929-0 and 2002, *The Wedge of Truth: Splitting the Foundations of Naturalism*, InterVarsity Press. ISBN 0-8308-2395-6

- C.S. Lewis, ed., 1996, "Miracles", Harper Collins. ISBN 0-06-065301-9

- Michael Rea, 2004, *World without Design: The Ontological Consequences of Naturalism*, Oxford University Press. ISBN 0-19-924761-7

- Victor Reppert, 2003, *C.S. Lewis's Dangerous Idea: In Defense of the Argument from Reason*, InterVarsity Press. ISBN 0-8308-2732-3

- Mark Steiner, 2002, *The Applicability of Mathematics as a Philosophical Problem*, Harvard University Press. ISBN 0-674-00970-3

16.9 Notes

[1] Schafersman 1996.

[2] Stone 2008, p. 2 "Personally, I place great emphasis on the phrase "in principle," since there are many things that science does not now explain. And perhaps we need some natural piety concerning the ontological limit question as to why there is anything at all. But the idea that naturalism is a polemical notion is important."

[3] Papineau 2007.

[4] Plantinga 2010

[5] Scott, Eugenie C. (1996). "Creationism, Ideology, and Science". In Gross, Levitt, and Lewis. *The Flight From Science and Reason*. The New York Academy of Sciences,. pp. 519–520.

[6] Carrier 2005, p. 26

[7] "Since philosophy is at least implicitly at the core of every decision we make or position we take, it is obvious that correct philosophy is a necessity for scientific inquiry to take place." (A.Sergei 2000)

[8] Scott, Eugenie C. (2008). "Science and Religion, Methodology and Humanism". NCSE. Retrieved 20 March 2012.

[9] (Strahler 1992, p. 3)

[10] (Gould 1987, p. 120)

[11] Gould 1987, p. 119

[12] (Gould 1965, pp. 223–228)

[13] (Gould 1984, p. 11)

[14] Simpson 1963, pp. 24–48

[15] Hooykaas 1963, p. 38

[16] Schafersman, Steven D. (1996). "Naturalism is Today An Essential Part of Science". Section "The Origin of Naturalism and Its Relation to Science". Certainly most philosophical naturalists today are materialists[...]

[17] Carrier, Richard (2010-08-09). "Free Preview". *Sense and Goodness without God: A Defense of Metaphysical Naturalism*. Bloomington, Indiana: AuthorHouse. Retrieved 2013-12-25.

[18] Carrier 2005, pp. 166–68

[19] Richard Carrier, [The Argument from Biogenesis: Probabilities Against a Natural Origin of Life], *Biology and Philosophy* 19.5 (November 2004), pp. 739-64.

[20] Carrier 2005, pp. 168–176, 326–327

[21] Rosenberg, Alexander (2009). "The Disenchanted Naturalist's Guide to Reality". *On the Human Forum*. National Humanities Center (United States). Archived from the original on 26 February 2012.

[22] Richard Carrier, On Defining Naturalism as a Worldview, *Free Inquiry* 30.3 (April/May 2010), pp. 50-51.

[23] Stace, W.T, Mysticism and Philosophy. N.Y.: Macmillan, 1960; reprinted, Los Angeles: Jeremy P. Tarcher, 1987.

[24] Carrier 2005, pp. 53–54

[25] Hankinson, R. J. (1997). *Cause and Explanation in Ancient Greek Thought*. Oxford University Press. p. 125. ISBN 978-0-19-924656-4.

[26] David Papineau, "The Rise of Physicalism" in *Physicalism and its Discontents*, Cambridge (2011). URL:http://ebooks.cambridge.org/ebook.jsf?bid=CBO9780511570797

[27] Williams, Sally (July 4, 2007). "The God curriculum". London: The Telegraph. Retrieved 2008-12-26.

[28] Schafersman, Steven D. (1996). "Naturalism is Today An Essential Part of Science". Section "The Origin of Naturalism and Its Relation to Science". Naturalism did not exist as a philosophy before the nineteenth century, but only as an occasionally adopted and non-rigorous method among natural philosophers. It is a unique philosophy in that it is not ancient or prior to science, and that it developed largely due to the influence of science.

[29] Schafersman, Steven D. (1996). "Naturalism is Today An Essential Part of Science". Section "The Origin of Naturalism and Its Relation to Science". Naturalism is almost unique in that it would not exist as a philosophy without the prior existence of science. It shares this status, in my view, with the philosophy of existentialism.

[30] Papineau, David (2007). "Naturalism". In Edward N. Zalta. *Stanford Encyclopedia of Philosophy*. It thus gives rise to a particularly strong form of ontological naturalism, namely the physicalist doctrine that any state that has physical effects must itself be physical.

[31] Lowder, Jeffery Jay (March 1999). "The Empirical Case for Metaphysical Naturalism". *Internet Infidels Newsletter*.

[32] "Argument from Physical Minds".

[33] Sagan, C.; Head, T. (2006). *Conversations with Carl Sagan*. Literary Conversations Series. University Press of Mississippi. p. 14. ISBN 9781578067367. LCCN 2005048747.

[34] Carrier 2005, p. 257

[35] "Gifford Lecture Series - Warrant and Proper Function 1987-1988".

[36] Plantinga, Alvin (11 April 2010). "Evolution, Shibboleths, and Philosophers — Letters to the Editor". The Chronicle of Higher Education. ...I do indeed think that evolution functions as a contemporary shibboleth by which to distinguish the ignorant fundamentalist goats from the informed and scientifically literate sheep.
According to Richard Dawkins, 'It is absolutely safe to say that, if you meet somebody who claims not to believe in evolution, that person is ignorant, stupid, or insane (or wicked, but I'd rather not consider that).' Daniel Dennett goes Dawkins one (or two) further: 'Anyone today who doubts that the variety of life on this planet was produced by a process of evolution is simply ignorant—inexcusably ignorant.' You wake up in the middle of the night; you think, can that whole Darwinian story really be true? Wham! You are inexcusably ignorant.
I do think that evolution has become a modern idol of the tribe. But of course it doesn't even begin to follow that I think the scientific theory of evolution is false. And I don't.

[37] Plantinga, Alvin (1993). *Warrant and Proper Function*. Oxford: Oxford University Press. Chap. 11. ISBN 0-19-507863-2.

[38] Beilby, J.K. (2002). "Introduction by Alvin Plantinga". *Naturalism Defeated?: Essays on Plantinga's Evolutionary Argument Against Naturalism*. Reference, Information and Interdisciplinary Subjects Series. Ithaca: Cornell University Press. pp. 1–2, 10. ISBN 978-0-8014-8763-7. LCCN 2001006111.

[39] Fitelson, Branden; Elliott Sober (1998). "Plantinga's Probability Arguments Against Evolutionary Naturalism" (PDF). *Pacific Philosophical Quarterly*. **79** (2): 115–129. doi:10.1111/1468-0114.00053. Retrieved 2007-03-06.

[40] Carrier 2005, pp. 181–188

[41] The Last Superstition: A Refutation of the New Atheism, Ch 2, pp. 31-49

[42] The Last Superstition, Ch 2, pp. 53-72

[43] The Last Superstition: A Refutation of the New Atheism, Ch 2, pp. 44-46

[44] The Last Superstition, Ch 3, pp. 90-119

[45] The Last Superstition, Ch 3, pp. 91-102

[46] The Last Superstition, Ch 3, pp. 102-110

[47] The Last Superstition, Ch 3, pp. 110-118

[48] The Last Superstition, Ch 3, pp. 110-114

[49] The Last Superstition, Ch 3, pp. 83

16.10 References

Books

- Audi, Robert (1996). "Naturalism". In Borchert, Donald M. *The Encyclopedia of Philosophy Supplement*. USA: Macmillan Reference. pp. 372–374.

- Carrier, Richard (2005). *Sense and Goodness without God: A defense of Metaphysical Naturalism*. AuthorHouse. p. 444. ISBN 1-4208-0293-3.

- Gould, Stephen J. (1984). "Toward the vindication of punctuational change in catastrophes and earth history". In Bergren, W. A.; Van Couvering, J. A. *Catastrophes and Earth History*. Princeton, New Jersey: Princeton University Press.

- Gould, Stephen J. (1987). *Time's Arrow, Time's Cycle: Myth and Metaphor in the Discovery of Geological Time*. Cambridge, MA: Harvard University Press. p. 119.

- Danto, Arthur C. (1967). "Naturalism". In Edwords, Paul. *The Encyclopedia of Philosophy*. New York: The Macmillan Co. and The Free Press. pp. 448–450.

- Hooykaas, R. (1963). *The principle of uniformity in geology, biology, and theology* (2nd ed.). London: E.J. Brill.

- Kurtz, Paul (1990). *Philosophical Essays in Pragmatic Naturalism*. Prometheus Books.

- Lacey, Alan R. (1995). "Naturalism". In Honderich, Ted. *The Oxford Companion to Philosophy*. Oxford University Press. pp. 604–606.

- Post, John F. (1995). "Naturalism". In Audi, Robert. *The Cambridge Dictionary of Philosophy.* Cambridge University Press. pp. 517–518.

- Rea, Michael (2002). *World Without Design: The Ontological Consequences of Naturalism.* Oxford University Press. ISBN 0-19-924760-9.

- Sagan, Carl (2002). *Cosmos.* Random House. ISBN 978-0-375-50832-5.

- Simpson, G. G. (1963). "Historical science". In Albritton Jr., C. C. *Fabric of geology.* Stanford, California: Freeman, Cooper, and Company.

- Strahler, Arthur N. (1992). *Understanding Science: An Introduction to Concepts and Issues.* Buffalo: Prometheus Books.

- Stone, J.A. (2008). *Religious Naturalism Today: The Rebirth of a Forgotten Alternative.* G - Reference, Information and Interdisciplinary Subjects Series. State University of New York Press. p. 2. ISBN 978-0-7914-7537-9. LCCN 2007048682.

Journals

- Gould, Stephen J. (1965). "Is uniformitarianism necessary". *American Journal of Science.* **263**.

Web

- A., Kate; Sergei, Vitaly (2000). "Evolution and Philosophy: Science and Philosophy". Think Quest. Retrieved 19 January 2009.

- Papineau, David (2007). "Naturalism". In Edward N. Zalta. *Stanford Encyclopedia of Philosophy* (Spring 2007 ed.).

- Plantinga, Alvin (2010). "Religion and Science". Stanford Encyclopedia of Philosophy. Retrieved 3 November 2010.

- Schafersman, Steven D. (1996). "Naturalism is Today An Essential Part of Science". Retrieved 3 November 2010.

16.11 External links

- "Naturalism" in the *Internet Encyclopedia of Philosophy*

- "Naturalism" in the *Stanford Encyclopedia of Philosophy*

- "Naturalism in Legal Philosophy" in the *Stanford Encyclopedia of Philosophy*

- "Naturalism in the Philosophy of Mathematics" in the *Stanford Encyclopedia of Philosophy*

- "Physicalism" in the *Stanford Encyclopedia of Philosophy*

- "Naturalism" in the *Catholic Encyclopedia*

- Center for Naturalism

- Naturalism entry in The Skeptic's Dictionary

- Naturalism Library at the Secular Web

- Naturalism as a Worldview resource page by Richard Carrier

- A Defense of Naturalism by Keith Augustine (2001)

Chapter 17

Religion of Humanity

Positivist temple in Porto Alegre

Religion of Humanity (from French *Religion de l'Humanité* or *église positiviste*) is a secular religion created by Auguste Comte, the founder of positivist philosophy. Adherents of this religion have built chapels of Humanity in France and Brazil.[1]

In the US and Europe, Comte's ideas influenced others, and contributed to the emergence of ethical societies and "ethical churches", which led to the development of Ethical culture, congregational humanist, and secular humanist organisations.

17.1 Origins

Comte developed the *religion of humanity* for positivist societies in order to fulfill the cohesive function once held by traditional worship. The religion was developed after Comte's passionate platonic relationship with Clotilde de Vaux, whom he idealised after her death. He became convinced that feminine values embodied the triumph of sentiment and morality. In a future science-based Positivist society there should also be a religion that would have power by virtue of moral force alone.[2] In 1849, he proposed a calendar reform called the "positivist calendar", in which months were named after history's greatest leaders, thinkers, and artists, and arranged in chronological order. Each day was dedicated to a thinker.

17.2 Tenets

According to Tony Davies, Comte's secular and positive religion was "a complete system of belief and ritual, with liturgy and sacraments, priesthood and pontiff, all organized around the public veneration of Humanity", referred to as the *Nouveau Grand-Être Suprême* (New Supreme Great Being). "This was later to be supplemented in a positivist trinity by the *Grand Fétish* (the Earth) and the *Grand Milieu* (Cosmic Space)".[3]

In *Système de politique positive* (1851–1854) Comte stated that the pillars of the religion are:

- **altruism**, leading to generosity and selfless dedication to others.
- **order**: Comte thought that after the French Revolution, society needed restoration of order.
- **progress**: the consequences of industrial and technical breakthroughs for human societies.

In *Catéchisme positiviste* (1851), Comte defined the Church of Humanity's seven sacraments:

- **Introduction**; (nomination and sponsoring)
- **Admission**; (end of education)
- **Destination**; (choice of a career)
- **Marriage**;
- **Retirement**; (age 63),
- **Separation**; (social extreme unction),

- **Incorporation**; (absorption into history) - 3 years after death.

17.3 Liturgy and priesthood

Erlon Jacques de Oliveira, Temple guardian in Porto Alegre

The Religion of Humanity was described by Thomas Huxley as "Catholicism minus Christianity".[2] In addition to a holy trinity of Humanity, the Earth and Destiny, it had a priesthood. Priests were *required* to be married, because of the ennobling influence of womanhood. They would conduct services, including Positivist prayer, which was "a solemn out-pouring, whether in private or in public, of men's nobler feelings, inspiring them with larger and more comprehensive thoughts." The purpose of the religion was to increase altruism, so that believers acted always in the best interests of humanity as a whole. The priests would be international ambassadors of altruism, teaching, arbitrating in industrial and political disputes, and directing public opinion. They should be scholars, physicians, poets and artists. Indeed all the arts, including dancing and singing should be practiced by them, like bards in ancient societies.

This required long training. They began training from the age of twenty-eight, studying in positivist schools. From thirty-five to forty-two a priest served in an apprentice position as teacher and ritualist. Only at the age of forty-two could he become a full priest. They earned no money and could not hold offices outside the priesthood. In this way their influence was purely spiritual and moral. The High Priest of Humanity was to live in Paris, which would replace Rome as the centre of religion.[2]

17.4 Influence

Davies argues that Comte's austere and "slightly dispiriting" philosophy of humanity - viewed as alone in an indifferent universe (which can only be explained by "positive" science) - "was even more influential in Victorian England than the theories of Charles Darwin or Karl Marx".[3]

The *system* was ultimately unsuccessful but, along with Darwin's *On the Origin of Species*, it influenced the proliferation of various Secular Humanist organizations in the 19th century, especially through the work of secularists such as George Holyoake and Richard Congreve. Although Comte's English followers, including George Eliot and Harriet Martineau, for the most part rejected the full panoply of his system, they liked the idea of a religion of humanity and his injunction to "vivre pour altrui" ("live for others", from which comes the word "altruism").

Profound criticism came from John Stuart Mill who advocated Comte but dismissed his Religion of Humanity in a move towards a differentiation between the (good) early Comte, the author of *The Course in Positive Philosophy* and the (problematic) late Comte, who authored the Religion of Humanity.[4]

17.4.1 Religion of Humanity in Brazil

Comtean Positivism was relatively popular in Brazil. In 1881 Miguel Lemos and Raimundo Teixeira Mendes organized the "Positivist Church of Brazil." In 1897 the "Temple of Humanity" was created.[5] The services at the Temple could go on for up to four hours and that, combined with a certain moral strictness, led to some decline during the Republican period.[6] Nevertheless it had appeal with the military class as Benjamin Constant joined the group before breaking with it because he deemed Mendes and Lemos as too fanatical. Cândido Rondon's conversion proved more solid as he remained an orthodox Positivist and a member of the faith long after the church's importance waned.[7] Although declined, the church still survives in Brazil. The national flag of Brazil bears the "Ordem e Progresso" ("Order and Progress"), inspired by Comte's motto of positivism: "L'amour pour principe et l'ordre pour base; le progrès pour but" ("Love as a principle and order as the basis; progress as the goal").[8]

17.5 Other examples

There are more examples of Religion of Humanity started by positivists, and there are several authors who have given the epithet to the religion they support, whatever the religion. In India Baba Faqir Chand established Manavta Mandir (Temple of Humanity) to spread his religion of humanity with scientific attitude as explained by David C. Lane in a book *The Unknowing Sage*. Comte influenced the thought of Victorian secularists George Holyoake (coiner of the term "secularism") and Richard Congreve.

17.6 See also

17.7 References

[1] "Où peut-on visiter un temple positiviste ? (Where Are Positivist Shrines to be Seen?)" (in French). Retrieved 2007-10-30.

[2] Rollin Chambliss, *Social Thought: From Hammurabi to Comte*, Dryden Press, New York, 1954, p.424.

[3] Davies, Tony. *Humanism*, The New Critical Idiom. Drakakis, John, series editor. University of Stirling, UK. Routledge, 1997, p.28-29 ISBN 0-415-11052-1

[4] See the central passage of Mill's *Auguste Comte and Positivism*: John Stuart Mill on Auguste Comte's Religion of Humanity ()

[5] Latin American Thought: Philosophical Problems and Arguments By Susana Nuccetelli: Page 184

[6] The Human Tradition in Modern Brazil By Peter M. Beattie: Pages 112-113

[7] Stringing Together a Nation: Candido Mariano da Silva Rondon and the... By Todd A. Diacon: pgs 83-84

[8] Bandeiras e significados Historianet. Retrieved on 2010-10-09. (in Portuguese).

17.8 External links

- English language site for Brazil's "Religion of Humanity"

- Olaf Simons, *The Religion of Humanity* (a structured collection of transcripts from English translations of Comte's major publications on the topic)

Chapter 18

International Humanist and Ethical Union

The **International Humanist and Ethical Union (IHEU)** is an umbrella organisation of humanist, atheist, rationalist, secular, skeptic, freethought and Ethical Culture organisations worldwide.[2] British philosopher and biologist Julian Huxley (also the first director of UNESCO) presided over the founding Congress of the IHEU in Amsterdam, 1952; Dutch philosopher and politician Jaap van Praag became its first chairman until 1975.

The IHEU works "to build and represent the global Humanist movement that defends human rights and promotes Humanist values world-wide."[3]

In 2002, the IHEU General Assembly unanimously adopted the Amsterdam Declaration 2002, which presents as the "official defining statement of World Humanism".[4] The Happy Human is the official symbol of the IHEU.

IHEU holds a World Humanist Congress usually every three years. The next is to be held in Miami, United States, in 2020.[5]

18.1 Humanism as a life stance

In 2002 at the IHEU's 50th anniversary World Humanist Congress, delegates unanimously passed a resolution known as the Amsterdam Declaration 2002, an update of the original Amsterdam Declaration (1952).[6]

The Amsterdam Declaration defines Humanism as a "lifestance" that is "ethical", "rational", supportive of "democracy and human rights", insisting "that personal liberty must be combined with social responsibility"; it is "an alternative to dogmatic religion"; it values "artistic creativity and imagination" and is aimed at living lives of "fulfillment" through the powers of "free inquiry", "science" and "creative imagination".[7]

In addition to the Amsterdam Declaration's "official statement of World Humanism", the IHEU provides a "Minimum Statement on Humanism":[8]

> Humanism is a democratic and ethical life stance, which affirms that human beings have the right and responsibility to give meaning and shape to their own lives. It stands for the building of a more humane society through an ethic based on human and other natural values in the spirit of reason and free inquiry through human capabilities. It is not theistic, and it does not accept supernatural views of reality.

Member Organisations of the IHEU are required according to IHEU's membership regulations to have objects that are "consistent" with this understanding of Humanism.[9]

18.1.1 Other major resolutions

In 2010, in an "unprecedented alliance"[10] of the IHEU, the European Humanist Federation and Catholics for Choice, launched the Brussels Declaration, a secular response to a proposed Berlin Declaration, under which the amended EU Constitution would have made references to "God" and the "Christian roots of Europe".[11]

At World Humanist Congress 2011, in Norway, the IHEU General Assembly adopted The Oslo Declaration on Peace, which concludes: "We urge each of our member organizations and Humanists globally to work for a more peaceful culture in their own nations and urge all governments to prefer the peaceful settlement of conflicts over the alternative of violence and war."[12]

At World Humanist Congress 2014, in the United Kingdom, the IHEU General Assembly adopted The Oxford Declaration on Freedom of Thought and Expression, which asserts: "Freedom of thought implies the right to develop, hold, examine and manifest our beliefs without coercion, and to express opinions and a worldview whether religious or non-religious, without fear of coercion. It includes the right to change our views or to reject beliefs previously held, or previously ascribed. Pressure to conform to ideologies of the state or to doctrines of religion is a tyranny."[13]

18.2 Organisation

18.2.1 Founding in 1952

Five Humanist organisations, the American Ethical Union, American Humanist Association, British Ethical Union (later the British Humanist Association and now Humanists UK), Vienna Ethical Society and the Dutch Humanist league hosted the founding congress of the IHEU in Amsterdam, 22–27 August 1952.[14][15] On the last day of the congress five resolutions were passed, which included a statement of the fundamentals of "modern, ethical Humanism", a resolution which would come to be known as the Amsterdam Declaration (1952).[14]

18.2.2 Current structure

IHEYO logo.

The IHEU is a democratic organisation,[2] the Board of which is elected by representatives of the Member Organisations at annual General Assemblies.[16] The President as of 2015 is Andrew Copson[17] (who is also the Chief Executive of Humanists UK as of 2010).[18] The IHEU headquarters is in London, sharing an office with Humanists UK.

Representatives of IHEU Member Organisations ratify new memberships annually during a General Assembly. Following the 2017 General Assembly, the IHEU listed its membership as 139 Member Organisations from 53 countries[19] from a variety of non-religious traditions.

A staff of four is headed by the current Chief Executive, Gary McLelland, and the IHEU maintains delegations to the United Nations Human Rights Council in Geneva, the United Nations in New York, and the Council of Europe in Strasbourg.[20]

The IHEU is an international NGO with Special Consultative Status with the United Nations, General Consultative Status at the Council of Europe, Observer Status with the African Commission on Human and Peoples' Rights, and maintains operational relations with UNESCO.

IHEU has a wing for people aged up to 35 called the International Humanist and Ethical Youth Organisation (IHEYO).

The name of the IHEU may soon change, as delegates at the organization's 2017 General Assembly passed a resolution "mandating the Board to oversee a transition to a revised identity for the organization".[21]

18.3 Strategy and activities

The aim of the IHEU is to "build, support and represent the global humanist movement, defending human rights, particularly those of non-religious people, and promoting humanist values world-wide".[22] As a campaigning NGO the IHEU aims "to influence international policy through representation and information, to build the humanist network, and let the world know about the worldview of Humanism."[3]

18.3.1 The Freedom of Thought Report

In 2012 the IHEU began publishing an annual report on "discrimination against humanists, atheists and the non-religious" called The Freedom of Thought Report.[23]

The report centres around a "Country Index" with a textual entry for every sovereign state.[24]

Each country is measured against a list of 64 boundary conditions, which are categorised into four thematic categories ("Constitution and government", "Education and children's rights", "Family, community, society, religious courts and tribunals", and "Freedom of expression, advocacy of humanist values") at five levels of overall "severity" ("Free and equal", "Mostly satisfactory", "Systemic discrimination", "Severe discrimination" and "Grave violations").[25] The 64 boundary conditions include for example: "'Apostasy' or conversion from a specific religion is outlawed and punishable by death", which is placed at the worst level of severity, and under the category "Freedom of expression", and: "There is state funding of at least some religious schools", which is a middle severity condition, under the category "Education and children's rights". The data from the report is freely available under a Creative Commons license.[26]

Cover of the downloadable 2016 edition of the IHEU Freedom of Thought Report - Key Countries Edition

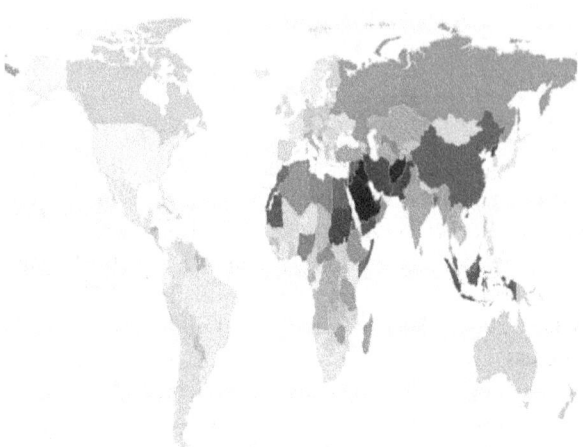

This composite map overlays the results from four separate categories of assessment in the IHEU Freedom of Thought Report, as to how countries discriminate against non-religious people. Countries block-filled in darker, redder colors are rated more severely in the report, while lighter, greener shades are more "free and equal".

Findings of the Freedom of Thought Report

In 2017, the report found that 30 countries meet at least one boundary condition at the most severe level ("Grave violations"), and a further 55 countries met at least one boundary condition in the next most severe level ("Severe discrimination").[26]

Responses to the Freedom of Thought Report

The various annual editions of the Freedom of Thought Report have been reported in the media under headlines such as: "How the right to deny the existence of God is under threat globally" (the Independent, UK);[27] "Most countries fail to respect rights of atheists – report" (Christian Today); and "Stephen Fry's mockery of religion could land him the death penalty in these countries" (the Washington Post).[28] The report has received coverage in the national media of countries that are severely criticised, for example "Malaysia's free thought, religious expression under 'serious assault', study shows" (the Malay Mail).[29]

Forewords and prefaces to the various annual editions of report have been written by then-United Nations Special Rapporteurs on Freedom of Religion or Belief, Heiner Bielefeldt, in 2012; two victims of "blasphemy" accusations, Kacem El Ghazzali and Alber Saber in 2013; human rights defenders Gulalai Ismail and Agnes Ojera in 2014; humanist activist and survivor of an anti-secularist machete attack in Bangladesh, Rafida Ahmed Bonya (2015); and United Nations Special Rapporteurs on Freedom of Religion or Belief, Ahmed Shaheed, in 2016. In 2015 and 2016 the annual edition of the Freedom of Thought Report was launched at the European Parliament in Brussels hosted by the European Parliamentary Intergroup on Freedom of Religion or Belief and Religious Tolerance chaired by Dennis de Jong MEP.[30][31]

In his foreword to the first edition of the Freedom of Thought Report, Heiner Bielefeldt wrote: "As a universal human right, freedom of religion or belief has a broad application. However, there seems to be little awareness that this right also provides a normative frame of reference for atheists, humanists and freethinkers and their convictions, practices and organizations. I am therefore delighted that for the first time the Humanist community has produced a global report on discrimination against atheists. I hope it will be given careful consideration by everyone concerned with freedom of religion or belief."[32]

At a panel event at the European Parliament for the launch of the 2015 edition, Bielefeldt said he "unambiguously welcomed" the report and reiterated with regard to "freedom of religion or belief" that it is "only a kind of short-hand", and "Formulations such as "religious freedom" obfuscate the scope of this human right which covers the identity-shaping, profound convictions and conviction-based practices of human beings broadly."[33]

The report was the subject of a question in the UK Parliament in 2013, to which David Lidington MP responded for the government asserting, "Our freedom of religion or belief policy is consistent with the key message of the International Humanist and Ethical Union's (IHEU) report: that international human rights law exists to protect the rights of individuals to manifest their beliefs, not to protect the beliefs themselves. The report records a sharp increase in the number of prosecutions for alleged criticism of religion by atheists on social media. Protecting freedom of expression online is a priority for the British Government and we have consistently argued against attempts to create a new international standard in order to protect religions from criticism."[34]

18.3.2 Focus of advocacy and campaigns

Recurring themes of the IHEU's advocacy and campaigns work include LGBTI rights and women's rights, sexual and reproductive health and rights, laws against blasphemy and apostasy, caste-based discrimination, slavery, and advocacy of secularism.

Persecuted non-religious individuals

Individuals persecuted for expressing their non-religious views (actual or perceived) have frequently been the subject of IHEU campaigns. Some prominent cases include:

- In the 1990s IHEU was instrumental in highlighting the threats against Taslima Nasrin who lives in exile from Bangladesh, and who also acted as a representative of the IHEU at UNESCO.[35]

- The IHEU and Amnesty International led the campaign in 2004 to try to obtain the release of Younus Shaikh[36] who was accused of "blasphemy" in Pakistan.

- In 2013 the IHEU urged the authorities in Egypt to ensure the safety of Alber Saber after he was accused of "offending religion" for allegedly linking to the YouTube video "Innocence of Muslims".[37]

- In 2014 the IHEU blew the whistle on the case of Mubarak Bala from Nigeria, who was detained in a psychiatric hospital after he talked openly about being an atheist.[38][39] He was freed following international media coverage.[40]

- In 2017, after a government minister in Malaysia said members of an atheist meetup group would be "hunted down"[41], the IHEU called for respect of the atheists' human rights[42], and the organization's condemnation of the minister's remarks was reported in Malaysian media.[43]

The IHEU delegation at the United Nations Human Rights Council has repeatedly raised the imprisonment and corporal punishment of Raif Badawi for "insulting religion",[44] and Waleed Abulkhair for "disrespecting the authorities",[45] both in Saudi Arabia.

The IHEU similarly highlights cases where individuals are accused of "apostasy", such as the blogger Mohamed Cheikh Ould Mkhaitir currently on death row in Mauritania,[46] and the poet Ashraf Fayadh currently imprisoned in Saudi Arabia.[47] In June 2016 at the 32nd session of the Human Rights Council the IHEU's delegate took the unusual step of reading one of Ashraf Fayadh's poems during General Debate.[48]

Bangladesh machete murders

See also: Attacks by Islamic extremists in Bangladesh

The IHEU complained that fundamentalists linked to the government were "terrorising" secular activists, including individuals in connection with IHEU Member Organisations, as far back as 2006.[49] However, a series of machete attacks primarily targeting secular and atheist bloggers and freethinkers in Bangladesh has been especially severe since 2013, and the IHEU has campaigned persistently in response[50] and highlighted the murders at the UN Human Rights Council.[51]

The IHEU responded in 2013 to the murder of blogger and activist Ahmed Rajib Haider and the machete attack on his friend Asif Mohiuddin, and highlighted the subsequent arrest and imprisonment of Mohiuddin and others for "hurting religious sentiments".[52][53][54]

When author and prominent leader of the Bengali freethought movement Avijit Roy was murdered, 26 February 2015, the IHEU revealed that he had been advising them on the situation in Bangladesh; the IHEU's Director of Communications commented, "This loss is keenly felt by freethinkers and humanists in South Asia and around the world. He was a colleague in humanism and a friend to all who respect human rights, freedom, and the light of reason."[55]

Following the murder of Washiqur Rahman Babu (or Oyasiqur Rhaman), 30 March 2015, the IHEU republished some of his final writings.[56]

Following the murder of Ananta Bijoy Das, 12 May 2016, the IHEU leaked parts of the letter Bijoy Das had recently received from Sweden rejecting his visa application, despite his having being invited to the country by Swedish PEN.

The IHEU highlighted "the failures of the Bangladeshi authorities to bring to justice the individuals and to break the networks behind this string of targeted killings", and also criticised Sweden's rejection of his visa application, commenting, "We call on all countries to recognise the legitimacy and sometimes the urgency and moral necessity of asylum claims made by humanists, atheists and secularists who are being persecuted for daring to express those views."[57]

Following the murder of Niladri Chattopadhyay Niloy (or Niloy Chatterjee, also known by his pen name Niloy Neel), 7 August 2015, the IHEU again attacked the government and authorities, saying, "Apparent failure to pursue the most obvious lines of inquiry even when initial arrests are made, and media manipulation resulting in conflicting stories, further makes reportage difficult and police operations opaque."[58]

A coordinated attack against two separate publishing houses in Dhaka, 31 October 2016, killed the publisher Faisal Arefin Dipon and seriously injured the publisher Ahmedur Rashid Chowdhury.[59] The IHEU later published an interview with Chowdhury about the attack and his escape to Norway.[60]

In August 2015 the IHEU coordinated a joint open letter in English and Bangla by a coalition of "Bloggers, free speech campaigners, humanist associations, religious and ex-Muslim groups"[61][62] calling on the president and prime minister of Bangladesh to "ensure the safety and security of those individuals whose lives are threatened by Islamist extremists... instruct the police to find the killers, not to harass or blame the victims... disassociate yourself publicly from those who call for death penalties against non-religious Bangladeshis..." and repeal the laws under which secular bloggers faced arrest and imprisonment.[61]

Following the murder of a student and secular activist Nazimuddin Samad, 6 April 2016,[63] and then the murder of university lecturer Professor Rezaul Karim Siddique, 23 April 2016, the IHEU's president Andrew Copson said "Unless the government [of Bangladesh] immediately begins to defend the right to speak and write freely, without adding the unprincipled and anti-secular qualifications that it keeps applying to freedom of expression, then very soon the only voices that will be heard will be those of murderous extremists."[64]

The IHEU, along with the IHEU Member Organisation the Dutch Humanist Association, and Hague Peace Projects, organised a "solidarity book fair" in The Hague, 26 February 2016, to coincide with the annual Ekushey Book Fair in Dhaka.[65]

The range of targets for these attacks began to broaden in the later part of 2015 and throughout 2016 to more often include minority religious individuals and foreigners, culminating in the July 2016 Dhaka attack in Gulshan Thana.

End Blasphemy Laws campaign

In January 2015, in part as a response to the Charlie Hebdo shooting, the IHEU alongside other transnational secular groups the European Humanist Federation and Atheist Alliance International and a two-hundred strong organisational coalition, founded the End Blasphemy Laws Campaign.[66][67] End Blasphemy Laws is "campaigning to repeal "blasphemy" and related laws worldwide."[68]

Other campaigns

The "First World Conference on Untouchability" was organised by the IHEU in London, June 2009.[69] Anticipating the event, the BBC News quoted then-Exeutive Director of the IHEU Babu Gogineni as saying that legal reforms alone would not end caste discrimination: "There are Dalit politicians in India, but nothing has changed. The answer is to educate Dalits and empower them."[70] The event was preceded by questions in the UK Parliament[71] and guests included Lord Desai and Lord Avebury from the UK House of Lords; Binod Pahadi, Member of the Constituent Assembly, Nepal; and Tina Ramirez, US Congressional Fellow on International Religious Freedom.[69] The Second World Conference on Untouchability was held in Kathmandu, in April 2014.[72]

In 2013 the IHEU criticised the US-based Appeal of Conscience Foundation for awarding their "World Statesman Award" to then-president of Indonesia Susilo Bambang Yudhoyono; IHEU argued that the award "is a slap in the face to prisoners of conscience across the world. While Alexander Aan suffers in an Indonesian jail for posting his beliefs to Facebook, his jailer will be honored in New York as a champion of freedom of belief."

In 2014 the IHEU as part of a "coalition of secular groups" led a campaign around the hashtag "#TwitterTheocracy" to protest the social media website Twitter's implementation of tools blocking "blasphemous" tweets in Pakistan.[73][74]

18.4 Historical dates and figures

18.4.1 Chairs and presidents

18.4.2 Awards

The IHEU makes a number of regular and occasional special awards.[76]

18.4. HISTORICAL DATES AND FIGURES

Previous IHEU President Sonja Eggerickx.

Current IHEU President Andrew Copson.

International Humanist Award

The International Humanist Award recognises outstanding achievements and contributions to the progress and defence of Humanism.

- 1970: Barry Commoner (United States of America), environmentalist professor

- 1974: Harold John Blackham (UK), founding member IHEU, IHEU secretary (1952-1966)

- 1978: Vithal Mahadeo Tarkunde (India), former judge of the Bombay High Court

- 1982: Kurt Partzsch (Germany), former Minister for Social Affairs

- 1986: Arnold Clausse (Belgium), professor emeritus of education

- 1986: The Atheist Centre (India), for pioneering social reform activities

- 1988: Andrei Sakharov (USSR), nuclear physicist, developer of the hydrogen bomb for the Soviet military, and winner of the Nobel Prize for Peace

- 1990: Alexander Dubček (Czechoslovakia), leader of Czechoslovakia during the "Prague Spring" of 1968

- 1992: Pieter Admiraal (Netherlands), a Dutch anaesthetist, and euthanasia advocate

- 1999: Professor Paul Kurtz (USA), writer and founder of the Committee for Skeptical Inquiry

- 2002: Amartya Sen (India), economist, social theorist, Master of Trinity College (Cambridge), and winner of the 1998 The Bank of Sweden Prize in Economic Sciences in Memory of Alfred Nobel

- 2005: Jean-Claude Pecker (France), astronomer

- 2008: Philip Pullman (UK), best-selling author of children's literature, including "His Dark Materials" trilogy

- 2011: Sophie in 't Veld, (Netherlands) MEP and vice-chair of the European Parliament Committee on Civil Liberties, and PZ Myers (USA), biology professor at University of Minnesota Morris, and author of the Pharyngula blog

- 2014: Gulalai Ismail (Pakistan), the founder and chair of Aware Girls, a charity which promotes the developmental and human rights of young women in Pakistan[77] and Wole Soyinka (Nigeria), Nobel Prize-winning author[78]

Distinguished Service to Humanism Award

The Distinguished Service to Humanism Award recognises the contributions of Humanist activists to International Humanism and to organised Humanism.

- 1988: Corliss Lamont (United States of America); Indumati Parikh (India); Mathilde Krim (United States)

- 1990: Jean Jacques Amy (Belgium)

- 1992: Indumati Parikh (India); Vern Bullough (USA); Nettie Klein, also volunteer IHEU secretary general (1982-1996)

- 1996: Jim Herrick (UK); James Dilloway

- 1999: Abe Solomon; Paul Postma

- 2002: Phil Ward

- 2005: Barbara Smoker (UK); Marius Dées de Stério

- 2007: Keith Porteous Wood (UK)
- 2008: Roy W Brown (UK)
- 2011: V B Rawat (India); Narendra Nayak (India); David Pollock (UK)
- 2012: Margaretha Jones (United States of America)
- 2013: Josh Kutchinsky (UK)
- 2014: Robbi Robson (UK)
- 2015: Hope Knutsson[79] (Iceland)
- 2016: Sonja Eggerickx[80] (Belgium)
- 2017: Leo Igwe[81] (Nigeria)

Other Awards

- 1978: Special Award for Service to World Humanism: Harold John Blackham; Jaap van Praag; Sidney Scheuer {also IHEU treasurer (1952-1987)}
- 1988: Humanist Laureate Award: Betty Friedan; Herbert Hauptman; Steve Allen
- 1988: Humanist of the Year Award: Henry Morgentaler
- 1992: Distinguished Human Rights Award: Elena Bonner
- 1996: Humanist Awards: Shulamit Aloni; Taslima Nasrin; Xiao Xuehui
- 2008: Lifetime Achievement Award: Levi Fragell[82] (Norway)
- 2017: Distinguished Services to Anti-Superstition Award: Narendra Dhabolkar[81] (India)

18.5 See also

- World Humanist Congress
- World Humanist Day

18.6 Footnotes

[1] http://iheu.org/about/staff-and-representatives/

[2] "About IHEU". IHEU. Retrieved 2007-11-12.

[3] "IHEU | About IHEU". *iheu.org*. Retrieved 2016-10-10.

[4] "Amsterdam Declaration 2002". IHEU. Retrieved 2013-02-27.

[5] "General Assembly | IHEU". *IHEU*. Retrieved 2017-08-18.

[6] Vandebrake, Mark. *Freethought resource guide: a directory of information, literature, art, organizations, & internet sites related to secular humanism, skepticism, atheism, & agnosticism*. Austin, Texas: CreateSpace. pp. Appendix A. ISBN 9781475020359.

[7] "IHEU | The Amsterdam Declaration". *iheu.org*. Retrieved 2016-10-10.

[8] "IHEU | What is Humanism?". *iheu.org*. Retrieved 2016-10-10.

[9] "Regulations on membership categories and dues". IHEU. Retrieved 2016-10-10.

[10] Choice, Catholics for a Free. "Säkuläre Werte für Europa: Brüsseler Erklärung zu Würde, Gleichheit und Freiheit /PR Newswire UK/". *www.prnewswire.co.uk*. Retrieved 2016-10-10.

[11] http://humanistfederation.eu/ckfinder/userfiles/files/position/Secular%20europe/The%20Brussels%20Declaration.pdf

[12] "IHEU | The Oslo Declaration on Peace". *iheu.org*. Retrieved 2016-10-10.

[13] IHEU, *Oxford Declaration on Freedom of Thought and Expression*, 12 August 2014. Retrieved 18 August 2014

[14] "1850-1952: The road to the founding congress". IHEU. Archived from the original on 11 August 2013. Retrieved 2 November 2013.

[15] Kurtz, Paul (2001). *Skepticism and humanism : the new paradigm*. New Brunswick, NJ [u.a.]: Transaction Publishers. p. 259. ISBN 0765800519.

[16] "IHEU | General Assembly". *iheu.org*. Retrieved 2016-10-10.

[17] "Andrew Copson elected new President of the International Humanist and Ethical Union". *British Humanist Association*. 2015-06-01. Retrieved 2016-10-10.

[18] "Senior Staff". 2012-04-27. Retrieved 2016-08-12.

[19] "IHEU | Our members". *iheu.org*. Retrieved 2017-08-18.

[20] "IHEU | Staff and Representatives". *iheu.org*. Retrieved 2016-10-10.

[21] "Tears and joy at IHEU General Assembly 2017". *IHEU*. 2017-08-10. Retrieved 2017-08-18.

[22] "IHEU | Vision and mission". *iheu.org*. Retrieved 2016-08-12.

[23] "IHEU | New global report on discrimination against the non-religious". *iheu.org*. Retrieved 2016-10-10.

[24] "Country Index". *Freedom of Thought Report*. Retrieved 2017-08-18.

[25] "The Ratings System". *Freedom of Thought Report*. Retrieved 2017-08-18.

[26] "Open Data". *Freedom of Thought Report*. Retrieved 2017-08-18.

[27] "The right to deny the existence of God is under threat". *The Independent*. 2014-12-09. Retrieved 2016-10-10.

[28] "Stephen Fry's mockery of religion could land him the death penalty in these countries". *Washington Post*. Retrieved 2016-10-10.

[29] "Malaysia's free thought, religious expression under 'serious assault', study shows". 2015-12-21. Retrieved 2016-10-10.

[30] "FoRB Free Thought Report Intergroup meeting | Religious Freedom | European Parliament Intergroup". Retrieved 2017-08-18.

[31] "New report shows: Persecution of adherents of non-theistic and atheist beliefs in the world on the rise | Religious Freedom | European Parliament Intergroup". Retrieved 2017-08-18.

[32] "Freedom of Thought Report | Documenting discrimination against the non-religious around the world". *freethoughtreport.com*. Retrieved 2016-10-10.

[33] "IHEU | IHEU's Freedom of Thought Report 2015… "Why should we pamper Saudi Arabia for oil?"". *iheu.org*. Retrieved 2016-10-10.

[34] Westminster, Department of the Official Report (Hansard), House of Commons,. "House of Commons Hansard Written Answers for 14 Oct 2013 (pt 0005)". *www.publications.parliament.uk*. Retrieved 2016-10-10.

[35] "IHEU | Taslima Nasrin's Visit to India". *iheu.org*. Retrieved 2016-10-10.

[36] "Ethical approach to a humane cause". The Hindu. January 28, 2004.

[37] "IHEU | IHEU urges calm and safeguarding of Alber Saber while on bail". *iheu.org*. Retrieved 2016-10-10.

[38] "Nigeria atheist Bala 'deemed mentally ill' in Kano state". *BBC News*. 2014-06-25. Retrieved 2017-08-18.

[39] Smith, David; correspondent, Africa (2014-06-25). "Nigerian man is locked up after saying he is an atheist". *The Guardian*. ISSN 0261-3077. Retrieved 2017-08-18.

[40] "Nigeria atheist Bala freed from Kano psychiatric hospital". *BBC News*. 2014-07-04. Retrieved 2017-08-18.

[41] "Government minister wants to 'hunt down' atheists in Malaysia". *The Independent*. 2017-08-09. Retrieved 2017-08-18.

[42] "IHEU deplores backlash and "hunt" against atheists in Malaysia". *IHEU*. 2017-08-08. Retrieved 2017-08-18.

[43] "Minister violating human rights by 'hunting' atheists, says humanist group". 2017-08-09. Retrieved 2017-08-18.

[44] "IHEU | Search Results: raif badawi". *iheu.org*. Retrieved 2016-10-10.

[45] "IHEU | Search Results: waleed". *iheu.org*. Retrieved 2016-10-10.

[46] "IHEU | IHEU condemns death sentence for "apostasy", handed to writer in Mauritania". *iheu.org*. Retrieved 2016-10-10.

[47] "IHEU | Death for "apostasy" must not stand! Free Ashraf Fayadh". *iheu.org*. Retrieved 2016-10-10.

[48] "IHEU | IHEU reads poem banned in Saudi for 'apostasy' to delegates at UN". *iheu.org*. Retrieved 2016-10-10.

[49] "IHEU | Bangladesh: secular intellectuals terrorised by Islamic fundamentalists". *iheu.org*. Retrieved 2016-10-10.

[50] "IHEU | Search Results: bangladesh". *iheu.org*. Retrieved 2016-10-10.

[51] "IHEU | At Human Rights Council, IHEU raises plight of Bangladeshi Atheist bloggers". *iheu.org*. Retrieved 2016-10-10.

[52] "IHEU | Arrests of "atheist bloggers" shows Bangladesh authorities are "walking into a trap set by fundamentalists"". *iheu.org*. Retrieved 2016-10-10.

[53] "IHEU | Prominent atheist blogger remains in danger pending blasphemy trial in Bangladesh". *iheu.org*. Retrieved 2016-10-10.

[54] "IHEU | Call to action: Defend the bloggers of Bangladesh". *iheu.org*. Retrieved 2016-10-10.

[55] "IHEU | Humanists appalled at the murder of secular activist and writer Avijit Roy". *iheu.org*. Retrieved 2016-10-10.

[56] "IHEU | "No, I will not write about war crimes, Islamic extremism, the country or politics anymore"". *iheu.org*. Retrieved 2016-10-10.

[57] "IHEU | Third atheist writer hacked to death in Bangladesh this year". *iheu.org*. Retrieved 2016-10-10.

[58] "IHEU | Niloy Neel – Fourth atheist activist murdered in a year in Bangladesh". *iheu.org*. Retrieved 2016-10-10.

[59] "IHEU | Coordinated machete attack on publishers of secular authors in Bangladesh". *iheu.org*. Retrieved 2016-10-10.

[60] "IHEU | Tutul: the Survivor – An interview with the target of a Bangladesh machete attack". *iheu.org*. Retrieved 2016-10-10.

[61] "IHEU | Huge alliance protests government response to Bangladesh blogger murders". *iheu.org*. Retrieved 2016-10-10.

[62] "Global free speech campaigners protest against blogger killings in Bangladesh". *bdnews24.com*. Retrieved 2016-10-10.

[63] "IHEU | Atheist student Nazimuddin Samad killed in Bangladesh". *iheu.org*. Retrieved 2016-10-10.

[64] "IHEU | Academic hacked to death in Bangladesh". *iheu.org*. Retrieved 2016-10-10.

[65] "IHEU | Europe stands with Bangladeshi atheists at Solidarity Book Fair in The Hague". *iheu.org*. Retrieved 2016-10-10.

[66] "End Blasphemy Laws campaign launched by international coalition". *The Irish Times*. Retrieved 2016-10-10.

[67] "IHEU | End Blasphemy Laws – a new campaign". *iheu.org*. Retrieved 2016-10-10.

[68] "About | End Blasphemy Laws". *end-blasphemy-laws.org*. Retrieved 2016-10-10.

[69] "IHEU | Global initiative against untouchability launched in London". *iheu.org*. Retrieved 2016-10-10.

[70] "BBC NEWS | UK | Is caste prejudice still an issue?". *news.bbc.co.uk*. Retrieved 2016-10-10.

[71] "IHEU | Parliamentary questions precede first World Conference on Untouchability in London, June 9 & 10". *iheu.org*. Retrieved 2016-10-10.

[72] "IHEU | IHEU supports Second World Conference on Untouchability". *iheu.org*. Retrieved 2016-10-10.

[73] "IHEU | #TwitterTheocracy campaign after social network blocks "blasphemy" in Pakistan". *iheu.org*. Retrieved 2016-10-10.

[74] "Twitter censorship targeted with hashtag activism - Tech Digest". *Tech Digest*. 2014-06-10. Retrieved 2016-10-10.

[75] "Andrew Copson elected new President of the International Humanist and Ethical Union". *British Humanist Association*. Retrieved 12 June 2015.

[76] "IHEU | IHEU Awards". *iheu.org*. Retrieved 2016-10-10.

[77] "Gulalai Ismail wins International Humanist of the Year Award". *British Humanist Association*. Retrieved 12 June 2015.

[78] "Wole Soyinka wins International Humanist Award". *British Humanist Association*. Retrieved 12 June 2015.

[79] "Achievements of Hope Knutsson and Siðmennt celebrated at 25th anniversary celebration".

[80] "New directions, diversity and determination at IHEU General Assembly 2016 | IHEU". *IHEU*. 2016-05-23. Retrieved 2017-08-18.

[81] "Our Distinguished Services Awards Recipients - 2017". *IHEU*. 2017-08-17. Retrieved 2017-08-18.

[82] 2011-08-23 (2011-08-23). "IHEU Events". Iheu.org. Retrieved 2014-08-01.

18.7 External links

- Official website
- Freedom of Thought Report website
- End Blasphemy Laws campaign website

Chapter 19

Renaissance humanism

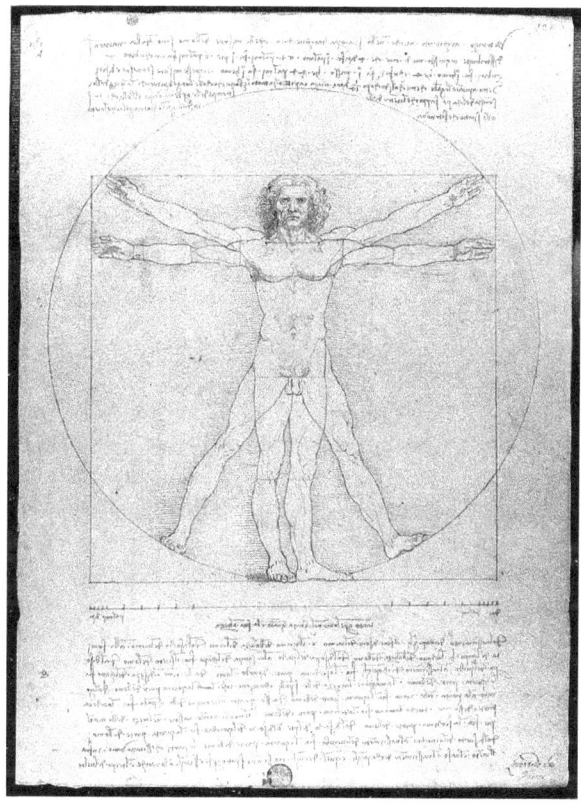

Leonardo da Vinci's Vitruvian Man *(c. 1490) shows the correlations of ideal human body proportions with geometry described by the ancient Roman architect Vitruvius in his* De Architectura. *Vitruvius described the human figure as being like the principal source of proportion among the Classical orders of architecture.*

Renaissance humanism is the study of classical antiquity, at first in Italy and then spreading across Western Europe in the 14th, 15th, and 16th centuries. The term *Renaissance humanism* is contemporary to that period — Renaissance (*rinascimento* "rebirth") and "humanist" (whence modern *humanism*; also *Renaissance humanism* to distinguish it from later developments grouped as humanism).[1]

Renaissance humanism was a response to the utilitarian approach and what came to be depicted as the "narrow pedantry" associated with medieval scholasticism.[2] Humanists sought to create a citizenry able to speak and write with eloquence and clarity and thus capable of engaging in the civic life of their communities and persuading others to virtuous and prudent actions. This was to be accomplished through the study of the "studia humanitatis", today known as the humanities: grammar, rhetoric, history, poetry, and moral philosophy.

According to one scholar of the movement,

> Early Italian humanism, which in many respects continued the grammatical and rhetorical traditions of the Middle Ages, not merely provided the old Trivium with a new and more ambitious name (*Studia humanitatis*), but also increased its actual scope, content and significance in the curriculum of the schools and universities and in its own extensive literary production. The *studia humanitatis* excluded logic, but they added to the traditional grammar and rhetoric not only history, Greek, and moral philosophy, but also made poetry, once a sequel of grammar and rhetoric, the most important member of the whole group.[3]

Humanism was a pervasive cultural mode and not the program of a small elite, a program to revive the cultural legacy, literary legacy, and moral philosophy of classical antiquity. There were important centres of humanism in Florence, Naples, Rome, Venice, Genoa, Mantua, Ferrara, and Urbino.

19.1 Origin

Some of the first humanists were great collectors of antique manuscripts, including Petrarch, Giovanni Boccaccio, Coluccio Salutati, and Poggio Bracciolini. Of the four, Petrarch was dubbed the "Father of Humanism" because of his devotion to Greek and Roman scrolls. Many worked for the Catholic Church and were in holy orders, like Petrarch, while others were lawyers and chancellors of Italian cities, and thus had access to book copying workshops, such as Petrarch's disciple Salutati, the Chancellor of Florence.

In Italy, the humanist educational program won rapid acceptance and, by the mid-15th century, many of the upper classes had received humanist educations, possibly in addition to traditional scholasticist ones. Some of the highest officials of the Catholic Church were humanists with the resources to amass important libraries. Such was Cardinal Basilios Bessarion, a convert to the Catholic Church from Greek Orthodoxy, who was considered for the papacy, and was one of the most learned scholars of his time. There were several 15th-century and early 16th-century humanist Popes[4] one of whom, Aeneas Silvius Piccolomini (Pope Pius II), was a prolific author and wrote a treatise on *The Education of Boys*.[5] These subjects came to be known as the humanities, and the movement which they inspired is shown as humanism.

The migration waves of Byzantine Greek scholars and émigrés in the period following the Crusader sacking of Constantinople and the end of the Byzantine Empire in 1453 greatly assisted the revival of Greek and Roman literature and science via their greater familiarity with ancient languages and works.[6][7] They included Gemistus Pletho, George of Trebizond, Theodorus Gaza, and John Argyropoulos.

Italian humanism spread northward to France, Germany, the Low Countries, and England with the adoption of large-scale printing after the end of the era of incunabula (or books printed prior to 1501), and it became associated with the Protestant Reformation. In France, preeminent humanist Guillaume Budé (1467–1540) applied the philological methods of Italian humanism to the study of antique coinage and to legal history, composing a detailed commentary on Justinian's Code. Budé was a royal absolutist (and not a republican like the early Italian *umanisti*) who was active in civic life, serving as a diplomat for François I and helping to found the Collège des Lecteurs Royaux (later the Collège de France). Meanwhile, Marguerite de Navarre, the sister of François I, was a poet, novelist, and religious mystic[8] who gathered around her and protected a circle of vernacular poets and writers, including Clément Marot, Pierre de Ronsard, and François Rabelais.

19.2 Paganism and Christianity in the Renaissance

Many humanists were churchmen, most notably Pope Pius II (Aeneas Silvius Piccolomini), Sixtus IV, and Leo X,[9][10] and there was often patronage of humanists by senior church figures.[11] Much humanist effort went into improving the understanding and translations of Biblical and early Christian texts, both before and after the Protestant Reformation, which was greatly influenced by the work of non-Italian, Northern European figures such as Desiderius Erasmus, Jacques Lefèvre d'Étaples, William Grocyn, and Swedish Catholic Archbishop in exile Olaus Magnus.

The Cambridge Dictionary of Philosophy describes the rationalism of ancient writings as having tremendous impact on Renaissance scholars:

> Here, one felt no weight of the supernatural pressing on the human mind, demanding homage and allegiance. Humanity—with all its distinct capabilities, talents, worries, problems, possibilities—was the center of interest. It has been said that medieval thinkers philosophised on their knees, but, bolstered by the new studies, they dared to stand up and to rise to full stature.[12]

Inevitably, the rediscovery of classical philosophy and science would eventually challenge traditional religious beliefs. In 1417, for example, Poggio Bracciolini discovered the manuscript of Lucretius, *De rerum natura*, which had been lost for centuries and which contained an explanation of Epicurean doctrine, though at the time this was not commented on much by Renaissance scholars, who confined themselves to remarks about Lucretius's grammar and syntax.[13] Lorenzo Valla, however, puts a defense of epicureanism in the mouth of one of the interlocutors of one of his dialogues.[14] Valla's defense, or adaptation, of Epicureanism was later taken up in *The Epicurean* by Erasmus, the "Prince of humanists:"

> If people who live agreeably are Epicureans, none are more truly Epicurean than the righteous and godly. And if it is names that bother us, no one better deserves the name of Epicurean than the revered founder and head of the Christian philosophy Christ, for in Greek *epikouros* means "helper." He alone, when the law of Nature was all but blotted out by sins, when the law of Moses incited to lists rather than cured them, when Satan ruled in the world unchallenged, brought timely aid to perishing humanity. Completely mistaken, therefore, are those who talk in their foolish fashion about Christ's having been sad and gloomy in character and calling upon us to follow a dismal mode of life. On the contrary, he alone shows the most enjoyable life of all and the one most full of true pleasure.[15]

This passage exemplifies the way in which the humanists saw pagan classical works, such as the philosophy of Epicurus, as being in harmony with their interpretation of Christianity.

Renaissance Neo-Platonists such as Marsilio Ficino (whose translations of Plato's works into Latin were still used into the 19th century) attempted to reconcile Platonism with Christianity, according to the suggestions of early Church fathers Lactantius and Saint Augustine. In this spirit, Pico della Mirandola attempted to construct a syncretism of all religions (he was not a humanist but an Aristotelian trained in Paris), but his work did not win favor with the church authorities.

Historian Steven Kreis expresses a widespread view (derived from the 19th-century Swiss historian Jacob Burckhardt), when he writes that:

> The period from the fourteenth century to the seventeenth worked in favor of the general emancipation of the individual. The city-states of northern Italy had come into contact with the diverse customs of the East, and gradually permitted expression in matters of taste and dress. The writings of Dante, and particularly the doctrines of Petrarch and humanists like Machiavelli, emphasized the virtues of intellectual freedom and individual expression. In the essays of Montaigne the individualistic view of life received perhaps the most persuasive and eloquent statement in the history of literature and philosophy.[16]

Two noteworthy trends in Renaissance humanism were Renaissance Neo-Platonism and Hermeticism, which through the works of figures like Nicholas of Kues, Giordano Bruno, Cornelius Agrippa, Campanella and Pico della Mirandola sometimes came close to constituting a new religion itself. Of these two, Hermeticism has had great continuing influence in Western thought, while the former mostly dissipated as an intellectual trend, leading to movements in Western esotericism such as Theosophy and New Age thinking.[17] The "Yates thesis" of Frances Yates holds that before falling out of favour, esoteric Renaissance thought introduced several concepts that were useful for the development of scientific method, though this remains a matter of controversy.

Though humanists continued to use their scholarship in the service of the church into the middle of the sixteenth century and beyond, the sharply confrontational religious atmosphere following the Protestant reformation resulted in the Counter-Reformation that sought to silence challenges to Catholic theology,[18] with similar efforts among the Protestant denominations. However, a number of humanists joined the Reformation movement and took over leadership functions, for example, Philipp Melanchthon, Ulrich Zwingli, John Calvin, and William Tyndale.

With the Counter Reformation initiated by the Council of Trent (1545-1563), positions hardened and a strict Catholic orthodoxy based on Scholastic philosophy was imposed. Some humanists, even moderate Catholics such as Erasmus, risked being declared heretics for their perceived criticism of the church.[19]

The historian of the Renaissance Sir John Hale cautions against too direct a linkage between Renaissance humanism and modern uses of the term humanism: "Renaissance humanism must be kept free from any hint of either "humanitarianism" or "humanism" in its modern sense of rational, non-religious approach to life ... the word "humanism" will mislead ... if it is seen in opposition to a Christianity its students in the main wished to supplement, not contradict, through their patient excavation of the sources of ancient God-inspired wisdom"[20]

19.3 Humanists

Main article: List of Renaissance humanists

19.4 See also

- Thomas More
- Greek scholars in the Renaissance
- Humanist Latin
- Legal humanists
- New Learning

19.5 Notes

[1] The term *la rinascita* (rebirth) first appeared, however, in its broad sense in Giorgio Vasari's *Vite de' più eccellenti architetti, pittori, et scultori Italiani* (The Lives of the Artists, 1550, revised 1568) Panofsky, Erwin. *Renaissance and Renascences in Western Art*, New York: Harper and Row, 1960. "The term *umanista* was used in fifteenth-century Italian academic slang to describe a teacher or student of classical literature and the arts associated with it, including that of rhetoric. The English equivalent 'humanist' makes its appearance in the late sixteenth century with a similar meaning. Only in the nineteenth century, however, and probably for the first time in Germany in 1809, is the attribute transformed into a substantive: *humanism*, standing for devotion to the literature of ancient Greece and Rome, and the

humane values that may be derived from them" Nicholas Mann "The Origins of Humanism", *Cambridge Companion to Humanism*, Jill Kraye, editor [Cambridge University Press, 1996], p. 1–2). The term "Middle Ages" for the preceding period separating classical antiquity from its "rebirth" first appears in Latin in 1469 as *media tempestas*.

[2] Craig W. Kallendorf, introduction to *Humanist Educational Treatises*, edited and translated by Craig W. Kallendorf (Cambridge, Massachusetts and London England: The I Tatti Renaissance Library, 2002) p. vii.

[3] Paul Oskar Kristeller, *Renaissance Thought II: Papers on Humanism and the Arts* (New York: Harper Torchbooks, 1965), p. 178. See also Kristeller's *Renaissance Thought I*, "Humanism and Scholasticism In the Italian Renaissance", *Byzantion 17* (1944–45), pp. 346–74. Reprinted in *Renaissance Thought* (New York: Harper Torchbooks), 1961.

[4] They include Innocent VII, Nicholas V, Pius II, Sixtus IV, Alexander VI, Julius II and Leo X. Innocent VII, patron of Leonardo Bruni, is considered the first humanist Pope. See James Hankins, *Plato in the Italian Renaissance* (New York: Columbia Studies in the Classical Tradition, 1990), p. 49; for the others, see their respective entries in Sir John Hale's *Concise Encyclopaedia of the Italian Renaissance* (Oxford University Press, 1981).

[5] See *Humanist Educational Treatises*, (2001) pp. 126–259. This volume (pp. 92–125) contains an essay by Leonardo Bruni, entitled "The Study of Literature", on the education of girls.

[6] Byzantines in Renaissance Italy

[7] Greeks in Italy

[8] She was the author of *Miroir de l'ame pecheresse* (*The Mirror of a Sinful Soul*), published after her death, among other devotional poetry. See also "Marguerite de Navarre: Religious Reformist" in Jonathan A. Reid, *King's sister--queen of dissent: Marguerite of Navarre (1492-1549) and her evangelical network* (*Studies in medieval and Reformation traditions, 1573-4188*; v. 139). Leiden; Boston: Brill, 2009. (2 v.: (xxii, 795 p.) ISBN 978-90-04-17760-4 (v. 1), 9789004177611 (v. 2)

[9] Löffler, Klemens (1910). "Humanism". *The Catholic Encyclopedia*. **VII**. New York: Robert Appleton Company. pp. 538–542.

[10] See note two, above.

[11] Davies, 477

[12] "Humanism". *The Cambridge Dictionary of Philosophy, Second Edition*. Cambridge University Press. 1999. p.397 quotation:

> The unashamedly humanistic flavor of classical writings had a tremendous impact on Renaissance scholar.

[13] Only in 1564 did French commentator Denys Lambin (1519–72) announce in the preface to the work that "he regarded Lucretius's Epicurean ideas as 'fanciful, absurd, and opposed to Christianity". Lambin's preface remained standard until the nineteenth century. (See Jill Kraye's essay, "Philologists and Philosophers" in the *Cambridge Companion to Renaissance Humanism* [1996], p. 153.) Epicurus's unacceptable doctrine that pleasure was the highest good "ensured the unpopularity of his philosophy" (Kraye [1996] p. 154.)

[14] Charles Trinkhaus regards Valla's "epicureanism" as a ploy, not seriously meant by Valla, but designed to refute Stoicism, which he regarded together with epicureanism as equally inferior to Christianity. See Trinkaus, *In Our Image and Likeness* Vol. 1 (University of Chicago Press, 1970), pp. 103–170

[15] John L. Lepage (5 December 2012). *The Revival of Antique Philosophy in the Renaissance*. Palgrave Macmillan. p. 111. ISBN 978-1-137-28181-4.

[16] Kreis, Steven (2008). "Renaissance Humanism". Retrieved 2009-03-03.

[17] Plumb, 95

[18] "Rome Reborn: The Vatican Library & Renaissance Culture: Humanism". The Library of Congress. 2002-07-01. Retrieved 2009-03-03.

[19] "Humanism". *Encyclopedic Dictionary of Religion*. **F–N**. Corpus Publications. 1979. p. 1733. ISBN 0-9602572-1-7.

[20] Hale, 171. See also Davies, 479-480 for similar caution.

19.6 Further reading

- Bolgar, R. R. *The Classical Heritage and Its Beneficiaries: from the Carolingian Age to the End of the Renaissance*. Cambridge, 1954.

- Cassirer, Ernst. *Individual and Cosmos in Renaissance Philosophy*. Harper and Row, 1963.

- Cassirer, Ernst (Editor), Paul Oskar Kristeller (Editor), John Herman Randall (Editor). *The Renaisssance Philosophy of Man*. University of Chicago Press, 1969.

- Cassirer, Ernst. *Platonic Renaissance in England*. Gordian, 1970.

- Celenza, Christopher S. *The Lost Italian Renaissance: Humanism, Historians, and Latin's Legacy*. Baltimore: Johns Hopkins University Press. 2004 ISBN 978-0-8018-8384-2

- Erasmus, Desiderius. "The Epicurean". In *Colloquies*.

- Garin, Eugenio. *Science and Civic Life in the Italian Renaissance*. New York: Doubleday, 1969.

- Garin, Eugenio. *Italian Humanism: Philosophy and Civic Life in the Renaissance*. Basil Blackwell, 1965.

- Garin, Eugenio. *History of Italian Philosophy*. (2 vols.) Amsterdam/New York: Rodopi, 2008. ISBN 978-90-420-2321-5

- Grafton, Anthony. *Bring Out Your Dead: The Past as Revelation*. Harvard University Press, 2004 ISBN 0-674-01597-5

- Grafton, Anthony. *Worlds Made By Words: Scholarship and Community in the Modern West*. Harvard University Press, 2009 ISBN 0-674-03257-8

- Hale, John. *A Concise Encyclopaedia of the Italian Renaissance*. Oxford University Press, 1981, ISBN 0-500-23333-0.

- Kallendorf, Craig W, editor. *Humanist Educational Treatises*. Cambridge, Mass.: The I Tatti Renaissance Library, 2002.

- Kraye, Jill (Editor). *The Cambridge Companion to Renaissance Humanism*. Cambridge University Press, 1996.

- Kristeller, Paul Oskar. *Renaissance Thought and Its Sources*. Columbia University Press, 1979 ISBN 978-0-231-04513-1

- Pico della Mirandola, Giovanni. *Oration on the Dignity of Man*. In Cassirer, Kristeller, and Randall, eds. *Renaissance Philosophy of Man*. University of Chicago Press, 1969.

- Skinner, Quentin. *Renaissance Virtues: Visions of Politics: Volume II*. Cambridge University Press, [2002] 2007.

- McManus, Stuart M. "Byzantines in the Florentine Polis: Ideology, Statecraft and Ritual during the Council of Florence". *Journal of the Oxford University History Society*, 6 (Michaelmas 2008/Hilary 2009).

- Melchert, Norman (2002). *The Great Conversation: A Historical Introduction to Philosophy*. McGraw Hill. ISBN 0-19-517510-7.

- Nauert, Charles Garfield. *Humanism and the Culture of Renaissance Europe (New Approaches to European History)*. Cambridge University Press, 2006.

- Plumb, J. H. ed.: *The Italian Renaissance* 1961, American Heritage, New York, ISBN 0-618-12738-0 (page refs from 1978 UK Penguin edn).

- Rossellini, Roberto. *The Age of the Medici*: Part 1, *Cosimo de' Medici*; Part 2, *Alberti* 1973. (Film Series). Criterion Collection.

- Symonds, John Addington. *The Renaissance in Italy*. Seven Volumes. 1875-1886.

- Trinkaus, Charles (1973). "Renaissance Idea of the Dignity of Man". In Wiener, Philip P. *Dictionary of the History of Ideas*. ISBN 0-684-13293-1. Retrieved 2009-12-02.

- Trinkaus, Charles. *The Scope of Renaissance Humanism*. Ann Arbor: University of Michigan Press, 1983.

- Wind, Edgar. *Pagan Mysteries in the Renaissance*. New York: W.W. Norton, 1969.

- Witt, Ronald. "In the footsteps of the ancients: the origins of humanism from Lovato to Bruni." Leiden: Brill Publishers, 2000

19.7 External links

- Humanism 1: An Outline by Albert Rabil, Jr.

- "Rome Reborn: The Vatican Library & Renaissance Culture: Humanism". The Library of Congress. 2002-07-01

Chapter 20

Renaissance humanism in Northern Europe

Renaissance Humanism came much later to Germany and Northern Europe in general than to Italy, and when it did, it encountered some resistance from the scholastic theology which reigned at the universities. Humanism may be dated from the invention of the printing press about 1450. Its flourishing period began at the close of the 15th century and lasted only until about 1520, when it was absorbed by the more popular and powerful religious movement, the Reformation, as Italian Humanism was superseded by the papal counter-Reformation. Marked features distinguished the new culture north of the Alps from the culture of the Italians. The university and school played a much more important part than in the South according to Catholic historians. The representatives of the new scholarship were teachers; even Erasmus taught in Cambridge and was on intimate terms with the professors at Basel. During the progress of the movement new universities sprang up, from Basel to Rostock. Again, in Germany, there were no princely patrons of arts and learning to be compared in intelligence and munificence to the Renaissance popes and the Medici. Nor was the new culture here exclusive and aristocratic. It sought the general spread of intelligence, and was active in the development of primary and grammar schools. In fact, when the currents of the Italian Renaissance began to set toward the North, a strong, independent, intellectual current was pushing down from the flourishing schools conducted by the Brethren of the Common Life. In the Humanistic movement, the German people was far from being a slavish imitator. It received an impulse from the South, but made its own path.

In the North, Humanism entered into the service of religious progress. German scholars were less brilliant and elegant, but more serious in their purpose and more exact in their scholarship than their Italian predecessors and contemporaries. In the South, the ancient classics absorbed the attention of the literati. It was not so in the North. There was no consuming passion to render the classics into German as there had been in Italy. Nor did Italian literature, with its often relaxed moral attitude, find imitators in the North. Boccaccio's Decameron was first translated

Albrecht Dürer, self-portrait, 1500

into German by the physician, Henry Stainhowel, who died in 1482. North of the Alps, attention was chiefly centred on the Old and New Testaments. Greek and Hebrew were studied, not with the purpose of ministering to a cult of antiquity, but to reach the fountains of the Christian system more adequately. In this way, preparation was made for the work of the Protestant Reformation. This focus on translation was a feature of the Christian humanists who helped to launch the new, post-scholastic era, among them Erasmus and Luther. In so doing, they also placed biblical texts above any human or institutional authority, an approach that emphasised the role of the reader in understanding a text for him or herself. Closely allied to the late medieval shift of scholarship from the monastery to the university, Christian humanism engendered a new freedom of expression, even though some of its proponents opposed that freedom of expression elsewhere, such as in their censure of the Anabaptists.

What was true of the scholarship of Germany was also true of its art. The painters, Albrecht Dürer, who was born and died at Nuremberg, 1471–1528, Lucas Cranach the Elder, 1472–1553, and for the most part Hans Holbein the Younger, 1497–1543, took little interest in mythology, apart from Cranach's nudes, and were persuaded by the Reformation, though most continued to take commissions for traditional Catholic subjects. Dürer and Holbein had close contacts with leading humanists. Cranach lived in Wittenberg after 1504 and painted portraits of Martin Luther, Philip Melanchthon and other leaders of the German Reformation. Holbein made frontispieces and illustrations for Protestant books and painted portraits of Erasmus and Melanchthon.

20.1 Italian roots of the humanism in Germany

If any one individual more than another may be designated as the connecting link between the learning of Italy and Germany, it is Aeneas Sylvius. By his residence at the court of Frederick III and at Basel, as one of the secretaries of the council, he became a well-known character north of the Alps long before he was chosen pope. The mediation, however, was not effected by any single individual. The fame of the Renaissance was carried over the pathways of trade which led from Northern Italy to Augsburg, Nuremberg, Konstanz and other German cities. The visits of Frederick III and the campaigns of Charles VIII and the ascent of the throne of Naples by the princes of Aragon carried Germans, Frenchmen and Spaniards to the greater centres of the peninsula. A constant stream of pilgrims travelled to Rome and the Spanish popes drew to the city throngs of Spaniards. As the fame of Italian culture spread, scholars and artists began to travel to Venice, Florence and Rome, and caught the inspiration of the new era.

To the Italians, Germany was a land of barbarians. They despised the German people for their rudeness and intemperance in eating and drinking. Aeneas was impressed by the beauty of Vienna, though it was quite small when compared to the greatest Italian cities.[1] However, he found that the German princes and nobles cared more for horses and dogs than for poets and scholars and loved their winecellars better than the muses. Campanus, a witty poet of the papal court, who was sent as legate to the Diet of

Pope Pius II

Regensburg (1471) by Pope Paul II, and afterwards was made a bishop by Pope Pius II, abused Germany for its dirt, cold climate, poverty, sour wine and miserable fare. He lamented his unfortunate nose, which had to smell everything, and praised his ears, which understood nothing.Johannes Santritter, himself being a German living in Italy, admitted that Italy was slightly ahead of Germany in the humanities. However, he also contended that many Italians were jealous of German science and technology, which he considered superior taking the examples of the printing press and the work of the astronomer Johannes Regiomontanus.

Such impressions were soon offset by the sound scholarship which arose in Germany and the Netherlands. And, if Italy contributed to Germany an intellectual impulse, Germany sent out to the world the printing press, the most important agent in the history of intellectual culture since the invention of the alphabet.

20.2 Universities

Main article: Medieval university

Before the first swell of the new movement was felt, the older German universities were already established: University of Vienna in 1365, University of Heidelberg in 1386, University of Cologne in 1388, University of Erfurt in 1392, University of Würzburg in 1402, University of Leipzig in 1409 and University of Rostock in 1419. During the last half of the 15th century, there were quickly added to this list universities at Greifswald and Freiburg 1457, Trier 1457, Basel 1459, Ingolstadt 1472, Tübingen and Mainz 1477, and Wittenberg 1502. Ingolstadt lost its distinct existence by incorporation in the University of Munich, 1826, and Wittenberg by removal to Halle.

Most of these universities had the four faculties, although the popes were slow to give their assent to the sanction of the theological department, as in the case of Vienna and Rostock, where the charter of the secular prince authorized their establishment. Strong as the religious influences of the age were, the social and moral habits of the students were by no means such as to call for praise. Parents, Luther said, in sending their sons to the universities, were sending them to destruction, and an act of the Leipzig university, dating from the close of the 15th century, stated that students came forth from their homes obedient and pious, but "how they returned, God alone knew."1061e to university archives and library.

20.3 Education

The theological teaching was ruled by the Schoolmen, and the dialectic method prevailed in all departments. In clashing with the scholastic method and curricula, the new teaching met with many a repulse, and in no case was it thoroughly triumphant till the era of the Reformation opened. Erfurt may be regarded as having been the first to give the new culture a welcome. In 1466, it received Peter Luder of Kislau, who had visited Greece and Asia Minor, and had been previously appointed to a chair in Heidelberg, 1456. He read on Virgil, Jerome, Ovid and other Latin writers. There Agricola studied and there Greek was taught by Nicolas Marschalck, under whose supervision the first Greek book printed in Germany issued from the press, 1501. There John of Wesel taught. It was Luther's alma mater and, among his professors, he singled out Trutvetter for special mention as the one who directed him to the study of the Scriptures.

Heidelberg, chartered by the elector Ruprecht I and Pope Urban VI, showed scant sympathy with the new movement. However, the elector-palatine, Philip, 1476–1508, gathered at his court some of its representatives, among them Reuchlin. Ingolstadt for a time had Reuchlin as professor and, in 1492, Conrad Celtes was appointed professor of poetry and eloquence.

In 1474, a chair of poetry was established at Basel. Founded by Pius II, it had among its early teachers two Italians, Finariensis and Publicius. Sebastian Brant taught there at the close of the century and among its notable students were Reuchlin and the Reformers, Leo Jud and Zwingli. In 1481, Tübingen had a stipend of oratoria. Here Gabriel Biel taught till very near the close of the century. The year after Biel's death, Heinrich Bebel was called to lecture on poetry. One of Bebel's distinguished pupils was Philip Melanchthon, who studied and taught in the university, 1512–1518. Reuchlin was called from Ingolstadt to Tübingen, 1521, to teach Hebrew and Greek, but died a few months later.

Leipzig and Cologne remained inaccessible strongholds of scholasticism, till Luther appeared, when Leipzig changed front. The last German university of the Middle Ages, Wittenberg, founded by Frederick the Wise and placed under the patronage of the Virgin Mary and St. Augustine, acquired a worldwide influence through its professors, Luther and Melanchthon. Not till 1518, did it have instruction in Greek, when Melanchthon, soon to be the chief Greek scholar in Germany, was called to one of its chairs at the age of 21. According to Luther, his lecture-room was at once filled brimful, theologians high and low resorting to it.

As seats of the new culture, Nuremberg and Strasbourg occupied, perhaps, even a more prominent place than any of the university towns. These two cities, with Basel and Augsburg, had the most prosperous German printing establishments. At the close of the 15th century, Nuremberg, the fountain of inventions, had four Latin schools and was the home of Albrecht Dürer the painter and his friend Willibald Pirkheimer, a patron of learning.

Popular education, during the century before the Reformation, was far more advanced in Germany than in other nations. Apart from the traditional monastic and civic schools, the Brothers of the Common Life had schools at Zwolle, Deventer, 's-Hertogenbosch and Liège in the Low Countries. All the leading towns had schools. The town of Schlettstadt(Selestat) in Alsace was noted as a classical centre. Here, Thomas Platter found Hans Sapidus teaching, and he regarded it as the best school he had found. In 1494, there were five pedagogues in Wesel, teaching reading, writing, arithmetic and singing. One Christmas the clergy of the place entertained the pupils, giving them each cloth for a new coat and a piece of money as begun with the 4th class.

Among the noted schoolmasters was Alexander Hegius, who taught at Deventer for nearly a quarter of a century, till his death in 1498. At the age of 40 he was not ashamed to sit at the feet of Agricola. He made the classics central in education and banished the old text-books. Trebonius, who taught Luther at Eisenach, belonged to a class of worthy men. The penitential books of the day called upon parents to be diligent in keeping their children off the streets and sending them to school.

20.3.1 See also

- Humanist Library of Sélestat

20.4 Leaders of humanism

Rudolph Agricola

The leading Northern humanists included Rudolph Agricola, Reuchlin and Erasmus. Agricola, whose original name was Roelef Huisman, was born near Groningen, 1443, and died 1485. He enjoyed the highest reputation in his day as a scholar and received unstinted praise from Erasmus and Melanchthon. He has been regarded as doing for Humanism in Germany what was done for Italy by Petrarch, the first life of whom, in German, Agricola prepared. After studying in Erfurt, Louvain and Cologne, Agricola went to Italy, spending some time at the universities in Pavia and Ferrara. He declined a professor's chair in favor of an appointment at the court of Philip of the Palatinate in Heidelberg. He made Cicero and Quintilian his models. In his last years, he turned his attention to theology and studied Hebrew. Like Pico della Mirandola, he was a monk. The inscription on his tomb in Heidelberg stated that he had studied what is taught about God and the true faith of the Saviour in the books of Scripture.

Johannes Trithemius by Tilman Riemenschneider

Another Humanist was Jacob Wimpheling, 1450–1528, of Schlettstadt, who taught in Heidelberg. He was inclined to be severe on clerical abuses but, at the close of his career, wanted to substitute for the study of Virgil and Horace, Sedulius and Prudentius. The poetic Sebastian Brant, 1457–1521, the author of the *Ship of Fools*, began his career as a teacher of law in Basel. Mutianus Rufus, in his correspondence, went so far as to declare that Christianity is as old as the world and that Jupiter, Apollo, Ceres and Christ are only different names of the one hidden God.

A name which deserves a high place in the German literature of the last years of the Middle Ages is John Trithemius, 1462–1505, abbot of a Benedictine convent at Sponheim, which, under his guidance, gained the reputation of a learned academy. He gathered a library of 2,000 volumes and wrote a patrology, or encyclopaedia of the Fathers, and a catalogue of the renowned men of Germany. Increasing differences with the convent led to his resignation in 1506, when he decided to take up the offer of the Lord Bishop of Würzburg, Lorenz von Bibra (bishop from 1495 to 1519), to become abbot of the *Schottenkloster* in Würzburg. He remained there until the end of his life. Prelates and nobles visited him to consult and read the Latin and Greek authors he had collected. These men and others contributed their part to that movement of which Reuchlin and Erasmus were the chief lights and which led to the Protestant Reformation.

20.5 See also

- Northern Renaissance
- German Renaissance
- Renaissance in the Low Countries

20.6 References

[1] *History of Vienna*, Jean-Paul Bled

20.7 Sources

This article is copied from

- Philip Schaff *History of the Christian Church*, Volume VI, **1882**

20.8 Literature

- Marco Heiles: *Topography of German humanism 1470-1550. An approach*

20.9 External links

- Topography of German humanism 1470-1550

Chapter 21

Humanism in France

Humanism in France found its way from Italy, but did not become a distinct movement until the 16th century was well on its way.

On the completion of the Hundred Years' War between France and England, the intellectual currents of humanism began to start. In 1464, Peter Raoul composed for the Duke of Burgundy a history of Troy. At that time the French still regarded themselves as descendants of Hector. If we except Paris, none of the French universities took part in the movement. Individual writers and printing-presses at Paris, Lyon, Rouen and other cities became its centres and sources. William Fichet and Robert Gaguin are usually looked upon as the first French Humanists. Fichet introduced "the eloquence of Rome" at Paris and set up a press at the Sorbonne. He corresponded with Bessarion and had in his library volumes of Petrarch, Guarino of Verona and other Italians. Gaguin copied and corrected Suetonius in 1468 and other Latin authors. Poggio's Jest-book and some of Valla's writings were translated into French. In the reign of Louis XI, who gloried in the title "the first Christian king", French poets celebrated his deeds. The homage of royalty took in part the place among the literary men of France that the cult of antiquity occupied in Italy.

Greek, which had been completely forgotten in France, had its first teachers in Gregory Tifernas, who reached Paris, 1458, John Lascaris, who returned with Charles VIII, and Hermonymus of Sparta, who had Reuchlin and Budaeus (known variously as William Budaeus (English), Guillaume Budé (French) and Guilielmus Budaeus (Latin)) among his scholars. An impetus was given to the new studies by the Italian, Aleander, afterwards famous for his association with Martin Luther at Worms. He lectured in Paris, 1509, on Plato and issued a *Latino-Greek lexicon*. In 1512 his pupil, Vatable, published the Greek grammar of Chrysoloras. Budaeus, perhaps the foremost Greek scholar of his day, founded the Collège de France, 1530, and finally induced Francis I to provide for instruction in Hebrew and Greek. The University of Paris at the close of the 14th century was sunk into a low condition and Erasmus bitterly complained of the food, the morals and the intellectual standards of the college of Montague which he attended. Budaeus urged the combination of the study of the Scriptures with the study of the classics and exclaimed of the Gospel of John, "What is it, if not the almost perfect sanctuary of the truth!"

François Rabelais

Jacques Lefèvre d'Étaples studied in Paris, Pavia, Padua and Cologne and, for longer or shorter periods, tarried in the greater Italian cities. He knew Greek and some Hebrew. From 1492–1506 he was engaged in editing the works of Aristotle and Raymundus Lullus and then, under the protection of Guillaume Briçonnet, bishop of Meaux, he turned his attention to theology. It was his purpose to offset the *Sentences* of Peter the Lombard by a system of theology giving only what the Scriptures teach. In 1509, he published the *Psalterum quintuplex*, a combination of five Latin versions of the Psalms, including a revision and a commentary by his own hand. In 1512, he issued a revised Latin translation of the Pauline Epistles with commentary. In this work, he asserted the authority of the Bible and the doctrine of justification by faith, without appreciating, however, the far-reaching significance of the latter opinion. Three years after the appearance of Luther's New Testament, Lefevre's French translation appeared, 1523. It was made from the Vulgate, as was his translation of the Old Testament, 1528. In 1522 and 1525, appeared his commentaries on the four Gospels and the Catholic Epistles. The former was put on the Index by the Sorbonne. The opposition to the free spirit of inquiry and to the Reformation, which the Sorbonne stirred up and French royalty adopted, forced him to flee to Strassburg and then to the liberal court of Margaret of Angoulême.

Among those who came into contact with Lefevre were Farel and Calvin, the Reformers of Geneva. In the meantime Clément Marot, 1495–1544, the first true poet of the French literary revival, was composing his French versification of the Psalms and of Ovid's *Metamorphoses*. The Psalms were sung for pleasure by French princes and later for worship in Geneva and by the Huguenots. When Calvin studied the humanities and law at Bourges, Orléans and Paris, about 1520, he had for teachers Cordier and L'Etoile, the canonists, and Melchior Wolmar, teacher of Greek, whose names the future Reformer records with gratitude and respect. He gave himself passionately to Humanistic studies and sent to Erasmus a copy of his work on Seneca's *Clemency*, in which he quoted frequently from the ancient classics and the Fathers. Had he not adopted the new religious views, it is possible he would now be known as an eminent figure in the history of French Humanism.

21.1 See also

Jest book

21.2 Further reading

- Tzvetan Todorov. *The Imperfect Garden: The Legacy of Humanism*. Princeton University Press. 2001. (text which focuses on the main views of French humanism)

21.3 External links and references

- Philip Schaff *History of the Christian Church*, Volume VI, 1882

- (in French) La Renaissance artistique et humaniste en France

Chapter 22

Center for Inquiry

Front entrance of Center For Inquiry Transnational

The **Center for Inquiry** (**CFI**) is a nonprofit educational organization. Its primary mission is to foster a secular society based on science, reason, freedom of inquiry, and humanist values.[1] CFI has headquarters in the United States and a number of locations around the world.

Center for Inquiry focuses on two primary subject areas:[2]

- Investigation of Paranormal and Fringe Science Claims through the Committee for Skeptical Inquiry

- Religion, Ethics, and Society through the Council for Secular Humanism

CFI is also active in promoting a scientific approach to medicine and health. The organization has been described as a think tank[3][4] and as a non-governmental organization.[5][6]

In January 2016, the Richard Dawkins Foundation for Reason and Science announced that it was merging with the Center for Inquiry, with Robyn Blumner as the CEO of the combined organizations.[7][8][9][10][11]

22.1 History

Philosopher Paul Kurtz (left) and author Martin Gardner at a CSICOP executive council meeting in 1979

The Center for Inquiry was established in 1991 by philosopher and author Paul Kurtz. It brought together two organizations: the Committee for the Scientific Investigation of Claims of the Paranormal[12] (CSICOP) and the Council for Secular Humanism[13] (CSH). CSICOP and CSH had previously operated in tandem but were now formally affiliated under one umbrella.

22.1.1 Expansion

By 1995 CFI had expanded into a new headquarters in Amherst, New York, and in 1996 opened its first branch office in Los Angeles, CFI West currently named CFI Los Angeles.[14] In the same year, CFI founded the Campus Freethought Alliance, organizing college students around its areas of interest.

By 1997 CFI had begun expanding its efforts internationally through an association with Moscow State University.

Between 2002 and 2003 CFI opened two new branches in New York City[15] and Tampa, Florida[16] in addition to expanding its west coast branch into a new building in Holly-

CFI Lecture Hall

wood, California. Located on Hollywood Boulevard, CFI Los Angeles also became home to the Steve Allen Theater, named after the former *Tonight Show* host and CFI supporter. In 2004, CFI continued to expand into cities across the United States with the creation of a network of community organizations called CFI Communities.[17]

In 2005 CFI once again expanded its Amherst headquarters with a new research wing. Additionally, CFI was granted special consultative status with the United Nations the same year.[18]

Since 2006 CFI has been expanding rapidly with a series of new branches in cities across North America and around the world. These include new Centers for Inquiry in Toronto, London, Washington, D.C., Indianapolis, Grand Rapids, Michigan, and Austin, Texas. The branch in Washington is headquarters to CFI's Office of Public Policy, which represents CFI's interests on Capitol Hill.

Their former affiliated organizations, the Council for Secular Humanism and the Committee for Skeptical Inquiry, ceased to exist as independent organizations, and have become programs of Center for Inquiry, since January 2015.[19]

Logo before its merger with the Richard Dawkins Foundation.

In January 2016, CFI announced that it was merging with the Richard Dawkins Foundation for Reason and Science, with Robyn Blumner as the CEO of the combined organizations.[7][8][9][10][20]

22.1.2 Departure of founder

According to Paul Kurtz, in June 2009, being at odds with new CEO Ronald Lindsay, Kurtz was voted out as chairman. Kurtz has described the direction of CFI under Lindsay as "angry atheism" in contrast to his affirmative humanist philosophical approach.[21] According to Ronald Lindsay,"Paul Kurtz voluntarily resigned from his positions with CFI and all its affiliates, including his position as editor-in-chief of Free Inquiry."[22] The Center for Inquiry Board Statement from 2010, thanks Kurtz for his "decades of service" and claims that "Much of CFI's success is due to Paul Kurtz's inspiration and leadership." The release states that with Kurtz's encouragement, new leadership was sought out, with the goal of transitioning Kurtz away from the CEO position. The Board according to CFI prior to 2010 had become concerned with Kurtz's "day-to-day management of the organization. In June 2008, the board appointed Dr. Ronald A. Lindsay president & CEO; in June 2009, the board elected Richard Schroeder chairman, with Dr. Kurtz moving to chairman emeritus." In May 2010, the Board accepted Kurtz's resignation from CFI.[23]

22.2 Paranormal and fringe science claims

Joe Nickell, Research Fellow at the Committee for Skeptical Inquiry

Through the Committee for Skeptical Inquiry (CSI), and its journal, *Skeptical Inquirer* magazine, published by the Center for Inquiry, CSI evaluates claims of the paranormal (phenomena allegedly beyond the range of normal scientific explanations), such as psychic phenomena, ghosts, communication with the dead, and alleged extraterrestrial visitations. It also explores the fringes and borderlands of the sciences, attempting to separate strictly evidence-based research from pseudosciences.

CSICOP was, alongside magician and prominent skeptic James Randi, sued by TV celebrity Uri Geller in the 1990s over claims made in the International Herald Tribune. The case ran for several years with Geller ordered to pay costs and other charges, and was ultimately settled in 1995.[24]

22.2.1 The Independent Investigations Group

IIG "Power Balance" testing exercise

Main article: Independent Investigations Group

The Independent Investigations Group, a volunteer group based at CFI Los Angeles, undertakes experimental testing of fringe claims.[25] It offers a cash prize (as of 2014 this has a value of USD 100,000) for successful demonstration of supernatural effects.[26] The IIG Awards (known as "Iggies") are presented for "scientific and critical thinking in mainstream entertainment". IIG has investigated, amongst other things, power bracelets, psychic detectives and a 'telepathic wonder dog'.

22.3 Religion, ethics, and society

Logo of the Council for Secular Humanism.

The Center promotes critical inquiry into the foundations and social effects of the world religions. Since 1983, initially through its connection with Committee for the Scientific Examination of Religion, it has focused on such issues as fundamentalism in Christianity and Islam, humanistic alternatives to religious ethics, and religious sources of political violence. It has taken part in protests against religious persecution around the world[27] and opposes religious privilege, for example benefits for clergy in the US Tax Code.[28]

CFI actively supports secular interests, such as secular state education.[29][30] It organizes conferences, such as *Women In Secularism* [31][32] and a conference focused on freethought advocate Robert Ingersoll.[33] CFI has provided meeting and conference facilities to other skeptical organisations, for example an atheist of color conference on social justice.[34][35]

CFI also undertakes atheist education and support activities,[36] for example sending freethought books to prisoners as part of its *Freethought Books Project*.[37]

CFI is active in advocating free speech,[38] and in promoting secular government.[39] It speaks against institutional religion in the armed forces.[40]

Free Inquiry is published by the Center for Inquiry, in association with the Council for Secular Humanism.

22.4 Publications

The results of research and activities supported by the Center and its affiliates are published and distributed to the public in seventeen separate national and international magazines, journals, and newsletters. Among them are CSH's *Free Inquiry* and *Secular Humanist Bulletin*,[41] and CSI's *Skeptical Inquirer*, CFI's *American Rationalist*.[42] The *Scientific Review of Alternative Medicine*, The *Scientific Review of Mental Health Practice*[43] and *Philo*, a journal covering philosophical issues, are no longer being published.

CFI has produced the weekly radio show and podcast, Point of Inquiry since 2005. Episodes are available free for down-

Tom Flynn, editor of Free Inquiry

load from iTunes. Current host, as of June 2017, is Paul Fidalgo. Notable guests have included Steven Pinker, Neil deGrasse Tyson and Richard Dawkins.

22.5 Projects and programs

22.5.1 Center for Inquiry On Campus

CFI Student Conference 2013 - Amherst, NY. Center, Eddie Tabash, a director of CFI

CFI On Campus[44] (originally the Campus Freethought Alliance) is a program launched by the Council for Secular Humanism in 1996 by Derek Araujo and others in order to reach out to university and high school students. The Center for Inquiry On Campus provides funding, speakers or debaters, literature, and other promotional and educational resources to student groups that affiliate, and supports over 200 campus groups around the world.

Center for Inquiry On Campus is directed by Debbie Goddard, who is also the director of African Americans for Humanism.[45][46] CFI on Campus employs a staff of organizers who help CFI student groups to advance their aims at their respective schools.

22.5.2 Skeptic's Toolbox

A lecture given by Ray Hyman at Skeptic's Toolbox 2012

Main article: Skeptic's Toolbox

The Skeptic's Toolbox is an annual four-day workshop workshop at the University of Oregon, Eugene sponsored by CFI[47] devoted to scientific skepticism. It was formed by psychologist and now-retired University of Oregon professor Ray Hyman, has been held every August since 1992. The workshop focuses on educating people to be better critical thinkers, and involves a central theme. The attendees are broken up into groups and given tasks that they must work on together and whose results they must present in front of the entire group on the last day.

22.5.3 Center for Inquiry Libraries

The Center for Inquiry Libraries[48] began as a small collection of books located in the offices of CSICOP in the late 1970s. When the first expansion of the Center for Inquiry building was completed in 1995, the library was prominently featured. The building opened on June 9, 1995, with such luminaries as Leon Jaroff, Herbert Hauptman, Stan Lundine, and Kendrick Frazier attending, and Steve Allen, prominent supporter of CFI, spoke at the opening ceremony.[49]

Gordon Stein was the Libraries' first director and acquired a large number of rare materials. Timothy Binga has been the Director of Libraries since 1996, and has been instrumental in the cataloging and organization of the large amount of materials acquired.

CFI's Libraries were created along the same lines as the organization; CSICOP and the Council for Secular Humanism had their own libraries, and there were a number of

CFI Library

shared libraries and collections as well. Highlights of the various collections include materials from Martin Gardner (some papers and books), Steve Allen (bound notebooks of clippings, notes, letters, and tipped-in pamphlets organized by subject), Martin T. Orne collection of books, papers, and case notes, and books and papers of noted philosophers Abraham Edel, Paul Edwards, Patrick Romanell, and Joseph Blau.[50]

It total, the books number around 70,000 volumes; this includes the world's foremost collections on skepticism, humanism, and freethought. Also, there are world-class collections on science, philosophy, American Philosophical Naturalism, the occult and paranormal, atheism, and other items related to the mission of the Center for Inquiry. In addition, there are archives, a reference section, periodicals, microfilm, and AV materials.

The Rare Book Room contains a signed Elizabeth Cady Stanton autobiography, first editions of such works as *The Age of Reason* by Thomas Paine, *Reason: The Only Oracle of Man* by Ethan Allen, many signed works by Robert Green Ingersoll, hard copies of *The Truth Seeker* (the newspaper of record for the Golden Age of Freethought), and a collection of Little Blue Books.

CFI Libraries are a member of OCLC[51], the Western New York Library Resources Council (WNYLRC)[51], New York Heritage and the Digital Public Library of America.

22.5.4 Secular Rescue

The Center for Inquiry has an emergency fund called Secular Rescue, formerly known as the Freethought Emergency Fund.[52] The fund is used to help freethought activists whose lives are under threat by Islamic radicals linked to Al Qaeda.[53]

22.5.5 Office of Public Policy

The Office of Public Policy (OPP) is the Washington D.C. political arm of the Center for Inquiry. The OPP's mandate is to lobby Congress and the Administration on issues related to science and secularism. This includes defending the separation of church and state, promoting science and reason as the basis of public policy, and advancing secular values.[54]

The OPP publishes position statements on its subjects of interest. Examples have included acupuncture, climate change, contraception and intelligent design.[55] The Office is an active participant in legal matters, providing experts for Congress testimony and amicus briefs in Supreme Court cases.[56] It publishes a list of bills it considers of interest as they pass through the U.S. legislative process.[57]

22.5.6 "Science and the Public" Master of Education program

In partnership with the Graduate School of Education at the State University of New York at Buffalo, CFI offers an accredited Master of Education program in Science and the Public, available entirely online.[58] Aimed at students preparing for careers in research, science education, public policy, science journalism, or further study in sociology, history and philosophy of science, science communication, education, or public administration, the program explores the methods and outlook of science as they intersect with public culture, scientific literacy, and public policy.

22.5.7 Skeptics and Humanist Aid and Relief Effort

The Skeptics and Humanist Aid and Relief Effort (previously the name began with the phrase "Secular Humanist") provides "an alternative for those who wish to contribute to charitable efforts without the intermediary of a religious organization in times of great need."[59] As of January 2010, all funds are being directed to the group Doctors Without Borders to aid the survivors of the 2010 Haiti earthquake. Previous relief efforts have included aid for survivors of the 2008 Sichuan earthquake and the October 2007 California wildfires.[60]

22.6 Past projects and programs

The following projects and programs are no longer active.

22.6.1 Camp Inquiry

The Center for Inquiry organized an annual summer camp for children called *Camp Inquiry*,[61] focusing on scientific literacy, critical thinking, naturalism, the arts, humanities, and humanist ethical development.[62] Camp Inquiry has been described as "a summer camp for kids with questions"[63] where spooky stories were followed by "reverse engineering sessions" as the participants were encouraged to determine the cause of an apparently supernatural experience. Camp Inquiry has been criticised as "Jesus Camp in reverse"; its organisers countered that the camp is not exclusive to atheist children and that campers are encouraged to draw their own conclusions based on empirical and critical thinking.

22.6.2 CFI Institute

The Center for Inquiry Institute[64] offered undergraduate level online courses, seminars, and workshops in critical thinking and the scientific outlook and its implications for religion, human values, and the borderlands of science. In addition to transferable undergraduate credit through the University at Buffalo system, CFI offered a thirty credit-hour Certificate of Proficiency in Critical Inquiry. The three-year curriculum plan offered summer sessions at the main campus at the University at Buffalo in Amherst.

22.6.3 Medicine and health

The Commission for Scientific Medicine and Mental Health (CSMMH)[65] stimulated critical scientific scrutiny of New Age medicine and the schools of psychotherapy. It supported naturalistic addiction recovery practices through Secular Organizations for Sobriety. CFI challenges the claims of alternative medicine[66] and advocates a scientific basis for healthcare.[67][68] CSMMH papers have covered topics such as pseudoscience in autism treatments[69] and in psychiatry.[70]

22.6.4 Naturalism Research Project

CFI also ran the Naturalism Research Project, a major effort to develop the theoretical and practical applications of philosophical naturalism. As part of this project, CFI's libraries, research facilities, and conference areas were available to scientists and scholars to advance the understanding of science's methodologies and conclusions about naturalism.[71]

Activities of the Naturalism Research Project included lectures and seminars by visiting fellows and scholars; academic conferences; and support CFI publications of important research. Among the central issues of naturalism include the exploration of varieties of naturalism; problems in philosophy of science; the methodologies of scientific inquiry; naturalism and humanism; naturalistic ethics; planetary ethics; and naturalism and the biosciences.[72]

22.7 CFI organization and locations

CFI's Rare Book Room, located at their Amherst, NY Headquarters

CFI is a nonprofit body registered as a charity in the United States.[73] It has 17 locations in the U.S., and has 16 international branches or affiliated organizations.[74] The organization has Centers For Inquiry in Amherst, New York (its headquarters), Los Angeles, New York City, Tampa Bay, Washington, D.C., Indiana, Austin, Chicago, San Francisco and Michigan.[75]

22.7.1 International activities

CFI has branches, representation or affiliated organizations in countries around the world.[75] It organizes its international activities under the banner *Center For Inquiry Transnational*. In addition, CFI holds consultative status to the United Nations as an NGO under the UN Economic and Social Council.[6] The Center participates in UN Human Rights Council debates, for example a debate on the subject of female genital mutilation during 2014.[76]

22.7.2 University exchange programs

CFI Moscow operates an exchange program where Russian students and scholars are able to visit CFI headquarters in Amherst and participate in a summer institute each year. Additional international programs exist in Germany (Rossdorf), France (Nice), Spain (Bilbao), Poland

(Warsaw), Nigeria (Ibadan), Uganda (Kampala), Kenya (Nairobi), Nepal (Kathmandu), India (Pune) (Hyderabad), Egypt (Cairo), China (Beijing), New Zealand (Auckland), Peru (Lima), Argentina (Buenos Aires), Senegal (Dakar), Zambia (Lusaka), and Bangladesh (Dacca).[77]

22.7.3 Centre for Inquiry Canada

Main article: Centre for Inquiry Canada

CFI Canada (CFIC) is the Canadian branch of CFI Transnational, headquartered in Toronto, Ontario, Canada. Justin Trottier served as National Executive Director from 2007-2011. Originally established and supported in part by CFI Transnational, CFI Canada has become an independent Canadian national organization with several provincial branches. CFI Canada has branches in Halifax, Montreal, Ottawa, Toronto, Saskatoon, Calgary, Okanagan (Kelowna) and Vancouver.

22.8 Affiliate organizations

- Centre for Inquiry Canada
- Centre for Inquiry UK[78][79]
- Committee for Skeptical Inquiry (CSI)
- Committee for the Scientific Examination of Religion (CSER)
- Commission for Scientific Medicine and Mental Health Practice (CSMMH)[65]
- Institute for the Secularisation of Islamic Society (ISIS)

22.9 In the media

CFI participates in media debates on science, health,[80] religion and its other areas of interest. Its "Keep Healthcare Safe and Secular" campaign promotes scientifically sound healthcare.[68][81] It has been an outspoken critic of dubious and unscientific healthcare practices, and engages in public debate on the merit and legality of controversial medical techniques. In 2014, CEO Ron Lindsay publicly criticized Stanislaw Burzynski's controversial Texas cancer clinic.[82]

CFI campaigns for a secular society, for example in opposing the addition of prayer text on public property.[83] The Center supports secular and free speech initiatives.[84]

On November 14, 2006 the CFI opened its Office of Public Policy in Washington, DC and issued a declaration "In Defense of Science and Secularism", which calls for public policy to be based on science rather than faith.[85] The next day the Washington Post ran an article about it entitled "Think Tank Will Promote Thinking".[4]

In 2011, video expert James Underdown of IIG and CFI Los Angeles did an experiment for "Miracle Detective" Oprah Winfrey Network which replicated exactly the angelic apparition that people claim cured a 14-year-old severely disabled child at Presbyterian Hemby Children's Hospital in Charlotte, North Carolina. The "angel" was sunlight from a hidden window, and the girl remained handicapped.[86]

22.9.1 Wyndgate Country Club and Richard Dawkins, 2011

During Richard Dawkins' October 2011 book tour, Center for Inquiry - the tour's sponsor - signed a contract with Wyndgate Country Club in Rochester Hills, Michigan, as the venue site. After seeing an interview with Dawkins on *The O'Reilly Factor*, an official at the club cancelled Dawkins' appearance. Dawkins said that the country club official accepted Bill O'Reilly's "twisted" interpretation of his book *The Magic of Reality* without having read it personally.[87][88] Sean Faircloth said that cancelling the reading "really violates the basic principles of America ... The Civil Rights Act ... prohibits discrimination based on race or religious viewpoint. ... [Dawkins has] published numerous books ... to explain science to the public, so it's rather an affront, to reason in general, to shun him as they did."[89] CFI Michigan executive director Jeff Seaver stated that "This action by The Wyndgate illustrates the kind of bias and bigotry that nonbelievers encounter all the time."[90][91] Following the cancellation, protests and legal action by CFI against the Wyndgate Country Club were pursued.[92][93] In 2013 this case was settled in favor of the Center For Inquiry.[94]

22.9.2 CSH actions against faith-based initiatives

In 2007, CSH sued the Florida Department of Corrections (DOC) to block the use of state funds in contracts to faith-based programs for released inmates, claiming that this use is prohibited under the "No Aid" provision or Blaine amendment of the Florida constitution. The initial decision found in favor of the DOC but, on appeal, the case was remanded in 2010 on just the issue of the unconstitutionality of appropriating state funds for this purpose.[95]

While this case was in progress, after the appellate finding, Republican legislators began an effort to amend the Florida constitution to remove the language of the Blaine amendment, succeeding in 2011 to place the measure on the 2012 ballot as amendment 8.[96][97] The ballot measure failed.[97][98]

In 2015, CHS (now CFI) and the state (along with its co-defendants) both filed for summary judgement. The court granted the state's motion in January, 2016, allowing the contested contracting practice to continue.[99] After consideration, CFI announced in February, 2016, that it would not appeal.[100][101]

22.9.3 Heckled at the UN

A CFI representative was repeatedly interrupted and heckled whilst presenting the Center's position on censorship at the UN Human Rights Council.[38] CFI advocated free speech, and opposed the punishment by Saudi authorities of Raif Badawi for running an Internet forum, whom they accused of atheism and liberalism. The Saudi delegation objected repeatedly to CFI's statement. CFI drew support from American, Canadian, Irish and French delegates.

22.9.4 Blasphemy Day

Main article: Blasphemy Day

Blasphemy Rights Day International encourages individuals and groups to openly express their criticism of or outright contempt for religion. It was founded in 2009 by the Center for Inquiry.[102] A student contacted the Center for Inquiry in Amherst, New York to present the idea, which CFI then supported. Ronald Lindsay, president and CEO of the Center for Inquiry said regarding Blasphemy Day, "We think religious beliefs should be subject to examination and criticism just as political beliefs are, but we have a taboo on religion", in an interview with CNN.[103] It takes place every September 30 to coincide with the anniversary of the publications of the controversial Jyllands-Posten Muhammad cartoons.

Blasphemy Day and CFI's related Blasphemy Contests[104] started (in CFI's own words) "a firestorm of controversy".[104] The use of confrontational free speech has been a topic of debate within the Humanist movement [105] [106] and cited as an example of a wider move towards New Atheism and away from the more conciliatory approach historically associated with Humanism.[107][108]

22.10 See also

- Floris van den Berg

22.11 References

[1] "Home". Center for Inquiry. Retrieved 2014-03-01.

[2] "Program Areas". Center for Inquiry. Retrieved 2014-03-01.

[3] "Center Stage". Center for Inquiry. Retrieved 2014-03-03.

[4] Kaufman, Marc (2006-11-14). "Think Tank Will Promote Thinking". Washington Post. Retrieved 2014-03-03.

[5] Metacrock (2012-08-19). "Center for Inquiry, Jesus Project, Atheist Organization". Atheist Watch. Retrieved 2014-03-03.

[6] "UN". Center for Inquiry. Retrieved 2014-06-16.

[7] "An Important Announcement from CFI President Ronald A. Lindsay". Center for Inquiry. Retrieved 2016-01-24.

[8] "F.A.Q: The CFI/Dawkins Foundation Merger". Center for Inquiry. Retrieved 2016-01-24.

[9] "Merger creates largest atheist organization". WBFO. Retrieved 2016-01-24.

[10] "'Royal wedding' of atheist group, Richard Dawkins Foundation launches woman to top post". Religion News. Retrieved 2016-01-24.

[11] "Richard Dawkins Foundation for Reason & Science to Merge with Center for Inquiry". richarddawkins.net. January 20, 2016. Retrieved January 30, 2016.

[12] Smith, Cameron M. "CSI". Csicop.org. Retrieved 2014-03-01.

[13] "Council for Secular Humanism". Secularhumanism.org. Retrieved 2014-03-01.

[14] "CFI Los Angeles". Cfiwest.org. Archived from the original on 2014-03-05. Retrieved 2014-03-01.

[15] "New York City's Home for Reason & Science | CFI NYC". Center for Inquiry. Retrieved 2014-03-01.

[16] "CFI Tampa Home". Center for Inquiry. 2014-02-24. Retrieved 2014-03-01.

[17] "CFI Communities; Center for Inquiry". Center For Inquiry. 2009-09-08. Archived from the original on 2009-09-08. Retrieved 2014-06-16.

[18] "Center for Inquiry Awarded Special Consultative NGO Status by the United Nations.". *Free Inquiry*. **25** (6): 7. Oct–Nov 2005. ISSN 0272-0701.

22.11. REFERENCES

[19] "A Unified Center for Inquiry, Stronger Than Ever". Center for Inquiry. Retrieved 2014-05-11.

[20] "Richard Dawkins Foundation for Reason & Science to Merge with Center for Inquiry". centerforinquiry.net. January 21, 2016. Retrieved January 30, 2016.

[21] Oppenheimer, Mark (October 2, 2010). "Closer Look at Rift Between Humanists Reveals Deeper Divisions". *The New York Times*. Retrieved October 3, 2015.

[22] Erich Vieth (2010-10-22). "The Center for Inquiry responds to the claims made by Paul Kurtz". Retrieved 2017-06-25.

[23] "CFI Board accepts Paul Kurtz's resignation". *Center for Inquiry Board Statement*. Center For Inquiry. Retrieved 27 January 2016.

[24] Wikisource:Uri Geller vs. James Randi decision

[25] >"ClairAudient Test". Ustream. 2011-08-21.

[26] "The IIG $100,000 Challenge". Independent Investigations Group. Retrieved 2014-06-21.

[27] "Ted Cruz Joins Demonstrators in Front of White House; Calls on Obama to Help Imprisoned Sudanese Christian Woman". Christianpost.com. 2014-06-12. Retrieved 2014-06-29.

[28] "CFI Tells Federal Court: End Taxpayer Funding of Clergy Housing". Center for Inquiry. 2014-06-24. Retrieved 2014-06-29.

[29] "Supreme Court prayer ruling may spur new alliances". The Washington Post. Retrieved 2014-06-09.

[30] "Mass. High Court upholds "under God" in pledge - News - The Beacon - Acton, MA". Acton.wickedlocal.com. 2014-05-09. Archived from the original on 2014-06-09. Retrieved 2014-06-09.

[31] Center for Inquiry (2014-04-25). "Women in Secularism III: 2014 Conference in Alexandria, VA". Web.archive.org. Archived from the original on 2014-06-09. Retrieved 2014-06-09.

[32] "Why atheists should care about transgender issues: A conversation with Kayley Whalen | Faitheist". Chrisstedman.religionnews.com. Archived from the original on 2014-06-09. Retrieved 2014-06-09.

[33] "Celebrate Ingersoll! New Conference August 16–17". Center for Inquiry. 2014-05-23. Retrieved 2014-06-29.

[34] Hutchinson, Sikivu (2014-06-25). "Atheism has a big race problem that no one's talking about". The Washington Post. Retrieved 2014-06-29.

[35] "Moving Social Justice Conference: October '14 CFI-Los Angeles". Freethought Blogs. March 20, 2014. Archived from the original on 2014-08-11.

[36] Colette M. Jenkins (2014-05-17). "Northeast Ohio billboard campaign urges nonbelievers to 'come out of the closet' - Local". Ohio. Archived from the original on 2014-06-09. Retrieved 2014-06-09.

[37] "Secular Group to Send 'Freethought' Books to Prisoners as Alternative to Religion". Christianpost.com. 2014-01-14. Retrieved 2014-06-29.

[38] Rori Donaghy (2014-06-24). "Saudi Arabia attempts to silence NGO at Human Rights Council". *Middle East Eye*. Archived from the original on 2014-08-05. Retrieved 2014-06-29.

[39] "Atheist to deliver town's opening prayer". Troyrecord.com. Retrieved 2014-06-29.

[40] "To Expand Religious Freedom in the Military, Republicans Need to Win the Senate, Congressman Says". Christianpost.com. 2014-05-28. Retrieved 2014-06-29.

[41] "Secular Humanist Bulletin - Council for Secular Humanism". The Council For Secular Humanism. 2008-11-11. Retrieved 2014-06-16.

[42] *The American Rationalist* volume LVII May/June 2011, Number 3, ISSN 0003-0708.

[43] "Scientific Review of Mental Health Practice". Srmhp.org. Retrieved 2014-03-01.

[44] "Home - CFI On Campus | Organizing atheist, freethinking, skeptical, and secular humanist students and faculty worldwide". Centerforinquiry.net. Retrieved 2014-03-01.

[45] "Debbie Goddard | African Americans for Humanism". Aahumanism.net. Retrieved 2015-07-20.

[46] "Debbie Goddard Named CFI's Director of Outreach". Center for Inquiry. 2012-11-20. Retrieved 2014-03-01.

[47] Committee for Skeptical Inquiry. "The Skeptic's Toolbox: 2013 workshop in Eugene, OR". Skepticstoolbox.org. Retrieved 2014-03-01.

[48] "CFI Libraries". CFI Libraries. Retrieved 2014-03-01.

[49] "Center for Inquiry Celebrates Grand Opening". *CSICOP News*. 1995-06-09. Archived from the original on 2008-03-23. Retrieved 2008-09-20.

[50] "About the Collections". Retrieved 2017-07-09.

[51] "About Us". Retrieved 2017-07-09.

[52] "Secular Rescue, a program of the Center For Inquiry". Retrieved 2017-06-25.

[53] "Amid Death Threats from Islamists, CFI Brings Secular Activist Taslima Nasrin to Safety in U.S.". Center for Inquiry. Retrieved 2015-06-01.

[54] "CFI Office of Public Policy". Center for Inquiry. Retrieved 2014-03-01.

[55] "CFI Office of Public Policy". Center for Inquiry. Retrieved 2014-06-29.

[56] "Press release - Amicus brief to Supreme Court". Center For Inquiry. January 28, 2014. Retrieved February 6, 2014.

[57] "CFI Office of Public Policy". Center for Inquiry. Retrieved 2014-06-29.

[58] "Science and the Public, EdM". University at Buffalo. Retrieved 2017-07-09.

[59] "SHARE Opens Fund for Haiti Quake Relief". Center for Inquiry. 2010-01-14. Retrieved 2014-06-16.

[60] "Secular S.H.A.R.E. raises USD47,000 for Haiti in less than 24 hours". Center For Inquiry. 2010-01-15. Retrieved 2014-06-16.

[61] "Camp Inquiry — Holland, New York". Campinquiry.org. Retrieved 2014-03-01.

[62] "Camp Offers Training Ground For Little Skeptics". NPR. Retrieved 2014-06-29.

[63] "Angie McQuaig: Camp Inquiry: A Summer Camp for Kids With Questions". Huffingtonpost.com. Retrieved 2014-06-29.

[64] "Institute Catalogue". Center for Inquiry. Retrieved 2014-03-01.

[65] "CSMMH". Commission for Scientific Medicine and Mental Health. 2010-09-25. Archived from the original on 2010-09-25. Retrieved 2014-03-03.

[66] "The pseudoscience of homeopathy". The Washington Post. Retrieved 2014-06-09.

[67] "CFI and CSI Petition FDA to Take Action on Homeopathic Drugs". Center for Inquiry. 2011-08-30. Archived from the original on 2011-09-25. Retrieved 2014-06-16.

[68] "CFI Launches Campaign to Keep Religion and Pseudoscience Out of Health Care". Center for Inquiry. 2014-06-03. Retrieved 2014-06-29.

[69] Riggott, Julie. (Spring–Summer 2005). "PSEUDOSCIENCE IN AUTISM TREATMENT: ARE THE NEWS AND ENTERTAINMENT MEDIA HELPING OR HURTING?". 4 (1). Scientific Review of Mental Health Practice.: 55–58. 4p. Archived from the original on 2014-09-01. Professional organizations which issued position statements indicating that facilitated communication is not a scientifically valid technique; Most effective intervention for autism according to scientists; Symptoms of autism; Reason for the abundance of untested and ineffective therapies for autism

[70] Power Therapies and possible threats to the science of psychology and psychiatry. By: Devilly, Grant J. Australian & New Zealand Journal of Psychiatry. Jun 2005, Vol. 39 Issue 6, p437-445. 9p. This paper reviews a collection of new therapies collectively self-termed 'The Power Therapies', outlining their proposed procedures and the evidence for and against their use. These therapies are then put to the test for pseudoscientific practice. It is concluded that these new therapies have offered no new scientifically valid theories of action, show only non-specific efficacy, show no evidence that they offer substantive improvements to extant psychiatric care, yet display many characteristics consistent with pseudoscience.

[71] "Education". Center for Inquiry. n.d. Retrieved 2014-03-01.

[72] Shook, John (March–April 2007). "Center for Inquiry launches Naturalism Research Project". *Skeptical Inquirer*. **31** (2). – via HighBeam (subscription required)

[73] "About". Center for Inquiry. Retrieved 2014-03-02.

[74] "About center For Inquiry". 2013. Retrieved 2014-02-28.

[75] "About | Center for Inquiry". Center for Inquiry. 2014-04-04. Archived from the original on 2014-06-15. Retrieved 2014-06-16.

[76] "Le Conseil des droits de l'homme tient une reunion-débat de haut niveau sur la lute contre les mutilations génitales féminines - Communiqués de presse - Actualités - StarAfrica.com". Fr.starafrica.com. 2014-06-16. Retrieved 2014-06-29.

[77] "Center For Inquiry On Campus locations". Retrieved 2014-06-25.

[78] CFI UK is a section of the British Humanist Association

[79] "Centre for Inquiry UK". Centreforinquiry.org.uk. Retrieved 2014-06-16.

[80] "Dr. Oz's bad day on Capitol Hill". TheHill. 2014-05-14. Retrieved 2014-06-29.

[81] "The Fight To Take Back Our Health Care System From Junk Science". ThinkProgress. 2014-06-05. Retrieved 2014-06-29.

[82] Liz Szabo (2014-07-25). "Texas medical board charges controversial cancer doctor". USA Today. Archived from the original on 2014-08-11.

[83] Kimberly Winston (2014-06-06). "Senate approves prayer plaque for World War II monument". Deseret News. Retrieved 2014-06-29.

[84] "Coalition of US-Based Firms Urges Pakistan to End Social Media Censorship | NDTV Gadgets". Gadgets.ndtv.com. 2014-06-12. Retrieved 2014-06-29.

[85] Park, Bob (2006-11-17). "Freedom Of Science: in Defense of Science and Secularism" (blog). *What's New with Bob Park*.

[86] "Guardian Angel: Video Expert Re-Creation | Guardian Angel: Video Expert Re-Creation". Oprah Winfrey Network. 2011-01-20. Retrieved 2011-01-28.

[87] "Rochester Hills Country Club Cancels Richard Dawkins Appearance". Fox News. 2011-10-13. Retrieved 2011-10-15.

[88] "Atheist Richard Dawkins snubbed by Detroit area country club". *Detroit Free Press*. 2011-10-12. Retrieved 2011-10-15.

[89] "Rochester Hills Country Club Cancels Richard Dawkins Appearance". *MyFoxDetroit.com*. Fox Television Stations, Inc. Retrieved 2011-10-15.

[90] "Protest tonight against club's decision to cancel atheist's appearance". *The Detroit News*. 2011-10-12. Retrieved 2011-10-15.

[91] "Atheist Richard Dawkins Rejected by Detroit Country Club?". The Christian Post. 2011-10-13. Retrieved 2014-06-16.

[92] "UPDATE: Dawkins Event Banned - CFI to Pursue Legal Remedies". Center For Inquiry. 2011-10-14. Retrieved 2011-10-15.

[93] "COMPLAINT AND DEMAND FOR JURY" (PDF). *UNITED STATES DISTRICT COURT, EASTERN DISTRICT OF MICHIGAN*. Center for Inquiry. April 27, 2012.

[94] "Mich. club settles atheist discrimination suit". *The Wall Street Journal*. 2013-02-26. Archived from the original on 2013-03-01. Retrieved 2013-02-27.

[95] "COUNCIL FOR SECULAR HUMANISM INC v. McNEIL". *findlaw.com*. April 27, 2010.

[96] Postal, Leslie (April 20, 2011). "Bill asks voters to OK taxpayer funding of religious institutions". Orlando Sentinel. Retrieved May 16, 2016.

[97] Mazzei, Patricia (April 27, 2011). "Blaine amendment repeal passes Florida house". Tampa Bay Times. Retrieved May 16, 2016.

[98] "Florida religious freedom Amendment 8". *ballotpedia.org*. Retrieved May 16, 2016.

[99] Fl 2nd District Court (January 20, 2016). "CFI v Jones (2016)" (PDF). *becketfund.org*. Retrieved May 16, 2016.

[100] "Center for Inquiry will not appeal adverse decision in Florida lawsuit". CFI. February 8, 2016. Retrieved May 16, 2016.

[101] Bettis, Kara (February 23, 2016). "Atheists drop suit to block Christian prison ministry funding". New Boston Post. Retrieved May 16, 2016.

[102] "Penn Jillette Celebrates Blasphemy Day in "Penn Says"". Center for Inquiry. 2009-09-29. Retrieved 2013-09-30.

[103] Basu, Moni (September 30, 2009). "Taking aim at God on 'Blasphemy Day'". *CNN.com*.

[104] "CFI Announces Blasphemy Contest Winners". Center for Inquiry. 2009-11-16. Retrieved 2014-07-23.

[105] Moni Basu CNN (2009-09-30). "Taking aim at God on 'Blasphemy Day'". CNN.com. Retrieved 2014-07-23.

[106] Mark Oppenheimer (October 1, 2010). "Closer Look at Rift Between Humanists Reveals Deeper Divisions". New York Times.

[107] "A Bitter Rift Divides Atheists". NPR. 2009-10-19. Retrieved 2014-07-23.

[108] St. Louis, Missouri (2009-10-19). "Quick Thoughts on the NPR "Bitter Rift" Story". Center for Inquiry. Retrieved 2014-07-23.

22.12 External links

- Official website
- How Camp Inquiry introduces kids to the principles of humanism by Dr. Angie McQuaig
- Point of Inquiry *Point of Inquiry* – the radio show and podcast of the Center for Inquiry.

Chapter 23

A Secular Humanist Declaration

A Secular Humanist Declaration was an argument for and statement of support for democratic secular humanism. The document was issued in 1980 by the Council for Democratic and Secular Humanism ("CODESH"), now the Council for Secular Humanism ("CSH"). Compiled by Paul Kurtz, it is largely a restatement of the content of the American Humanist Association's 1973 Humanist Manifesto II, of which he was co-author with Edwin H. Wilson. Both Wilson and Kurtz had served as editors of *The Humanist*, from which Kurtz departed in 1979 and thereafter set about establishing his own movement and his own periodical. His Secular Humanist Declaration was the starting point for these enterprises.

23.1 Table of Contents

1. Free Inquiry
2. Separation of Church and State
3. The Ideal of Freedom
4. Ethics Based on Critical Intelligence
5. Moral Education
6. Religious Skepticism
7. Reason
8. Science and Technology
9. Evolution
10. Education

23.2 Signatories

Before the list of signatories, the declaration has the following disclaimer: "Although we who endorse this declaration may not agree with all its specific provisions, we nevertheless support its general purposes and direction and believe that it is important that they be enunciated and implemented. We call upon all men and women of good will who agree with us to join in helping to keep alive the commitment to the principles of free inquiry and the secular humanist outlook. We submit that the decline of these values could have ominous implications for the future of civilization on this planet."

23.2.1 United States

- George Abell (professor of astronomy, UCLA)
- John Anton (professor of philosophy, Emory University)
- Khoren Arisian (minister, First Unitarian Society of Minneapolis)
- Isaac Asimov (science fiction author)
- Paul Beattie (minister, All Souls Unitarian Church; president, Fellowship of Religious Humanism)
- H. James Birx (professor of anthropology and sociology, Canisius College)
- Brand Blanshard (professor emeritus of philosophy, Yale)
- Joseph L. Blau (Professor Emeritus of Religion, Columbia)
- Francis Crick (Nobel Prize Laureate, Salk Institute)
- Arthur Danto (professor of philosophy, Columbia University)
- Albert Ellis (executive director, Institute for Rational Emotive Therapy)
- Roy Fairfield (former professor of social science, Antioch)

- Herbert Feigl (professor emeritus of philosophy, University of Minnesota)

- Joseph Fletcher (theologian, University of Virginia Medical School)

- Sidney Hook (professor emeritus of philosophy, NYU, fellow at Hoover Institute)

- George Hourani (professor of philosophy, State University of New York at Buffalo)

- Walter Kaufmann (professor of philosophy, Princeton)

- Marvin Kohl (professor of philosophy, medical ethics, State University of New York at Fredonia)

- Richard Kostelanetz (writer, artist, critic)

- Paul Kurtz (Professor of Philosophy, State University of New York at Buffalo)

- Joseph Margolis (professor of philosophy, Temple University)

- Floyd Matson (professor of American Studies, University of Hawaii)

- Ernest Nagel (professor emeritus of philosophy, Columbia)

- Lee Nisbet (associate professor of philosophy, Medaille)

- George Olincy (lawyer)

- Virginia Olincy

- W. V. Quine (professor of philosophy, Harvard University)

- Robert Rimmer (novelist)

- Herbert Schapiro (Freedom from Religion Foundation)

- Herbert W. Schneider (professor emeritus of philosophy, Claremont College)

- B. F. Skinner (professor emeritus of psychology, Harvard)

- Gordon Stein (editor, The American Rationalist)

- George Tomashevich (professor of anthropology, Buffalo State University College)

- Valentin Turchin (Russian dissident; computer scientist, City College, City University of New York)

- Sherwin Wine (rabbi, Birmingham Temple, founder, Society for Humanistic Judaism)

- Marvin Zimmerman (professor of philosophy, State University of New York at Buffalo)

23.2.2 Canada

- Henry Morgentaler (physician, Montreal)

- Kai Nielsen (professor of philosophy, University of Calgary)

23.2.3 France

- Yves Galifret (executive director, Union Rationaliste)

- Jean-Claude Pecker (professor of astrophysics, College de France, Academie des Sciences)

23.2.4 Great Britain (i.e. Scotland, Wales and England)

- Sir A.J. Ayer (professor of philosophy, Oxford University)

- H.J. Blackham (former chairman, Social Morality Council and British Humanist Association)

- Bernard Crick (professor of politics, Birkbeck College, London University)

- Sir Raymond Firth (professor emeritus of anthropology, University of London)

- Jim Herrick (then editor of The Freethinker)

- Zhores A. Medvedev (Russian dissident; Medical Research Council)

- Dora Russell (Mrs. Bertrand Russell) (author)

- Lord Ritchie-Calder (president, Rationalist Press Association)

- Harry Stopes-Roe (senior lecturer in science studies, University of Birmingham; chairman, British Humanist Association)

- Nicolas Walter (editor, New Humanist)

- Baroness Barbara Wootton (Deputy Speaker, House of Lords)

23.2.5 India

- B. Shah (president, Indian Secular Society; director, Institute for the Study of Indian Traditions)
- V. M. Tarkunde (Supreme Court Judge, chairman, Indian Radical Humanist Association)

23.2.6 Israel

- Shulamit Aloni (lawyer, member of Knesset, head of Citizens Rights Movement)

23.2.7 Norway

- Alastair Hannay (professor of philosophy, University of Trondheim)

23.2.8 Yugoslavia

- Milovan Djilas (author, former vice president of Yugoslavia)
- Mihailo Marković (professor of philosophy, Serbian Academy of Sciences & Arts and University of Belgrade)
- Svetozar Stojanović (professor of philosophy, University of Belgrade)

23.3 See also

- Amsterdam Declaration 2002 - the defining statement of Humanism worldwide

23.4 External links

- A Secular Humanist Declaration

Chapter 24

Amsterdam Declaration

The **Amsterdam Declaration 2002** is a statement of the fundamental principles of modern Humanism passed unanimously by the General Assembly of the International Humanist and Ethical Union (IHEU) at the 50th anniversary World Humanist Congress in 2002. According to the IHEU, the declaration "is the official statement of World Humanism."

It is officially supported by all member organisations of the IHEU including:

- Humanistic Association Netherlands (Humanistisch Verbond)
- American Humanist Association
- British Humanist Association
- Humanist Canada
- Human-Etisk Forbund, the Norwegian Humanist Association
- Humanistischer Verband Deutschlands, the Humanist Association of Germany
- Council of Australian Humanist Societies
- Council for Secular Humanism
- Gay and Lesbian Humanist Association
- Humanist Association of Ireland
- Indian Humanist Union
- Philippine Atheists and Agnostics Society (PATAS)

A complete list of signatories can be found on the IHEU page (see references).

This declaration makes exclusive use of capitalized *Humanist* and *Humanism*, which is consistent with IHEU's general practice and recommendations for promoting a unified Humanist identity. To further promote Humanist identity, these words are also free of any adjectives, as recommended by prominent members of IHEU. Such usage is not universal among IHEU member organizations, though most of them do observe these conventions.

24.1 Humanist principles

(*see References for complete text*)

The official defining statement of World Humanism is:

- Humanism is ethical. It affirms the worth, dignity and autonomy of the individual and the right of every human being to the greatest possible freedom compatible with the rights of others. Humanists have a duty of care to all humanity including future generations. Humanists believe that morality is an intrinsic part of human nature based on understanding and a concern for others, needing no external sanction.

- Humanism is rational. It seeks to use science creatively, not destructively. Humanists believe that the solutions to the world's problems lie in human thought and action rather than divine intervention. Humanism advocates the application of the methods of science and free inquiry to the problems of human welfare. But Humanists also believe that the application of science and technology must be tempered by human values. Science gives us the means but human values must propose the ends.

- Humanism supports democracy and human rights. Humanism aims at the fullest possible development of every human being. It holds that democracy and human development are matters of right. The principles of democracy and human rights can be applied to many human relationships and are not restricted to methods of government.

- Humanism insists that personal liberty must be combined with social responsibility. Humanism ventures to build a world on the idea of the free person responsible to society, and recognizes our dependence and responsibility for the natural world. Humanism is undogmatic, imposing no creed upon its adherents. It is thus committed to education free from indoctrination.

- Humanism is a response to the widespread demand for an alternative to dogmatic religion. The world's major religions claim to be based on revelations fixed for all time, and many seek to impose their world-view on all of humanity. Humanism recognizes that reliable knowledge of the world and ourselves arises through a continuing process of observation, evaluation and revision.

- Humanism values artistic creativity and imagination and recognises the transforming power of art. Humanism affirms the importance of literature, music, and the visual and performing arts for personal development and fulfilment.

- Humanism is a lifestance aiming at the maximum possible fulfilment through the cultivation of ethical and creative living and offers an ethical and rational means of addressing the challenges of our time. Humanism can be a way of life for everyone everywhere.

The Amsterdam Declaration explicitly states that Humanism rejects dogma, and imposes no creed upon its adherents.

24.2 History

At the first World Humanist Congress in the Netherlands in 1952, the IHEU general assembly agreed a statement of the fundamental principles of modern Humanism - **The Amsterdam Declaration**.

At the 50th anniversary World Humanist Congress in 2002, the IHEU general assembly unanimously passed a resolution updating that declaration - "The Amsterdam Declaration 2002".

24.3 References

1. ^ "Capitalization [of *Humanism*] is not mandatory... It is recommended usage and the normal usage within IHEU"—Jeremy Webbs, IHEU webmaster, from a response to a Wikipedia editor inquiry, dated 2 March 2006.

2. ^ *Humanism is Eight Letters, No More*—endorsed by Harold John Blackham, Levi Fragell, Corliss Lamont, Harry Stopes-Roe and Rob Tielman.

24.4 External links

- Amsterdam Declaration 2002 - the IHEU general assembly unanimous resolution

Chapter 25

American Humanist Association

Warning: Page using Template:Infobox organization with unknown parameter "bgcolor" (this message is shown only in preview).
Warning: Page using Template:Infobox organization with unknown parameter "fgcolor" (this message is shown only in preview).

The **American Humanist Association** (**AHA**) is an educational organization in the United States that advances Secular Humanism, a philosophy of life that, without theism or other supernatural beliefs, affirms the ability and responsibility of human beings to lead personal lives of ethical fulfillment that aspire to the greater good of humanity.[1]

The American Humanist Association was founded in 1941 and currently provides legal assistance to defend the constitutional rights of secular and religious minorities,[2] actively lobbies Congress on church-state separation and other issues,[3] and maintains a grassroots network of 150 local affiliates and chapters that engage in social activism, philosophical discussion and community-building events.[4] The AHA has several publications, including the bi-monthly magazine *The Humanist*, a quarterly newsletter *Free Mind*, a peer-reviewed semi-annual scholastic journal *Essays in the Philosophy of Humanism*, and a daily online news site TheHumanist.com.[5]

25.1 Background

In 1927 an organization called the "Humanist Fellowship" began at a gathering in Chicago. In 1928 the Fellowship started publishing the *New Humanist* magazine. H.G. Creel was the first editor. The *New Humanist* was published from 1928 to 1936. By 1935 the Humanist Fellowship had become the "Humanist Press Association", the first national association of humanism in the United States.[6]

The first *Humanist Manifesto* was issued by a conference held at the University of Chicago in 1933. Signatories included the philosopher John Dewey, but the majority were ministers (chiefly Unitarian) and theologians. They identified humanism as an ideology that espouses reason, ethics, and social and economic justice.[7]

In July 1939 a group of Quakers, inspired by the 1933 Humanist Manifesto, incorporated under the state laws of California the Humanist Society of Friends as a religious, educational, charitable nonprofit organization authorized to issue charters anywhere in the world and to train and ordain its own ministry. Upon ordination these ministers were then accorded the same rights and privileges granted by law to priests, ministers, and rabbis of traditional theistic religions.[8]

25.2 History

Curtis Reese was a leader in the 1941 reorganization and incorporation of the "Humanist Press Association" as the American Humanist Association. Along with its reorganization, the AHA began printing *The Humanist* magazine. The AHA was originally headquartered in Yellow Springs, Ohio, then San Francisco, California, and in 1978 Amherst, New York.[6] Subsequently, the AHA moved to Washington, D.C..

In 1952 the AHA became a founding member of the International Humanist and Ethical Union (IHEU) in Amsterdam, Netherlands.[9] As an international coalition of Humanist organizations, the IHEU stands today as the only international umbrella group for Humanism.

The AHA was the first national membership organization to support abortion rights. Around the same time, the AHA joined hands with the American Ethical Union (AEU) to help establish the rights of nontheistic conscientious objectors to the Vietnam War. This time also saw Humanists involved in the creation of the first nationwide memorial societies, giving people broader access to cheaper alternatives than the traditional burial. In the late 1960s the AHA also secured a religious tax exemption in support of its celebrant program, allowing Humanist celebrants to legally officiate at weddings, perform chaplaincy functions, and in other ways enjoy the same rights as traditional clergy.

In 1991 the AHA took control of the Humanist Society, a religious Humanist organization that now runs the celebrant program. Since 1991 the organization has worked as an adjunct to the American Humanist Association to certify qualified members to serve in this special capacity as ministers. The Humanist Society's ministry prepares Humanist Celebrants to lead ceremonial observances across the nation and worldwide. Celebrants provide millions of Americans an alternative to traditional religious weddings, memorial services, and other life cycle events.[10] After this transfer, the AHA commenced the process of jettisoning its religious tax exemption and resumed its exclusively educational status. Today the AHA is recognized by the U.S. Internal Revenue Service as a nonprofit, tax exempt, 501(c)(3), publicly supported educational organization.

Membership numbers are disputed, but Djupe and Olson place it under 50,000.[11] The AHA has over 575,000 followers on Facebook and over 42,000 followers on Twitter.[12][13]

25.3 Adjuncts and affiliates

The AHA is also the supervising organization for various Humanist affiliates and adjunct organizations.

25.3.1 Black Humanist Alliance

The Black Humanist Alliance of the American Humanist Association was founded in 2016 as a pillar of its new "Initiatives for Social Justice."[14] Like the Feminist Humanist Alliance and the LGBT Humanist Alliance, the Black Humanist Alliance uses an intersectional approach to addressing issues facing the Black community. As its mission states, the BHA "concern ourselves with confronting expressions of religious hegemony in public policy," but is "also devoted to confronting social, economic, and political deprivations that disproportionately impact Black America due to centuries of culturally ingrained prejudices."[15]

25.3.2 Feminist Humanist Alliance

The Feminist Humanist Alliance (formerly the Feminist Caucus) of the American Humanist Association was established in 1977 as a coalition of both women and men within the AHA to work toward the advancement of women's rights and equality between the sexes in all aspects of society. Originally called the Women's Caucus, the new name was adopted in 1985 as more representative of all the members of the caucus and of the caucus' goals. Over the years, members of the Caucus have advocated for the passage of the Equal Rights Amendment and participated in various public demonstrations, including marches for women's and civil rights. In 1982, the Caucus established its annual Humanist Heroine Award, with the initial award being presented to Sonia Johnson. Other Humanist Heroines include Tish Sommers, Christine Craft, and Fran Hosken.[16] In 2012 the Feminist Caucus declared it would be organizing around two principal efforts: "Refocusing on passing the ERA" and "Promoting the Universal Declaration of Human Rights."[17]

In 2016, the Feminist Caucus, mirroring the Black Humanist Alliance and the LGBT Humanist Alliance, reorganized as the Feminist Humanist Alliance as a component of their larger "Initiatives for Social Justice."[14] As stated on its website, the "refinement in vision" emphasized "FHA's more active partnership with outreach programs and social justice campaigns with distinctly inclusive feminist objectives."[18] Its current goal is to provide a "movement powered by and for women, transpeople, and genderqueer people to fight for social justice. We are united to create inclusive and diverse spaces for activists and allies on the local and national level."[19]

25.3.3 LGBTQ Humanist Alliance

The LGBTQ Humanist Alliance (formerly LGBT Humanist Council) of the American Humanist Association is committed to advancing equality for lesbian, gay, bisexual, and transgender people and their families. The alliance "seeks to cultivate safe and affirming communities, promote humanist values, and achieve full equality and social liberation of LGBTQ persons."[20]

Paralleling the Black Humanist Alliance and the Feminist Humanist Alliance, the Council reformed in 2016 as the LGBTQ Humanist Alliance as a larger part of the AHA's "Initiatives for Social Justice."[14]

25.3.4 Humanist Charities

Humanist Disaster Recovery (formerly Humanist Charities) was established in 2005 and its purpose includes applying uniquely Humanist approaches to those in need and directing the generosity of American humanists to worthy disaster relief and development projects around the world. In 2011 Humanist Charities raised $5,000 from AHA members to donate to the Japan Earthquake Relief Fund.[21]

In September 2008 Humanist Charities raised over $2,500 for the Children of the Border project, a relief and development project to expand emergency medical service and

Official logo of Humanist Disaster Recovery

Official logo of the AHLC

health care for expectant mothers living in the Haitian border region of the Dominican Republic.[22]

In June 2014 Humanist Charities joined forces with the Foundation Beyond Belief's Humanist Crisis Reposonse to form Humanist Disaster Recovery. AHA's Executive Director Roy Speckhardt commented that, "This merger is a positive move that will grow the relief efforts of the humanist community. The end result will be more money directed to charitable activities, dispelling the false claim that nonbelievers don't give to charity." As such, the American Humanist Association agreed to promote and raise awareness of Humanist Disaster Recovery while the Foundation Beyond Belief managed its implementation.[23]

Between 2013-2016, Humanist Disaster Recovery has raised funds for victims of the Typhoon Haiyan in the Philippines, the Syrian Refugee Crisis, Refugee Children of the U.S. Border, Tropical Cyclone Sam, and the Nepal and Ecuadoran Earthquakes.[24] They also facilitate a volunteer network called the Humanist Disaster Recovery (HDR) Teams, who directly help communities in need.[25] The HDR Teams helped to rebuild homes in Columbia, South Carolina after the effects of Hurricane Joaquin.[26]

25.3.5 Appignani Humanist Legal Center

The American Humanist Association launched the Appignani Humanist Legal Center (AHLC) in 2006 to ensure that humanists' constitutional rights are represented in court. Through amicus activity, litigation, and legal advocacy, a team of cooperating lawyers, including Jim McCollum, Wendy Kaminer, and Michael Newdow, provide legal assistance by challenging perceived violations of the Establishment Clause.

- The AHLC's first independent litigation was filed on November 29, 2006, in the United States District Court for the Southern District of Florida. Attorney James Hurley, the AHLC lawyer serving as lead counsel, filed suit against the Palm Beach County Supervisor of Elections on behalf of Plaintiff Jerry Rabinowitz, whose polling place was a church in Delray Beach, Florida. The church featured numerous religious symbols, including signs exhorting people to "Make a Difference with God" and anti-abortion posters, which the AHLC claimed demonstrated a violation of the Establishment Clause. In the voting area itself, "Rabinowitz observed many religious symbols in plain view, both surrounding the election judges and in direct line above the voting machines. He took photographs that will be entered in evidence."[27] U.S. District Judge Donald M. Middlebrooks ruled that Jerry Rabinowitz did not have standing to challenge the placement of polling sites in churches, and dismissed the case.[28]

- In February 2014, AHA brought suit to force the removal of the Bladensburg Peace Cross, a war memorial honoring 49 residents of Prince George's County, Maryland, who died in World War I. AHA represented the plaintiffs, Mr. Lowe, who drives by the memorial "about once a month" and Fred Edwords, former AHA Executive director.[29][30] AHA argued that the presence of a Christian religious symbol on public property violates the First Amendment clause prohibiting government from establishing a religion. Town officials feel the monument to have historic and patriotic significant to local residents.[30][31] A member of the local American Legion Post said, "I mean, to me, it's like they're slapping the veterans in the face. I mean, that's a tribute to the veterans, and for some reason, I have no idea what they have against veterans. I mean, if it wasn't for us veterans they wouldn't have the right to do what they're trying to do."[32]

- In March 2014, a Southern California woman reluctantly removed a roadside memorial from near a freeway ramp where her 19-year-old son was killed after the AHA contacted the city council calling the cross on city-owned property a "serious constitutional violation".[33]

- AHLC represented an atheist family who claimed that the equal rights amendment of the Massachusetts constitution prohibits mandatory daily recitations of the Pledge of Allegiance because the anthem contains the phrase "under God." In November 2012 the Massachusetts Supreme Judicial Court permitted a direct appeal with oral arguments set for early 2013.[34] In May 2014, the Massachusetts Supreme Judicial Court ruled in a unanimous decision that the daily recitation of the phrase "under god" in the US Pledge of Allegiance does not violate the plaintiffs' equal protection rights under the Massachusetts Constitution.

- In February 2015 New Jersey Superior Court Judge David F. Bauman dismissed a lawsuit challenging the Pledge of Allegiance, ruling that "...the Pledge of Allegiance does not violate the rights of those who don't believe in God and does not have to be removed from the patriotic message."[35] In a twenty-one page decision, Bauman wrote, "Under (the association members') reasoning, the very constitution under which (the members) seek redress for perceived atheistic marginalization could itself be deem unconstitutional, an absurd proposition which (association members) do not and cannot advance here."[35]

25.4 Advertising campaigns

2008 Bus Campaign

The American Humanist Association has received media attention for its various advertising campaigns; in 2010, the AHA's campaign was said to be the more expensive than similar ad campaigns from the American Atheists and Freedom From Religion Foundation.[36]

In 2008 the AHA ran ads on buses in Washington, D.C. that proclaimed "Why believe in a god? Just be good for goodness' sake",[37] and since 2009 the organization has paid for billboard advertisements nationwide.[38] One such billboard, which stated "No God...No Problem" was repeatedly vandalized.[39]

In 2010 the AHA launched another ad campaign promoting Humanism, which *The New York Times* said was the "first (atheist campaign) to include spots on television and cable"[40] and was described by CNN as the "largest, most extensive advertising campaign ever by a godless organization".[41] The campaign featured violent or sexist quotes from holy books, contrasted with quotes from humanist thinkers, including physicist Albert Einstein, biologist Richard Dawkins, and anthropologist Carleton Coon, and was largely underwritten by Todd Stiefel, a retired pharmaceutical company executive.[40]

In late 2011 the AHA launched a holiday billboard campaign, placing advertisements in 7 different cities: Kearny, New Jersey; Washington, D.C.; Cranston, Rhode Island; Bastrop, Louisiana; Oregon City, Oregon; College Station, Texas and Rochester Hills, Michigan", cities where AHA states "atheists have experienced discrimination due to their lack of belief in a traditional god".[42] The organization spent more than $200,000 on their campaign which included a billboard reading "Yes, Virginia, there is no god."[43]

In November 2012, the AHA launched a national ad campaign to promote a new website, KidsWithoutGod.com, with ads using the slogans "I'm getting a bit old for imaginary friends"[44] and "You're Not The Only One."[45] The campaign included bus advertising in Washington, DC, a billboard in Moscow, Idaho, and online ads on the family of websites run by Cheezburger and Pandora Radio, as well as Facebook, Reddit, Google, and YouTube.[46] Ads were turned down for content by Disney, Time for Kids and National Geographic Kids.[47]

25.5 National Day of Reason

The National Day of Reason was created by the American Humanist Association and the Washington Area Secular Humanists in 2003. In addition to serving as a holiday for secularists, the National Day of Reason was created in response to the perceived unconstitutionality of the National Day of Prayer. According to the organizers of the National Day of Reason, the National Day of Prayer, "violates the First Amendment of the United States Constitution because it asks federal, state, and local government entities to set aside tax dollar supported time and space to engage in religious ceremonies".[48]

Several organizations associated with the National Day of Reason have organized food drives and blood donations, while other groups have called for an end to prayer invocations at city meetings.[49][50] Other organizations, such as the Oklahoma Atheists and the Minnesota Atheists, have organized local secular celebrations as alternatives to the National Day of Prayer.[51] Additionally, many individuals affiliated with these atheistic groups choose to protest the official National Day of Prayer.[52]

25.6 Reason Rally

In 2012, the American Humanist Association co-sponsored the Reason Rally, a national gathering of "humanists, atheists, freethinkers and nonbelievers from across the United States and abroad" in Washington, D.C.[53] The rally, held on the National Mall, had speakers such as Richard Dawkins, James Randi, Adam Savage, and student activist Jessica Ahlqvist. AHA Executive Director Roy Speckhardt also spoke. According to the Huffington Post, the event's attendance was between 8,000-10,000 while the Atlantic reported a crowd of nearly 20,000.[54][55]

The AHA also co-sponsored the 2016 Reason Rally, held at the Lincoln Memorial.[56]

25.7 Famous awardees

The American Humanist Association has named a "Humanist of the Year" annually since 1953. It has also granted other honors to numerous leading figures, including Salman Rushdie (Outstanding Lifetime Achievement Award in Cultural Humanism 2007), Oliver Stone (Humanist Arts Award, 1996), Katharine Hepburn (Humanist Arts Award 1985), John Dewey (Humanist Pioneer Award, 1954), Jack Kevorkian (Humanist Hero Award, 1996) and Vashti McCollum (Distinguished Service Award, 1991).

25.8 AHA's Humanists of the Year

The AHA website presents the list of the following Humanists of the Year:[57]

- Anton J. Carlson - 1953
- Arthur F. Bentley - 1954
- James P. Warbasse - 1955
- C. Judson Herrick - 1956
- Margaret Sanger - 1957
- Oscar Riddle - 1958
- Brock Chisholm - 1959
- Leó Szilárd - 1960
- Linus Pauling - 1961
- Julian Huxley - 1962
- Hermann J. Muller - 1963
- Carl Rogers - 1964
- Hudson Hoagland - 1965
- Erich Fromm - 1966
- Abraham H. Maslow - 1967
- Benjamin Spock - 1968
- R. Buckminster Fuller - 1969
- A. Philip Randolph - 1970
- Albert Ellis - 1971
- B.F. Skinner - 1972
- Thomas Szasz - 1973
- Joseph Fletcher - 1974
- Mary Calderone - 1974
- Henry Morgentaler - 1975
- Betty Friedan - 1975
- Jonas E. Salk - 1976
- Corliss Lamont - 1977
- Margaret E. Kuhn - 1978
- Edwin H. Wilson - 1979
- Andrei Sakharov - 1980
- Carl Sagan - 1981
- Helen Caldicott - 1982
- Lester A. Kirkendall - 1983
- Isaac Asimov - 1984
- John Kenneth Galbraith - 1985
- Faye Wattleton - 1986

- Margaret Atwood - 1987
- Leo Pfeffer - 1988
- Gerald A. Larue - 1989
- Ted Turner - 1990
- Lester R. Brown - 1991
- Kurt Vonnegut - 1992
- Richard D. Lamm - 1993
- Lloyd Morain - 1994
- Mary Morain - 1994
- Ashley Montagu - 1995
- Richard Dawkins - 1996
- Alice Walker - 1997
- Barbara Ehrenreich - 1998
- Edward O. Wilson - 1999
- William F. Schulz - 2000
- Stephen Jay Gould - 2001
- Steven Weinberg - 2002
- Sherwin T. Wine - 2003
- Daniel Dennett - 2004
- Murray Gell-Mann - 2005
- Steven Pinker - 2006
- Joyce Carol Oates - 2007
- Pete Stark - 2008
- PZ Myers - 2009
- Bill Nye - 2010
- Rebecca Goldstein - 2011
- Gloria Steinem - 2012
- Dan Savage - 2013
- Barney Frank - 2014
- Lawrence M. Krauss - 2015
- Jared Diamond - 2016
- Adam Savage - 2017

25.9 See also

- John Dewey
- Charles Francis Potter
- Bertrand Russell

25.10 References

[1] "About Humanism". Retrieved 2009-06-04.

[2] "AHLC mission statement". Retrieved 2012-03-22.

[3] "AHA Action Center". Retrieved 2012-03-22.

[4] "Local Group Information". Retrieved 2012-03-22.

[5] List of Publications americanhumanist.org (Retrieved 2011-10-01)

[6] Harris, Mark W., *The A to Z of Unitarian Universalism*, Scarecrow Press, 2009 ISBN 9780810863330

[7] Walter, Nicolas. *Humanism: What's in the Word?* (London: RPA/BHA/Secular Society Ltd, 1937), p.43.

[8] "Humanist Society's Early History". Retrieved 2012-03-28.

[9] "IHEU founding". Retrieved 2012-03-22.

[10] "Humanist Society's Services". Retrieved 2012-03-28.

[11] Djupe, Paul A. and Olsen, Laura R., "American Humanist Association", *Encyclopedia of American Religion and Politics", Infobase Publishing, 2014*

[12] "Security Check Required". *www.facebook.com*. Retrieved 2016-06-17.

[13] "American Humanist (@americnhumanist) | Twitter". *twitter.com*. Retrieved 2016-06-17.

[14] "Humanist Group Launches Initiatives for Racial Justice, Women's Equality and LGBTQ Rights". *American Humanist Association*. Retrieved 2016-06-17.

[15] "Mission - The Black Humanist Alliance - Menu". *blackhumanists.org*. Retrieved 2016-06-17.

[16] "Feminist Caucus Previous Work". Retrieved 2012-03-28.

[17] "The Feminist Caucus of the American Humanist Association". Retrieved 2012-09-30.

[18] "History - Feminist Humanist Alliance". *feministhumanists.org*. Retrieved 2016-06-16.

[19] "What We Do - Feminist Humanist Alliance". *feministhumanists.org*. Retrieved 2016-06-16.

[20] "About the LGBTQ Humanist Alliance". *LGBTQ Humanist Alliance*. Retrieved 2016-06-17.

[21] "Recent Projects". Retrieved 2012-03-28.

[22] "Humanist Charities Past Work". Archived from the original on January 13, 2013. Retrieved 2012-03-28.

[23] "Humanist Charities and Humanist Crisis Response Announce Merger". *American Humanist Association*. Retrieved 2016-06-16.

[24] "Humanist Disaster Recovery Drive". *Foundation Beyond Belief*. Retrieved 2016-06-16.

[25] "Humanist Disaster Recovery Teams". *Foundation Beyond Belief*. Retrieved 2016-06-16.

[26] "HDR Teams: South Carolina 2016". *Foundation Beyond Belief*. Retrieved 2016-06-16.

[27] Jones, Susan (2006-11-30). "'Humanists' Challenge Voting Booths in Churches". *crosswalk.com*. Retrieved 2012-03-28.

[28] "Voting in churches is constitutional, says Florida federal court.". *www.thefreelibrary.com*. 2009-09-01. Retrieved 2012-03-28.

[29] Brown, Matthew Hay. "Veterans' cross in Maryland at the center of national battle", Baltimore Sun, *May 25, 2014*

[30] Kuruvilla, Carol. "Humanists suing to tear down cross-shaped World War I memorial", *Daily News*, March 1, 2014

[31] Jacobs, Danny. "Bladensburg Peace Cross Sparks Legal War", *Daily Record*, March 1, 2014

[32] "Bladensburg residents argue over WWI memorial". *WJLA*. Retrieved 12 June 2015.

[33] "Mother Removes Cross Memorial After Dispute With Atheist Rights Group". *NBC Southern California*. Retrieved 12 June 2015.

[34] "SJC to hear case from atheist family". Retrieved 2012-11-18.

[35] "'Under God' is not discriminatory and will stay in pledge, judge says". *NJ.com*. Retrieved 12 June 2015.

[36] Laurie Goodstein, Atheist Groups Promote a Holiday Message: Join Us, *New York Times* (November 9, 2010).

[37] "'Why Believe in a God?' Ad Campaign Launches on D.C. Buses". *Fox News*. 2011-12-01.

[38] "American Humanist Association | 2009". Americanhumanist.org. Retrieved 2012-12-05.

[39] "Humanists replace billboard for the second time | News | KLEW CBS 3 - News, Weather and Sports - Lewiston, ID". Klewtv.com. Archived from the original on 2013-11-13. Retrieved 2012-12-05.

[40] Goodstein, Laurie (2010-11-09). "Atheists' Holiday Message: Join Us". *The New York Times*.

[41] "Humanists launch huge 'godless' ad campaign". *CNN*. 2010-11-09.

[42] "Humanists Launch "Naughty" Awareness Campaign". Americanhumanist.org. 2011-11-21. Retrieved 2012-12-05.

[43] "Ad Campaign Promoting Atheism Across U.S. Draws Ire and Protest - ABC News". Abcnews.go.com. 2010-12-05. Retrieved 2012-12-05.

[44] Duke, Barry (2012-11-14). "Getting too old for imaginary friends? American humanists have the answers". Freethinker.co.uk. Retrieved 2012-12-05.

[45] "Kids Without God ad campagin". Americanhumanist.org. 2012-11-13. Retrieved 2012-12-05.

[46] "National ad campaign promotes KidsWithoutGod.com on buses and online". Secular News Daily. 2012-11-14. Archived from the original on 2012-11-24. Retrieved 2012-12-05.

[47] "Atheist Ad Campaign Promotes Kids Without God; Already, Companies Are Refusing to Run Ads". Patheos.com. 2012-11-13. Retrieved 2012-12-05.

[48] National Day of Reason History

[49] "Positive Protest Against the Day of Prayer! (Center for Atheism, New York)". Retrieved 12 June 2015.

[50] Janet Zinc (May 6, 2010). "On National Day of Prayer, atheists renew call to end invocations at Tampa city meetings". *St. Petersburg Times*. Retrieved May 7, 2011.

[51] Minnesota Atheists Day of Reason Archived October 25, 2011, at the Wayback Machine.

[52] "National Day of Reason May 5, 2011". *WordPress.com*. Retrieved May 7, 2011.

[53] "American Humanist Association Sponsors Reason Rally, Largest Atheist Event in History". *American Humanist Association*. Retrieved 2016-06-17.

[54] "PHOTOS: Atheists Rally On National Mall For Political Change". *The Huffington Post*. 2012-03-24. Retrieved 2016-06-17.

[55] Woods, Benjamin Fearnow and Mickey. "Richard Dawkins Preaches to Nonbelievers at Reason Rally". *The Atlantic*. Retrieved 2016-06-17.

[56] "American Humanist Association to Co-Sponsor Reason Rally 2016, National Gathering of Humanists and Atheists". *American Humanist Association*. Retrieved 2016-06-17.

[57] "The Humanist of the Year". American Humanist Association. Archived from the original on 14 January 2013. Retrieved 1 May 2012.

25.11 External links

- Official website
- American Humanist Association at GuideStar
- "Edwin H. Wilson Papers of the American Humanist Association, 1913-1989". *Special Collections Research Center*. Carbondale: Southern Illinois University.
- Bladensburg War Memorial in Town of Bladensburg

Chapter 26

Humanists UK

Humanists UK,[1] known from 1967 until May 2017 as the **British Humanist Association** (**BHA**), is a charitable organisation which promotes Humanism and aims to represent "people who seek to live good lives without religious or superstitious beliefs" in the United Kingdom[2] by campaigning on issues relating to humanism, secularism, and human rights.

The charity also supports humanist and non-religious ceremonies in England and Wales, Northern Ireland, and the Crown dependencies[3] and maintains a national network of accredited celebrants for humanist funeral ceremonies, weddings, civil partnerships, and baby namings, in addition to a network of volunteers who provide likeminded support and comfort to non-religious people in hospitals and prisons. The current President of Humanists UK is Shappi Khorsandi and the Chief Executive is Andrew Copson. The association currently has 70 affiliated regional and special interest groups and claims a total of approximately 60,000 members and supporters.[4]

Humanists UK also has sections which run as staffed national humanist organisations in both Wales and Northern Ireland. Wales Humanists and Northern Ireland Humanists each have an advisory committee drawn from the membership and a development officer. Wales Humanists and Northern Ireland Humanists campaign on devolved issues in Cardiff and Belfast and work to expand the provision of humanist ceremonies, pastoral care, and support for teachers in those countries.[5][6]

26.1 Aims

The organisation's Articles of Association sets out its aims as:

- The advancement of Humanism, namely a non-religious ethical lifestance the essential elements of which are a commitment to human wellbeing and a reliance on reason, experience and a naturalistic view of the world.

- The advancement of education and in particular the study of and the dissemination of knowledge about Humanism and about the arts and science as they relate to Humanism.

- The promotion of equality and non-discrimination and the protection of human rights as defined in international instruments to which the United Kingdom is party, in each case in particular as relates to religion and belief.

- The promotion of understanding between people holding religious and non-religious beliefs so as to advance harmonious cooperation in society.[7]

The organisation also wishes to build itself as a sustainable and nationally-recognised organisation as a voice for non-religious people.[8]

26.2 History

The organisation was founded in 1896 by American Stanton Coit as the Union of Ethical Societies, which brought together existing ethical societies in Britain. In 1963 H. J. Blackham became the first Executive Director,[9] and the society became the British Humanist Association in 1967, during the Presidency of philosopher A.J. Ayer.

This transition followed a decade of discussions which nearly prompted a merger of the Union of Ethical Societies with the Rationalist Press Association and the South Place Ethical Society. In 1963 the discussions went as far as creating an umbrella Humanist Association of which Harold Blackham (later to become a President of the BHA) was the Executive Director. However, the BHA, the Rationalist Association and the South Place Ethical Society remain separate entities today and in 1967 the Union of Ethical Societies alone became the British Humanist Association.[10]

In the 1960s the BHA campaigned on the repeal of Sunday Observance laws and the reform of the 1944 Education

Act's clauses on religion in schools. More generally the BHA aimed to defend freedom of speech, support the elimination of world poverty and remove the privileges given to religious groups. It was claimed in 1977 that the BHA aimed "to make humanism available and meaningful to the millions who have no alternative belief."[10]

BHA supporters, including Andrew Copson and Polly Toynbee, taking part in a 'No Prayer Breakfast' event at the Labour Party Conference in 2012

At this time the BHA also supported a growing number of local communities, continuing today as a network of affiliated local humanist groups. A network of celebrants able to conduct non-religious funerals, weddings, naming ceremonies and same sex affirmations (before the law allowing gay civil partnerships) was also developed and continues today as Humanist Ceremonies.[11]

Social concerns persisted in the BHA's programme. The BHA was a co-founder of the Social Morality Council (now transmuted into the Norham Foundation), which brought together believers and unbelievers concerned with moral education and with finding agreed solutions to moral problems in society. The BHA was active in arguing for voluntary euthanasia and the right to obtain an abortion. It has always sought an "open society".

The BHA claimed that the rules on religious programming within the BBC constitute a "religious privilege"[12] and reserve particular criticism for the Thought for the Day slot on Radio 4's Today programme.[13] In April 2009 a "breakthrough" in the BHA's campaign saw Andrew Copson invited to participate as the first humanist representative in the BBC's new Standing Conference on Religion and Belief, replacing the Central Religious Advisory Committee.[14]

In May 2017, the organisation renamed itself Humanists UK. Its chief executive, Andrew Copson, said that the change followed "a long, evidence-driven process with focus groups of non-religious people across the UK and research involving over 4,000 of our supporters... Humanists UK represents not just a new logo, but a totally new, friendly look that captures the essence of humanism: open, inclusive, energetic, and modern, with people and their stories placed first and foremost...".[15]

26.3 Campaigns

26.3.1 Schools

The organisation opposes faith schools because "The majority of the evidence [...] points towards their being an unfair and unpopular part of our state education system which the majority of people in Britain want them phased out."[16] In addition, they claim that faith schools are "exclusive, divisive and counter intuitive to social cohesion" and blame religious admissions procedures for "creating school populations that are far from representative of their local populations in religious or socio-economic terms."[17]

While the organisation is opposed to faith schools receiving any state funding whatsoever, it supports the Fair Admissions Campaign which has a more limited scope because "it furthers our aims of ending religious discrimination and segregation in state schools; and secondly because we know how important this particular topic is."[18] The organisation campaigns for reform of Religious Education in the UK including a reformed subject covered by the national curriculum which is inclusive of non-religious viewpoints, such as "Belief and Values Education". They believe that "all pupils in all types of school should have the opportunity to consider philosophical and fundamental questions, and that in a pluralist society we should learn about each other's beliefs, including humanist ones".[19]

They also support humanist volunteers on the Standing Advisory Council on Religious Education which currently determines the Religious Education syllabus for each local authority. Educational issues have always featured prominently in BHA campaigns activities, including efforts to abolish daily worship in schools and to reform Religious Education so that it is objective, fair and balanced and includes learning about humanism as an alternative life stance.

The organisation opposes the teaching of creationism in schools. In September 2011, the BHA launched their "Teach evolution, not creationism" campaign,[20] which aimed to establish statutory opposition to creationism in the UK education system.[21] The Department for Education amended the funding agreement to allow the withdrawal of funding for free schools that teach creationism as established scientific fact.[22]

26.3.2 Constitutional reform

The organisation campaigns for a secular state, which it defines as "a state where public institutions are separate from religious institutions and treat all citizens impartially regardless of their religious or non-religious beliefs."[23] It points to issues such as the joint role of the British monarch (both Supreme Governor of the Church of England and Head of State), the reserved places for bishops in the House of Lords, the status of the Church of England (the officially established church[24]), and other "discriminations based on religion or belief within the system" such as those in education and Public Services.[25]

26.3.3 Ethical issues

Richard Dawkins accepting the Services to Humanism award at the British Humanist Association Annual Conference in 2012

The BHA has supported the rights for those who need assistance in ending their own lives, and lobbied parliament for a change in the law,[26] on behalf of Tony Nicklinson and Paul Lamb, in their 'Right to Die' legal cases.[27] In 2014, it intervened in a Supreme Court case in which the court stated it would rule again on a potential declaration of incompatibility between restrictions on the right to die and the Human Rights Act should Parliament fail to legislate decisively.[28]

Persistent campaigns include retaining the legality of abortion in Great Britain and securing its legalisation in Northern Ireland,[29] defending embryonic stem cell research for medical purposes,[30] challenging the state funding of homeopathy through the National Health Service,[31] and calling for consistent and humane law on the slaughter of animals.[32] It has also campaigned for 'opt-out' organ donor registers to improve the availability of life-saving organs in the UK; Wales became the first part of the UK to adopt such a register in 2015.[33]

The organisation also campaigns on marriage laws, demanding full equality for same-sex and humanist marriage ceremonies throughout the UK. The BHA had been providing same-sex wedding ceremonies for decades, and had strongly supported legalising same-sex marriage years in advance of eventual UK and Scottish legislation.[34][35] In 2013, it secured an amendment to the same sex marriage bill to require UK Government to consult on letting humanist celebrants conduct legal marriages. Though the consultation result strongly indicated that legalisation should go ahead, ministers has held off on using order-making powers to effect the change. It also campaigns for same-sex and humanist marriages in Northern Ireland.[36] In 2017, it supported a humanist couple to challenge Northern Ireland's refusal to give legal recognition to humanist marriages through the High Court in Belfast,[37] which resulted in legalisation of humanist marriages in Northern Ireland in June 2017.[38]

Many of its campaigns are based on free speech and human rights legislation and it has paid special attention to the Human Rights Act 1998.[39] In 2008, the blasphemy law was repealed, an issue over which the BHA had long campaigned.[40][41][42][43] It has called for unification of existing anti-discrimination legislation and has contributed to the Discrimination Law Review which developed the Equality Act 2010.[44]

26.3.4 Public awareness

Ariane Sherine and BHA Vice President Richard Dawkins at the bus campaign launch

On 21 October 2008, the British Humanist Association lent its official support to *Guardian* journalist Ariane Sherine as she launched a fundraising drive to raise money for the UK's first atheist advertising campaign, the Atheist Bus Campaign. The campaign aimed to raise funds to place the slogan "There's probably no God. Now stop worrying and enjoy your life" on the sides of 30 London buses for four weeks in January 2009. Expecting to raise £5,500 over six months, the atheist author Richard Dawkins agreed to match donations up to £5,500 to make £11,000 total.[45] The campaign raised over £153,000,[46] enabling a nationwide advertising campaign to be launched on 6 January 2009.

On 8 January 2009 Christian Voice announced they had made an official complaint to the Advertising Standards Authority asserting that the Atheist Bus slogan broke rules on "substantiation and truthfulness".[47] In total the ASA received 326 complaints about the campaign, with many claiming that the wording was offensive to the religious,[48] however the BHA disagreed with the complaint and commented on the plausibility of the ASA making a claim as to the "probability of God's existence".[49] Robert Winston criticised the campaign as "arrogant".[50] The ASA ruled that the slogan was not in breach of advertising code.[48]

In 2011, the British Humanist Association campaigned to get atheists, agnostics and other non-believers to tick the "no religion" box in response to the optional religion question in the 2011 census (as opposed to writing in either a joke religion like "Jedi" or ticking the religion one grew up in). The BHA believed the question was worded in such a way as to increase the number of currently non-religious or nominally religious people who list the religion they grew up in rather than their current religious views, and thus the results would have been skewed to make the country seem more religious than it actually is. The BHA believes that this supposed overstatement of religious belief creates a situation where "public policy in matters of religion and belief will unduly favour religious lobbies and discriminate against people who do not live their lives under religion".[51]

Posters for the campaign which used the slogan "If you're not religious, for God's sake say so" were refused by companies owning advertising hoardings in railway stations following advice from the Advertising Standards Authority who believe the adverts had "the potential to cause widespread and serious offence".[52]

The Census results for England and Wales showed that 14.1 million people, about a quarter of the entire population (25%), stated they had no religion at all, a rise of 6.4 million over the decade. The BHA said the fall in the number of Christians was "astounding", and calculated that they could be in a minority by 2018.[53]

Set up in 2010, the Resolution Revolution campaign aims to "[recast] the tired old New Year resolution – so often about breaking a negative habit – as a pledge to do something positive for others".[54] Participation is open to all and not restricted to humanists or the non-religious.[55]

> "New Year is a time for renewal – but beyond diets and gyms, not just for ourselves. Resolution Revolution is a humanist social action initiative, turning good intentions outwards to others. The more people that get involved, even in a small way, the bigger the impact is. Spending cuts don't make a cohesive society, but generous actions do."
> — *Polly Toynbee*[56]

In 2014, the BHA launched two public awareness campaigns. The first, called "That's Humanism!", was an Internet-based campaign revolving around four videos on humanist responses to ethics, happiness, death, and the scientific method, as narrated by its distinguished supporter, Stephen Fry. The videos, which were widely shared on social media, were intended to introduce non-religious people who were humanist in their outlook to the existence of a community of like-minded people living their lives on the basis of reason and empathy.[57] The second campaign, called "Thought for the Commute", was a London Underground campaign featuring posters depicting humanist responses from Virginia Woolf, George Eliot, Bertrand Russell and A.C. Grayling to the question "What's it all for?" The campaign intended to be a positive introduction to Humanism for commuters, as well as to highlight the exclusion of humanist voices from BBC slots such as *Thought for the Day*. After announcing that it intended to replicate it in other UK cities,[58][59] the campaign moved to bus posters in Birmingham, Manchester and Liverpool for four weeks in November and December 2014, this time depicting humanist responses from Jim Al-Khalili, Jawaharlal Nehru, Natalie Haynes and Russell once again.[60]

26.4 Organisation

26.4.1 Presidents

- Shappi Khorsandi (from 2016)
- Jim Al-Khalili (2013–2016)
- Polly Toynbee (2007–2013)
- Linda Smith (2004–2006; died 27 February 2006)
- Claire Rayner (1999–2004)
- Hermann Bondi (1981–1999)
- James Hemming (1977–1980)
- H. J. Blackham (1974–1977)
- George Melly (1972–1974)
- Edmund Leach (1970–1972)
- A.J. Ayer (1965–1970)
- Julian Huxley (1963–1965)

In April 2011 it was announced that Professor A.C. Grayling would succeed Polly Toynbee as president of the BHA in July 2011.[61] However, in June the BHA announced that Professor Grayling had decided not to take up that position, because of what he described as "controversy generated by activities in another area of my public life." The BHA stated that Polly Toynbee would continue as President until a new appointment was made later in 2011;[62] she remained President for a further 18 months until in December 2012 it was announced that physicist Jim Al-Khalili would become President in January 2013.[63]

26.4.2 Staff

- Andrew Copson – Chief Executive
- Richy Thompson – Director of Public Affairs and Policy[64]
- Luke Donnellan – Head of Education
- Isabel Russo - Head of Ceremonies
- Simon O'Donoghue – Head of Pastoral Support
- Catriona McLellan – Director of Operations
- Andrew West – Web Manager[65]

The charity also contracts an officer on the ground both in Wales and Northern Ireland, who coordinate its national sections (or branches) Wales Humanists and Northern Ireland Humanists. These officers are in turn supported by national committees of volunteers whose advice supports the charity's service delivery in those countries, strategic litigation, and lobbying on devolved issues at the Northern Ireland Assembly and National Assembly for Wales.[5][6]

26.4.3 Humanist celebrants

Main article: Humanist celebrant

Humanist equivalents of otherwise religious celebrations are conducted by humanist celebrants, trained and accredited by the British Humanist Association across England, Wales and Northern Ireland,[66][67] while the Humanist Society Scotland performs similar ceremonies in Scotland.[68] Non-religious funerals are legal within the UK;[69] over 8,000 funerals are carried out by humanist celebrants in England and Wales each year.[70] Between 600 and 900 weddings and 500 baby namings per year are also conducted by BHA-accredited celebrants.[71] In England, Wales, and Northern Ireland, a Humanist wedding or partnership ceremony must be supplemented by a process of obtaining a civil marriage or partnership certificate through a Register Office to be legally recognised, but can be led by a Humanist celebrant.[72]

The humanist funeral for former Welsh First Minister Rhodri Morgan at the Welsh Assembly was conducted by a Humanists UK celebrant, and was the first national funeral in the United Kingdom to be led by a humanist celebrant.[73][74]

The June 2017 wedding of Laura Lacole and Eunan O'Kane was the first legally recognised humanist marriage in Northern Ireland. The status of subsequent humanist marriages in Northern Ireland was reserved pending further legal judgement.

26.4.4 Pastoral carers

Humanists UK maintains a network of roughly 150 trained and accredited volunteers in England, Wales, and Northern Ireland who go into hospitals, hospices, prisons, universities, and other settings to provide like-minded comfort and support to non-religious people during times of distress, much like a traditional religious chaplain. This network is known as the Non-Religious Pastoral Support Network. The project was initiated by data evidence which suggested that non-religious patients and inmates often refused support from a chaplain if they were themselves non-religious.[75] Since 2014, the National Offender Management Service has recognised the legal right of prisoners to access non-religious pastoral carers,[76] and since 2015, NHS England has recommended that every hospital in England offers a voluntary or employed non-religious carer.[75]

26.4.5 Young Humanists

Young Humanists is the organisation's youth wing, which launched early in 2015 with a number of events in cities across the UK.[77][78][79]

26.4.6 Patrons

Numerous prominent people from the worlds of science, philosophy, the arts, politics, and entertainment are publicly aligned with Humanists UK, including Professor Alice Roberts, Tim Minchin, Stephen Fry, Matty Healy, Sandi Toksvig, Philip Pullman, and Dan Snow.[80]

In the 20th century, key members of the BHA's "advisory council" included Karl Popper, Vanessa Redgrave, Harold Pinter, E M Forster, Bertrand Russell, John Maynard Smith, Harry Kroto, and Barbara Wootton.[81]

Young Humanists logo.

26.4.7 Affiliations

Humanists UK is a member organisation of the International Humanist and Ethical Union, International Humanist and Ethical Youth Organisation and of the European Humanist Federation.[82]

In September 2008, Humanists UK joined with religious organisations, teachers' unions, and other human rights campaigns groups to found the Accord coalition, a diverse coalition made up of groups that oppose religious segregation in education.[83]

Humanist Students is a national federation of atheist, humanist, secularist, and skeptic societies at universities and is part of Humanists UK. Its elected delegates traditionally represent Humanists UK at International Humanist and Ethical Youth Organisation events alongside Young Humanists.[84]

Humanists UK has traditionally worked closely with the British Pregnancy Advisory Service, which was founded by the President of Birmingham Humanists, sexologist Martin Cole, in 1968. Humanists UK was a founding member of the BPAS "We Trust Women" coalition, which campaigns for the full decriminalisation of abortion throughout the UK.[85]

The organisation supports a network of affiliated humanist groups throughout the UK and aims to encourage local campaigning, charity work, socialising, and events on a local level, and provides resources to assist the creation and running of such groups. Some of these groups are formally partnered with Humanists UK, which entitles them to added staff and promotional support, while others maintain a looser affiliate agreement. As of 2017, the number of partner groups stands at 47, with 15 affiliates.[86]

The charity has also sponsored philosophical debates.[87] at HowTheLightGetsIn Festival.

26.5 Annual award

From 2011 the BHA presented an annual award for special contributions to Humanism. It is known as the Humanist of the Year Award, having been known prior to 2014 as the Services to Humanism Award. The award is usually presented during the BHA annual conference, with an exception being made in 2014 when an international award was presented at the World Humanist Congress. Past winners are:

- 2011 Philip Pullman, Services to Humanism Award[88]

- 2012 Richard Dawkins, Services to Humanism Award[89]

- 2013 Terry Pratchett, Humanist of the Year Award[90]

- 2014 Gulalai Ismail, International Humanist of the Year Award[91]

- 2015 Alice Roberts, Humanist of the Year Award[92]

- 2016 Lord Dubs, Humanist of the Year Award[93][94]

26.6 Criticism

Bryan Appleyard has criticised both the British Humanist Association and the National Secular Society for their campaign that the Scouts' Oath of Allegiance is religious discrimination.[95][96] Similar views were expressed by Deborah Orr.[97] Terry Sanderson of the National Secular Society argued that the oath should be modified, as it has in the past allowed non-Christians to become Scouts, so that the non-religious can participate in Scouting without having to compromise their human rights.[98] Both the Scout Association and Girlguiding UK subsequently amended their oaths to accommodate nonreligious young people.

26.7 See also

- All Party Parliamentary Humanist Group

- Atheist, Humanist, and Secular Students
- Humanist Society Scotland
- LGBT Humanists UK
- Non-Prophet Week
- Rationalist Association
- Separation of church and state (UK)

26.8 References

[1] "British Humanist Association becomes Humanists UK", *Politics.co.uk*, 22 May 2017. Retrieved 23 May 2017

[2] "About Us: The British Humanist Association". Retrieved 13 May 2011.

[3] "Humanist Ceremonies". Wedding Guide UK. Retrieved 14 October 2013.

[4] "About Us". *Humanists UK*. Retrieved 3 July 2017.

[5] "Humanists celebrate vision of a more secular Northern Ireland". *Humanists UK*. 11 March 2017. Retrieved 3 July 2017.

[6] "Nearly a million Welsh adults have a humanist approach to life, YouGov research shows". *Humanists UK*. 30 November 2016. Retrieved 3 July 2017.

[7] BHA Articles of Association

[8] British Humanist Association: Our aims. Retrieved 2 November 2013

[9] "Would life be better if we knew all the answers?". *New Humanist*: 3. March–April 2009.

[10] "Our History since 1896". Retrieved 2 November 2013.

[11] "Humanist Ceremonies | Non-Religious Ceremonies & Celebrations". Humanism.org.uk. Retrieved 14 October 2013.

[12] http://www.humanism.org.uk/campaigns/broadcasting

[13] http://www.humanism.org.uk/campaigns/broadcasting/thought-for-the-day

[14] "'Breakthrough' in religious broadcasting as humanist appointed to consultative committee". Ekklesia. Retrieved 14 October 2013.

[15] "BHA becomes Humanists UK", *Humanists UK*, 22 May 2017. Retrieved 23 May 2017

[16] Bingham, John (30 September 2013). "Faith schools protests dragging children into ideological 'battleground' – bishop". *The Telegraph*. London. Retrieved 5 November 2013.

[17] "'Faith' schools » British Humanist Association". Humanism.org.uk. Retrieved 14 October 2013.

[18] "Fair Admissions Campaign". British Humanist Association. Retrieved 2 November 2013.

[19] "Religious Education » British Humanist Association". Humanism.org.uk. Retrieved 14 October 2013.

[20] "Teach evolution, not creationism!". Retrieved 2 November 2013.

[21] "BHA: Government changes rules to require Free Schools to teach evolution". *British Humanist Association*. politics.co.uk. 30 November 2012. Retrieved 2 November 2013.

[22] Doward, Jamie (15 January 2012). "Richard Dawkins celebrates a victory over creationists". *The Guardian*. London.

[23] "Constitutional reform » British Humanist Association". Humanism.org.uk. Retrieved 14 October 2013.

[24] "The History of The Church of England". Cofe.anglican.org. Retrieved 14 October 2013.

[25] "Public service reform » British Humanist Association". Humanism.org.uk. Retrieved 14 October 2013.

[26] "Right-to-die campaigners Nicklinson and Lamb lose battle". *BBC News*. 31 July 2013. Retrieved 5 November 2013.

[27] Jones, Nelson (14 May 2013). "Assisted dying isn't contested on religious grounds – it's about power, paternalism and control". *New Statesman*. Retrieved 5 November 2013.

[28] "Assisted dying". *British Humanist Association*. Retrieved 20 October 2016.

[29] "Sexual and reproductive rights » British Humanist Association". Humanism.org.uk. Retrieved 13 October 2016.

[30] "Human tissues". *British Humanist Association*. Retrieved 20 March 2013.

[31] "Homeopathy » British Humanist Association". humanism.org.uk. Retrieved 13 October 2016.

[32] "Animal welfare". *British Humanist Association*. Retrieved 20 March 2013.

[33] "Organ donation". *British Humanist Association*. Retrieved 13 October 2016.

[34] Lennard, Derek (4 April 2014). "Galha's journey to success". *HumanistLife*. British Humanist Association. Retrieved 12 August 2015.

[35] Copson, Andrew (21 June 2015). "Humanists have always been champions of LGBT rights". *PinkNews*. Retrieved 12 August 2015.

[36] "Marriage laws » British Humanist Association". Humanism.org.uk. Retrieved 13 October 2016.

[37] "Friday hearing in Northern Ireland couple's case for legal humanist marriage". *Humanists UK*. 23 May 2017. Retrieved 26 May 2017.

[38] "Success! Couple win challenge to lack of legal recognition of humanist marriages in Northern Ireland". *Humanists UK*. 9 June 2017. Retrieved 9 June 2017.

[39] "Meaning of 'Public Authority' » British Humanist Association". Humanism.org.uk. Retrieved 14 October 2013.

[40] "Blasphemy » British Humanist Association". Humanism.org.uk. Retrieved 14 October 2013.

[41] "New Commons push for an end to Britain's blasphemy laws". Ekklesia. Retrieved 14 October 2013.

[42] "UK | Q & A: Blasphemy law". BBC News. 18 October 2004. Retrieved 14 October 2013.

[43] Peter Tatchell (11 July 1977). "Blasphemy Law is dead | Rationalist Association". Newhumanist.org.uk. Retrieved 14 October 2013.

[44] "Working towards a Single Equality Act: The Government's Equalities Review and Discrimination Law Review". The British Humanist Association.

[45] "'No God' slogans for city's buses". *BBC News*. 21 October 2008. Retrieved 27 April 2010.

[46] British Humanist Association. "Atheist Bus Campaign Official is fundraising for British Humanist Association". Justgiving.com. Retrieved 14 October 2013.

[47] "'No God' campaign draws complaint". *BBC News*. 8 January 2009. Retrieved 27 April 2010.

[48] "Atheist ads 'not breaking code'". *BBC News*. 21 January 2009. Retrieved 2 November 2013.

[49] "Stephen Green challenges Atheist Bus adverts: BHA responds » British Humanist Association". Humanism.org.uk. 8 January 2009. Retrieved 14 October 2013.

[50] "The One Show – Backstage Blog: What did you think of the show? (03/02/09)". BBC. 3 February 2009. Retrieved 14 October 2013.

[51] McManus, John (4 March 2011). "Humanist religious question census campaign launched". BBC News. Retrieved 2 November 2013.

[52] Travis, Alan (4 March 2011). "Humanist census posters banned from railway stations". *The Guardian*. London. Retrieved 2 November 2013.

[53] Booth, Robert (12 December 2012). "Census reveals decline of Christianity – Guardian". London: Guardian. Retrieved 12 December 2012.

[54] Knowles, Joanne (6 December 2010). "A new twist on New Year's resolutions". *The Guardian*. London.

[55] McManus, John (30 December 2010). "Humanists call for new year resolutions to help others". *BBC News*. Retrieved 5 November 2013.

[56] "Resolution Revolution encourages us to make 'social resolutions' this New Year". Retrieved 14 May 2011.

[57] Selby, Jenn (25 March 2014). "Stephen Fry's humanist secret of happiness is the best thing you'll watch today". *The Independent*. Retrieved 13 October 2014.

[58] Lusher, Adam (22 September 2014). "British Humanist Association launches Tube poster campaign as antidote to Thought for the Day". *The Independent*. Retrieved 13 October 2014.

[59] Bingham, John (22 September 2014). "Where is my train and where are we all going? Humanists urge commuters to ponder meaning of life". Retrieved 13 October 2014.

[60] "'Thought for the Commute' bus poster campaign launches across Manchester, Liverpool, and Birmingham". *British Humanist Association*. 13 November 2014. Retrieved 29 December 2014.

[61] "Anthony Grayling named new British Humanist Association President » British Humanist Association". Humanism.org.uk. 4 April 2011. Retrieved 14 October 2013.

[62] "Anthony Grayling has decided not to take office as BHA President". 17 June 2011. Retrieved 2 November 2013.

[63] "Jim Al-Khalili named President-elect of British Humanist Association". *British Humanist Association*. 14 December 2012. Retrieved 2 November 2013.

[64] "Richy Thompson appointed new BHA Director of Public Affairs and Policy". *British Humanist Association*. Retrieved 12 March 2017.

[65] "Staff >> British Humanist Association". *British Humanist Association*. Retrieved 2 November 2013.

[66] "Humanist Ceremonies". *British Humanist Association*. Retrieved 2 November 2013.

[67] "Humani: The humanist association of Northern Ireland". Retrieved 2 November 2013.

[68] "Humanist Ceremonies and Celebrants". *Humanist Society Scotland*. Retrieved 2 November 2013.

[69] "BBC – Religions – Atheism: Funerals". Retrieved 2 November 2013.

[70] Matthew Engelke (14 May 2012). "What is a good death? Ritual, whether religious or not, still counts". *The Guardian*. Retrieved 2 November 2013.

[71] Feldman, Sally (September–October 2013). "Oh, Happy Day". *New Humanist*: 14.

[72] "Humanist Weddings and Partnership Celebrations". *British Humanist Association*. Retrieved 2 November 2013.

[73] "Public humanist funeral for Rhodri Morgan at National Assembly for Wales". *Humanists UK*. 31 May 2017. Retrieved 31 May 2017.

[74] "Rhodri Morgan funeral to be held at the Senedd, Cardiff". *BBC Wales News*. 31 May 2017. Retrieved 31 May 2017.

[75] "Humanist Pastoral Support". *Humanists UK*. Retrieved 3 July 2017.

[76] "Chaplaincy and pastoral support". *Humanists UK*. Retrieved 3 July 2017.

[77] Fuller, Alice (25 March 2015). "Gemeinschaft schaffen im Humanismus". *Diesseits* (in German). Humanistischer Verband Deutschlands. Retrieved 10 August 2015.

[78] "Launch". *Young Humanists website*. BHA. 2015. Retrieved 10 August 2015.

[79] Alice Fuller (7 October 2014). "Welcome and thanks". *Facebook*. Retrieved 10 August 2015.

[80] "Patrons". Retrieved 19 February 2015.

[81] "20th century Humanism". *Humanists UK*. Retrieved 25 August 2017.

[82] "Our Affiliations". Retrieved 2 November 2013.

[83] "Our Members". *Accord Coalition*. 18 August 2010. Retrieved 2 November 2013.

[84] "AHS >> What We Do". Retrieved 2 November 2013.

[85] "We Trust Women: The Coalition". *We Trust Women*. British Pregnancy Advisory Service. Retrieved 25 August 2017.

[86] "Local Groups". *British Humanist Association*. Retrieved 4 March 2017.

[87] http://iai.tv/video/a-touch-of-evil

[88] "Philip Pullman awarded for services to Humanism". *British Humanist Association*. Retrieved 7 March 2017.

[89] "Richard Dawkins awarded for services to Humanism". *British Humanist Association*. Retrieved 7 March 2017.

[90] "BHA mourns patron Terry Pratchett". *British Humanist Association*. Retrieved 7 March 2017.

[91] "Gulalai Ismail wins International Humanist of the Year Award". *British Humanist Association*. Retrieved 8 March 2017.

[92] "Alice Roberts wins Humanist of the Year at BHA Annual Conference 2015". *British Humanist Association*. Retrieved 7 March 2017.

[93] "Lord Dubs awarded Humanist of the Year 2016". *British Humanist Association*. Retrieved 7 March 2017.

[94] "Lord Alf Dubs awarded Humanist of the year 2016". *Politics*. Retrieved 7 March 2017.

[95] Petre, Jonathan (12 April 2008). "Scout's oath 'is religious discrimination'". *Daily Telegraph*. London.

[96] Appleyard, Bryan (1 February 2008). "Oh Grow Up!".

[97] Orr, Deborah (2 February 2008). "Labour promised social justice along with economic competence. It failed ...". *The Independent*. London.

[98] Sanderson, Terry (22 February 2012). "Scouting without God". *The Guardian*. London.

26.9 External links

- Humanists UK website
- Archives of the BHA at Bishopsgate Library

Chapter 27

Humanistischer Verband Deutschlands

Warning: Page using Template:Infobox organization with unknown parameter "bgcolor" (this message is shown only in preview).
Warning: Page using Template:Infobox organization with unknown parameter "fgcolor" (this message is shown only in preview).

The **Humanistischer Verband Deutschlands** (English: *Humanist Association of Germany*) is an organization to promote and spread a secular humanist worldview and an advocate for the rights of nonreligious people. It was founded 1993 in Berlin and counts about 20,000 members. Its president is the philosopher and political scientist Frieder Otto Wolf. The HVD is a member of the International Humanist and Ethical Union and the European Humanist Federation.

HVD headquarters in Berlin.

27.1 Aims

The HVD is committed to secularism, human rights, democracy, egalitarianism and mutual respect. The association works for an open and inclusive society with freedom of belief and speech, and for an end to the privileged position of religion and/or churches in law, education, broadcasting and wherever else it occurs.

27.2 Activities

The HVD is a provider of humanist and non-religious ceremonies in many regions,[1] and maintains more than three dozen Kindergartens in Berlin, Nuremberg, Fürth, Regensburg, Hanover and Braunschweig. It runs a humanistic primary school in Fürth,[2] and a second school is set to open its doors in Munich soon. In Berlin, the HVD is responsible for the optional school subject *Humanistische Lebenskunde*, with about 54,000 participants in 2013. The HVD supports a youth organization *Junge Humanistinnen & Humanisten in Deutschland* (JuHu).[3]

The association is a well-known consultant for Advance health care directives, runs hospices and offers consultation on family planning and abortion.

27.3 References

[1] http://www.humanismus.de/feierkultur

[2] http://www.humanistische-schule.de/

[3] Website of JuHu

27.4 External links

- HVD Website

Chapter 28

Humanist Society Scotland

Humanist Society Scotland (HSS) is a Scottish registered charity that promotes humanist views. It is a member of the European Humanist Federation and the International Humanist and Ethical Union.

28.1 History and aims

The **Humanist Society of Scotland** was formed in 1989 out of an association of local humanist groups around Scotland, the Society's objective is "to represent the views of people in Scotland who wish to lead good and worthwhile lives guided by reason and compassion rather than religion or superstition", and to provide a distinct Scottish voice in complement to the British Humanist Association.[2] In 2010, the Society reported having 7,000 members.[2] In 2011, reflective of this growth, it became a charitable company limited by guarantee, and in 2012 it dropped the "of" from its name to become Humanist Society Scotland.[2]

The Society also claims to have a representative role for the 28% of Scots (at the 2001 census) who identify themselves as having no religion. The Society believes that the wording of the census question tends to inflate the numbers of people identifying themselves as religious who were brought up in a tradition of religious belief but who either no longer believe or who have significant doubts. The Society has campaigned to persuade the Registrar General to amend the question for the 2011 census.

In April 2015, Gordon MacRae was appointed Chief Executive.[1]

The official symbol of the HSS is an adaptation of the Happy Human symbol which incorporates the Saltire. The Society also publishes a quarterly magazine, *Humanitie*.

28.2 Campaigns

The Society campaigns for a secular state in Scotland, and to abolish religious privilege. Its main efforts have concentrated on seeking to allow legal humanist weddings, and to secularise state education.

28.2.1 Weddings

In January 2001, the Society lodged a petition with the Scottish Parliament calling for the Marriage (Scotland) Act 1977 to be amended to allow legal humanist wedding ceremonies, alongside religious and civil ones.[3] Although the Act was not amended, section 12 of the Act allows the Registrar General for Scotland to authorise temporary additional celebrants.[4] In 2005, the Registrar agreed to authorise 12 celebrants from the Humanist Society, in part because of a concern that allowing legal religious weddings but not legal humanist ones might not be consistent with the right to "freedom of thought, conscience and religion", which includes non-religious belief, in Article 9 of the European Convention on Human Rights. The first legal humanist wedding took place at Edinburgh Zoo on 18 June 2005 between Karen Watts (from Ireland) and Martin Reijns (from the Netherlands).

Humanist weddings have since becoming increasingly popular and, in 2010, with over 70 celebrants authorised to conduct them 2,092 legal humanist weddings took place in Scotland, becoming the third most popular form of Wedding in Scotland after Registrars and the Church of Scotland. The Society organises training, mentoring and performance reviews of celebrants, and submits names of celebrants to the Registrar General annually for authorisation. Prior to the Marriage and Civil Partnership (Scotland) Act 2014, the Society performed a similar role for celebrants to conduct same-sex commitment ceremonies and weddings, although formal authorisation by the Registrar is not required for these ceremonies since they had no effect on the legal status of individuals concerned. Since the legalisation of same-sex marriages, both sets of same-sex and opposite-sex marriages are treated the same way.

In 2017 the society received official status from the Scottish Parliament as the first non-religious body that could solemnise weddings.[5] Scotland is currently the only part of the United Kingdom where a Humanist marriage is considered a legal wedding by the state.

28.2.2 Other issues

The Society supports both the End of Life Assistance (Scotland) Bill, introduced in the Scottish Parliament by Margo MacDonald MSP,[6] and the campaign for equal marriage in Scotland to allow same sex couples to be legally married as an alternative to civil partnerships as well as allowing opposite sex couples access to civil partnerships. In 2013 the group, along with the Edinburgh Secular Society, started a campaign against religious representation on council education committees in Scotland.[7]

28.3 References

[1] "Senior Management Team". *HSS website*. Humanist Society Scotland. Retrieved 5 September 2015.

[2] "About HSS". *Humanist Society Scotland*. Retrieved 17 May 2017.

[3] Petition to the Scottish Parliament to End Discrimination in the Marriage Law of Scotland

[4] Marriage (Scotland) Act 1977 (c.15)

[5] "Holyrood gives Humanist Society fantastic news over appointing celebrants". Aberdeen Evening Express. 14 February 2017. Retrieved 26 May 2017.

[6] End of Life Assistance (Scotland) Bill, Humanist Society Scotland website.

[7] "Campaigners call for end to religious 'interference' in schools". STV. 19 May 2013. Retrieved 20 August 2013.

28.4 External links

- Official website

Chapter 29

Norwegian Humanist Association

Warning: Page using Template:Infobox organization with unknown parameter "native_name HEF" (this message is shown only in preview).

The **Norwegian Humanist Association** (Norwegian: *Human-Etisk Forbund*, HEF) is one of the largest secular humanist associations in the world, with 84,300 members.[1] Those members constitute 1.7% of the national population of 5,038,100,[2] making the HEF by far the largest such association in the world in proportion to population.

Founded in 1956, the HEF is a member of the International Humanist and Ethical Union (IHEU). The Norwegian Humanist Association is an organization for people who base their ethics on human, not religious values. Most members are agnostics or atheists. HEF supports the following statement of the IHEU:

> Humanism is a democratic, non-theistic and ethical life stance which affirms that human beings have the right and responsibility to give meaning and shape to their lives and therefore reject supernatural views of reality.

Former HEF secretary general, Levi Fragell, was president of the IHEU (1988–2003) and is Chair of IHEU's Committee for Growth and Development.[3] In June, 2007 Åse Kleveland became chairman of the Board of the organization.

According to its bylaws, the organization works to ensure access to humanist ceremonies and spreading knowledge of humanism. The organization previously also worked for a separation of church and state (the Evangelical-Lutheran Church of Norway was the state church of Norway until 2012).[4] A civil confirmation organized by HEF has gained popularity among the Norwegian young during recent years. About 17 percent of Norwegian 15-year-olds are now taking part in HEF's civil confirmation.

On July 9, 2006 a prominent member of the Norwegian Humanist Association, Mr. Jens Brun-Pedersen, called for prime minister Jens Stoltenberg to advocate the separation of church and state. He argued that the second article of the constitution of Norway which defines the "Evangelical-Lutheran Religion" as "the official religion of the State" and the 12th article of the constitution which requires half of the ministers of the cabinet to be members of the state church is discriminatory, and that Norway can't criticise countries advocating sharia law when the constitution favours Lutheran members of society.[5]

The Norwegian Humanist Association hosted the 18th World Humanist Congress of the IHEU. The Congress was held at the Oslo Congress Center in Oslo, Norway, on the 12–14 August 2011. The theme was Humanism and Peace. HEF also hosted the 3rd in 1962 and the 9th in 1986.

A youth wing of the HEF, Norwegian Humanist Youth (*Humanistisk Ungdom*), was founded in August 2007. The current President of the Board is Anders Garbom Backe (for the period 2013-2014).

29.1 See also

- Norwegian Heathen Society

29.2 References

[1] "Trus- og livssynssamfunn utanfor Den norske kyrkja, 1. januar 2013". *Statistics Norway* (in Norwegian Bokmål). Retrieved 2014-03-17.

[2] "Population". *Statistics Norway*. Retrieved 2012-09-01.

[3] https://web.archive.org/web/20061003183246/http://www.iheu.org/node/1230. Archived from the original on October 3, 2006. Retrieved December 27, 2006. Missing or empty |title= (help)

[4] Finngeir Hiorth (1997): Church and State in Norway

[5] *Visionary or missionary?* - Jens Brun-Pedersen, Dagbladet July 9, 2006

29.3 External links

- Official website of the Norwegian Humanist Association
- Member magazine - Fri tanke
- Norwegian Humanist Youth

Chapter 30

Alternatives to the Ten Commandments

Several **alternatives to the Ten Commandments** have been promulgated by different persons and groups, which intended to improve on the lists of laws known as the Ten Commandments that appear in the Bible.

30.1 Examples

30.1.1 Christopher Hitchens

Christopher Hitchens was an English American author, columnist, essayist, orator, religious and literary critic, social critic and journalist.

His new Ten Commandments are:[1][2]

1. Do not condemn people on the basis of their ethnicity or their color.

2. Do not ever even think of using people as private property, or as owned, or as slaves.

3. Despise those who use violence or the threat of it in sexual relations.

4. Hide your face and weep if you dare to harm a child.

5. Do not condemn people for their inborn nature - why would God create so many homosexuals only in order to torture and destroy them?

6. Be aware that you, too, are an animal, and dependent on the web of nature. Try and think and act accordingly.

7. Do not imagine that you can escape judgment if you rob people with a false prospectus rather than with a knife.

8. Turn off that fucking cell phone - you can have no idea how unimportant your call is to us.

9. Denounce all jihadists and crusaders for what they are: psychopathic criminals with ugly delusions. And terrible sexual repressions.

10. Be willing to renounce any god or any faith if any holy commandments should contradict any of the above.

- In short: Don't swallow your moral code in tablet form.

30.1.2 Richard Dawkins

Richard Dawkins is an English ethologist, evolutionary biologist and author.

These are the alternative to the Ten Commandments, cited by Dawkins in his book *The God Delusion*:[3][4]

1. Do not do to others what you would not want them to do to you.

2. In all things, strive to cause no harm.

3. Treat your fellow human beings, your fellow living things, and the world in general with love, honesty, faithfulness and respect.

4. Do not overlook evil or shrink from administering justice, but always be ready to forgive wrongdoing freely admitted and honestly regretted.

5. Live life with a sense of joy and wonder.

6. Always seek to be learning something new.

7. Test all things; always check your ideas against the facts, and be ready to discard even a cherished belief if it does not conform to them.

8. Never seek to censor or cut yourself off from dissent; always respect the right of others to disagree with you.

9. Form independent opinions on the basis of your own reason and experience; do not allow yourself to be led blindly by others.

10. Question everything.

Dawkins uses these proposed commandments to make a larger point that "it is the sort of list that any ordinary, decent person today would come up with". He then adds four more of his own devising:

- Enjoy your own sex life (so long as it damages nobody else) and leave others to enjoy theirs in private whatever their inclinations, which are none of your business.

- Do not discriminate or oppress on the basis of sex, race or (as far as possible) species.

- Do not indoctrinate your children. Teach them how to think for themselves, how to evaluate evidence, and how to disagree with you.

- Value the future on a timescale longer than your own.

30.1.3 Bertrand Russell

Bertrand Russell was a British philosopher, logician, mathematician, historian, writer, social critic, political activist and Nobel laureate.

He formulated these ten commandments:[5]

1. Do not feel absolutely certain of anything.

2. Do not think it worth while to proceed by concealing evidence, for the evidence is sure to come to light.

3. Never try to discourage thinking for you are sure to succeed.

4. When you meet with opposition, even if it should be from your husband or your children, endeavour to overcome it by argument and not by authority, for a victory dependent upon authority is unreal and illusory.

5. Have no respect for the authority of others, for there are always contrary authorities to be found.

6. Do not use power to suppress opinions you think pernicious, for if you do the opinions will suppress you.

7. Do not fear to be eccentric in opinion, for every opinion now accepted was once eccentric.

8. Find more pleasure in intelligent dissent than in passive agreement, for, if you value intelligence as you should, the former implies a deeper agreement than the latter.

9. Be scrupulously truthful, even if the truth is inconvenient, for it is more inconvenient when you try to conceal it.

10. Do not feel envious of the happiness of those who live in a fool's paradise, for only a fool will think that it is happiness.

30.1.4 Bayer and Figdor's Ten Non-Commandments

As detailed in the book *Atheist Mind, Humanist Heart: Rewriting the Ten Commandments for the Twenty-first Century* by Lex Bayer and the Stanford Humanist Chaplain John Figdor, it is devoted to the subject of creating a secular alternative to the Ten Commandments and encouraging readers to formulate and discover their own list of beliefs.[6][7]

1. The world is real, and our desire to understand the world is the basis for belief.

2. We can perceive the world only through our human senses.

3. We use rational thought and language as tools for understanding the world.

4. All truth is proportional to the evidence.

5. There is no God.

6. We all strive to live a happy life. We pursue things that make us happy and avoid things that do not.

7. There is no universal moral truth. Our experiences and preferences shape our sense of how to behave.

8. We act morally when the happiness of others makes us happy.

9. We benefit from living in, and supporting, an ethical society.

10. All our beliefs are subject to change in the face of new evidence, including these.

30.1.5 The Atheists' New Ten Commandments

These are the ten winning beliefs of the Rethink Prize, a crowdsourcing competition to rethink the Ten Commandments. The contest drew more than 2,800 submissions from 18 countries and 27 U.S. states. Winners were selected by a panel of judges. [8][9]

1. Be open-minded and be willing to alter your beliefs with new evidence.

2. Strive to understand what is most likely to be true, not to believe what you wish to be true.

30.1. EXAMPLES

3. The scientific method is the most reliable way of understanding the natural world.
4. Every person has the right to control of their body.
5. God is not necessary to be a good person or to live a full and meaningful life.
6. Be mindful of the consequences of all your actions and recognize that you must take responsibility for them.
7. Treat others as you would want them to treat you, and can reasonably expect them to want to be treated. Think about their perspective.
8. We have the responsibility to consider others, including future generations.
9. There is no one right way to live.
10. Leave the world a better place than you found it.

30.1.6 Ten Indian Commandments

The Bird Clan of East Central Alabama has the Ten Indian Commandments.[10]

1. Remain close to the Great Spirit.
2. Show great respect for your fellow beings.
3. Give assistance and kindness wherever needed.
4. Be truthful and honest at all times.
5. Do what you know to be right.
6. Look after the well being of mind and body.
7. Treat the earth and all that dwell there on with respect.
8. Take full responsibility for your actions.
9. Dedicate a share of your efforts to the greater good.
10. Work together for the benefit of all man kind.

30.1.7 Selig Starr

Orthodox Jewish Rabbi Selig Starr's formulation of the Ten Modern Important Commandments was not intended to be an alternate to The Ten Commandments. Selig Starr's formulation of "Ten Modern Important Commandments:"[11]

1. Remember to embrace equally all the three fundamentally Jewish loves - love of God, Torah, and the Jewish People.

2. Remember not to minimize any one of [the above] in any way whatsoever.
3. Remember that time is the most precious element in your mental treasury; therefore, spend it very carefully.
4. Remember not to spend your spiritual harvest time more on one crop than on the others.
5. Remember that personal flattery is your worst enemy, while expert criticism is your best friend.
6. Remember that human behavior must be analysed and comprehended; some people are acting as spiders, while others [behave] like flies enwrapped in the deadly silken threads of the spiders. Avoid the company of either one of them.
7. Remember that six million of American Jews are waiting for your spiritual Orthodox guidance. Do not disappoint them.
8. Remember that you have been trained to fight two internal enemies, ignorance and confusion, the latter the greater.
9. Remember that our spiritual Orthodox survival depends solely on the ability of our leaders to rescue the wine while the barrel is broken, to watch over our Torah inheritance while the ghetto walls have been eliminated.
10. Remember that destiny has bestowed upon the incoming Jewish generation the greatest among the most precious blessings, and at the same time, imposed upon our selected Talmudic scholars the greatest responsibilities to be sincere servants of God, Torah, and Israel (as an independent state and everlasting people).

30.1.8 Ted Kaczynski

Kaczynski (also known as the "Unabomber"), an American domestic terrorist, mathematical prodigy and an advocate of a nature-centered form of anarchism, proposed the Six Principles:[12]

1. Do not harm anyone who has not previously harmed you, or threatened to do so.
2. You can harm others in order to forestall harm with which they threaten you, or in retaliation for harm that they have already inflicted on you.
3. One good turn deserves another: If someone has done you a favor, you should be willing to do her or him a comparable favor if and when he or she should need one.

4. The strong should have consideration for the weak.

5. Do not lie.

6. Abide faithfully by any promises or agreements that you make.

He considers the Six Principles to be an innate sense of fairness, present in all people even if they may be overrun by culture.

30.1.9 Summum

Summum is an informal gathering of people registered as a tax exempt organization in the state of Utah, U.S., in 1975.

Summum contradicts the historical Biblical account of the Ten Commandments by claiming that prior to returning with the Commandments, Moses descended from Mount Sinai with a first set of tablets inscribed with seven principles they call aphorisms.

According to the group, the seven principles are:[13][14]

1. SUMMUM is MIND, thought; the universe is a mental creation.

2. As above, so below; as below, so above.

3. Nothing rests; everything moves; everything vibrates.

4. Everything is dual; everything has an opposing point; everything has its pair of opposites; like and unlike are the same; opposites are identical in nature, but different in degree; extremes bond; all truths are but partial truths; all paradoxes may be reconciled.

5. Everything flows out and in; everything has its season; all things rise and fall; the pendulum swing expresses itself in everything; the measure of the swing to the right is the measure of the swing to the left; rhythm compensates.

6. Every cause has its effect; every effect has its cause; everything happens according to Law; Chance is just a name for Law not recognized; there are many fields of causation, but nothing escapes the Law of Destiny.

7. Gender is in everything; everything has its masculine and feminine principles; Gender manifests on all levels.

30.1.10 Hu Jintao

Eight Honors and Eight Shames by Hu Jintao is a set of moral concepts developed by former General Secretary Hu Jintao for the citizens of the People's Republic of China. They follow a Chinese tradition of making lists and trying to give guidance to the people in all regards of life. The state focus is not an error, but tries to bind people to the morality of the (communist) state and to further it's glory.

1. Love the country; do it no harm.

2. Serve the people; never betray them

3. Follow science; discard ignorance

4. Be diligent;not indolent.

5. Be united, help each other; make no gains at others' expense.

6. Be honest and trustworthy; do not sacrifice ethics for profit.

7. Be disciplined and law-abiding; not chaotic and lawless.

8. Live plainly, work hard; do not wallow in luxuries and pleasures.

30.2 See also

- Moral Code of the Builder of Communism

30.3 References

[1] Hitchens, Christopher, "The New Commandments", *Vanity Fair*, April 2010

[2] "Christopher Hitchens reading the Vanity Fair piece in video format". Commonsenseatheism.com. 2010-03-08. Retrieved 2012-12-10.

[3] Dawkins, Richard (2006). *The God Delusion*. Boston: Houghton Mifflin. p. 406. ISBN 0-618-68000-4.

[4] "The New Ten Commandments". Retrieved 2009-03-30.

[5] Bertrand Russell (1951). "A Liberal Decalogue". Panarchy – A Gateway to Selected Documents and Web Sites. Retrieved 2014-12-29.

[6] Kimberly Winston (November 20, 2014). "10 Commandments for atheists: a guide for nonbelievers who want to explore their values". The Washington Post. Retrieved February 15, 2015.

[7] "Atheist Mind, Humanist Heart". Retrieved February 15, 2015.

[8] Daniel Burke (December 20, 2014). "Behold, atheists' new Ten Commandments". CNN. Retrieved February 15, 2015.

[9] "The Rethink Prize". Retrieved February 15, 2015.

[10] "The Ten Indian Commandments". Retrieved 30 November 2016.

[11] Ohr Shmuel. Hebrew Theological College. Skokie, IL. 1996

[12] "Theodore J. Kaczynski. The Road to Revolution. Ripped by: TOTAL FREEDOM, 2009" (PDF). Retrieved 2012-12-09.

[13] "The Aphorisms of Summum and the Ten Commandments". Retrieved 30 November 2016.

[14] "Principles of Creation". Retrieved 30 November 2016.

30.4 External links

- 10 Humanist Commandments
- Eight I'd Really Rather You Didn'ts

Chapter 31

Maslow's hierarchy of needs

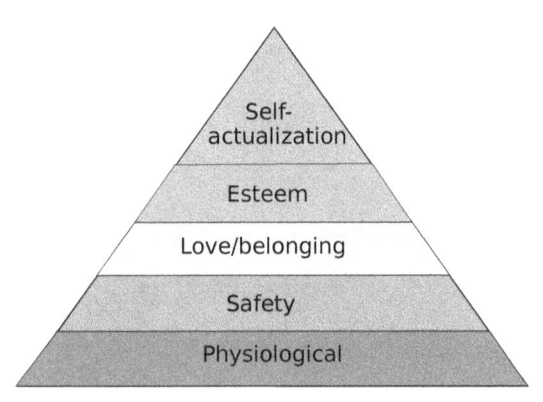

Maslow's hierarchy of needs, represented as a pyramid with the more basic needs at the bottom[1]

Maslow's hierarchy of needs is a theory in psychology proposed by Abraham Maslow in his 1943 paper "A Theory of Human Motivation" in *Psychological Review*.[2] Maslow subsequently extended the idea to include his observations of humans' innate curiosity. His theories parallel many other theories of human developmental psychology, some of which focus on describing the stages of growth in humans. Maslow used the terms "physiological", "safety", "belonging" and "love", "esteem", "self-actualization", and "self-transcendence" to describe the pattern that human motivations generally move through. The goal of Maslow's Theory is to attain the sixth level of stage: self transcendent needs.[3]

Maslow studied what he called exemplary people such as Albert Einstein, Jane Addams, Eleanor Roosevelt, and Frederick Douglass rather than mentally ill or neurotic people, writing that "the study of crippled, stunted, immature, and unhealthy specimens can yield only a cripple psychology and a cripple philosophy."[4]:236 Maslow studied the healthiest 1% of the college student population.[5]

Maslow's theory was fully expressed in his 1954 book *Motivation and Personality*.[4] The hierarchy remains a very popular framework in sociology research, management training[6] and secondary and higher psychology instruction.

31.1 Hierarchy

Maslow's hierarchy of needs is often portrayed in the shape of a pyramid with the largest, most fundamental levels of needs at the bottom and the need for self-actualization and self-transcendence at the top.[1][7]

The most fundamental and basic four layers of the pyramid contain what Maslow called "deficiency needs" or "d-needs": esteem, friendship and love, security, and physical needs. If these "deficiency needs" are not met – with the exception of the most fundamental (physiological) need – there may not be a physical indication, but the individual will feel anxious and tense. Maslow's theory suggests that the most basic level of needs must be met before the individual will strongly desire (or focus motivation upon) the secondary or higher level needs. Maslow also coined the term "metamotivation" to describe the motivation of people who go beyond the scope of the basic needs and strive for constant betterment.[8]

The human brain is a complex system and has parallel processes running at the same time, thus many different motivations from various levels of Maslow's hierarchy can occur at the same time. Maslow spoke clearly about these levels and their satisfaction in terms such as "relative", "general", and "primarily". Instead of stating that the individual focuses on a certain need at any given time, Maslow stated that a certain need "dominates" the human organism.[4] Thus Maslow acknowledged the likelihood that the different levels of motivation could occur at any time in the human mind, but he focused on identifying the basic types of motivation and the order in which they should be met.

31.1.1 Physiological needs

Physiological needs are the physical requirements for human survival. If these requirements are not met, the human body cannot function properly and will ultimately fail. Physiological needs are thought to be the most important; they should be met first.

Air, water, and food are metabolic requirements for survival in all animals, including humans. Clothing and shelter provide necessary protection from the elements. While maintaining an adequate birth rate shapes the intensity of the human sexual instinct, sexual competition may also shape said instinct.[2]

31.1.2 Safety needs

Once a person's physiological needs are relatively satisfied, their safety needs take precedence and dominate behavior. In the absence of physical safety – due to war, natural disaster, family violence, childhood abuse, etc. – people may (re-)experience post-traumatic stress disorder or transgenerational trauma. In the absence of economic safety – due to economic crisis and lack of work opportunities – these safety needs manifest themselves in ways such as a preference for job security, grievance procedures for protecting the individual from unilateral authority, savings accounts, insurance policies, disability accommodations, etc. This level is more likely to be found in children as they generally have a greater need to feel safe.

Safety and Security needs include:

- Personal security
- Financial security
- Health and well-being
- Safety net against accidents/illness and their adverse impacts

31.1.3 Social belonging

After physiological and safety needs are fulfilled, the third level of human needs is interpersonal and involves feelings of belongingness. This need is especially strong in childhood and it can override the need for safety as witnessed in children who cling to abusive parents. Deficiencies within this level of Maslow's hierarchy – due to hospitalism, neglect, shunning, ostracism, etc. – can adversely affect the individual's ability to form and maintain emotionally significant relationships in general, such as:

- Friendships
- Intimacy
- Family

According to Maslow, humans need to feel a sense of belonging and acceptance among their social groups, regardless whether these groups are large or small. For example, some large social groups may include clubs, co-workers, religious groups, professional organizations, sports teams, and gangs. Some examples of small social connections include family members, intimate partners, mentors, colleagues, and confidants. Humans need to love and be loved – both sexually and non-sexually – by others.[2] Many people become susceptible to loneliness, social anxiety, and clinical depression in the absence of this love or belonging element. This need for belonging may overcome the physiological and security needs, depending on the strength of the peer pressure.

31.1.4 Esteem

All humans have a need to feel respected; this includes the need to have self-esteem and self-respect. Esteem presents the typical human desire to be accepted and valued by others. People often engage in a profession or hobby to gain recognition. These activities give the person a sense of contribution or value. Low self-esteem or an inferiority complex may result from imbalances during this level in the hierarchy. People with low self-esteem often need respect from others; they may feel the need to seek fame or glory. However, fame or glory will not help the person to build their self-esteem until they accept who they are internally. Psychological imbalances such as depression can hinder the person from obtaining a higher level of self-esteem or self-respect.

Most people have a need for stable self-respect and self-esteem. Maslow noted two versions of esteem needs: a "lower" version and a "higher" version. The "lower" version of esteem is the need for respect from others. This may include a need for status, recognition, fame, prestige, and attention. The "higher" version manifests itself as the need for self-respect. For example, the person may have a need for strength, competence, mastery, self-confidence, independence, and freedom. This "higher" version takes precedence over the "lower" version because it relies on an inner competence established through experience. Deprivation of these needs may lead to an inferiority complex, weakness, and helplessness.

Maslow states that while he originally thought the needs of humans had strict guidelines, the "hierarchies are interrelated rather than sharply separated".[4] This means that esteem and the subsequent levels are not strictly separated; instead, the levels are closely related.

31.1.5 Self-actualization

Main article: Self-actualization

"What a man can be, he must be."[4]:91 This quotation forms the basis of the perceived need for self-actualization. This level of need refers to what a person's full potential is and the realization of that potential. Maslow describes this level as the desire to accomplish everything that one can, to become the most that one can be.[4]:92 Individuals may perceive or focus on this need very specifically. For example, one individual may have the strong desire to become an ideal parent. In another, the desire may be expressed athletically. For others, it may be expressed in paintings, pictures, or inventions.[4]:93 As previously mentioned, Maslow believed that to understand this level of need, the person must not only achieve the previous needs, but master them.

31.1.6 Self-transcendence

In his later years, Maslow explored a further dimension of needs, while criticizing his own vision on self-actualization.[9] The self only finds its actualization in giving itself to some higher goal outside oneself, in altruism and spirituality.[10] "Transcendence refers to the very highest and most inclusive or holistic levels of human consciousness, behaving and relating, as ends rather than means, to oneself, to significant others, to human beings in general, to other species, to nature, and to the cosmos" (*Farther Reaches of Human Nature*, New York 1971, p. 269).

31.2 Application to Nursing

Nurses can apply Maslow's hierarchy of basic needs in the assessment, planning, implementation, and evaluation of patient care. It helps the nurse identify unmet needs as they become health care needs, and allows the nurse to locate the patient on the health-illness continuum and to incorporate the human dimensions and health models into meeting needs.

31.2.1 The Human Dimensions and Basic Human Needs

All basic human needs are interrelated and may require nursing actions at more than one level at a given time. For example, in caring for a person coming into the emergency department with a heart attack, the nurse's immediate concern in the patient's physiologic needs (e.g., oxygen and pain relief). At the same time, safety needs (e.g., for ensuring that the person does not fall off the examining table) and love and belonging needs (e.g., for having a family member nearby if possible) are still major considerations.[11]

31.3 Research

Recent research appears to validate the existence of universal human needs, although the hierarchy proposed by Maslow is called into question.[12][13]

Following World War II, the unmet needs of homeless and orphaned children presented difficulties that were often addressed with the help of attachment theory, which was initially based on Maslow and others' developmental psychology work by John Bowlby.[14] Originally dealing primarily with maternal deprivation and concordant losses of essential and primal needs, attachment theory has since been extended to provide explanations of nearly all the human needs in Maslow's hierarchy, from sustenance and mating to group membership and justice.[15]

31.4 Criticism

31.4.1 Ranking

Global ranking

In their extensive review of research based on Maslow's theory, Wahba and Bridwell found little evidence for the ranking of needs that Maslow described or for the existence of a definite hierarchy at all.[16]

The order in which the hierarchy is arranged has been criticized as being ethnocentric by Geert Hofstede.[17] Maslow's hierarchy of needs fails to illustrate and expand upon the difference between the social and intellectual needs of those raised in individualistic societies and those raised in collectivist societies. The needs and drives of those in individualistic societies tend to be more self-centered than those in collectivist societies, focusing on improvement of the self, with self-actualization being the apex of self-improvement. In collectivist societies, the needs of acceptance and community will outweigh the needs for freedom and individuality.[18]

Ranking of sex

The position and value of sex on the pyramid has also been a source of criticism regarding Maslow's hierarchy. Maslow's hierarchy places sex in the physiological needs category along with food and breathing; it lists sex solely from an individualistic perspective. For example, sex is placed with other physiological needs which must be satisfied before a person considers "higher" levels of motivation. Some critics feel this placement of sex neglects the emotional, familial, and evolutionary implications of sex within the community, although others point out that this is true of all of the basic needs.[19][20] There are also people who do not want sex, such as some asexuals.[21][22][23]

Changes to the hierarchy by circumstance

The higher-order (self-esteem and self-actualization) and lower-order (physiological, safety, and love) needs classification of Maslow's hierarchy of needs is not universal and may vary across cultures due to individual differences and availability of resources in the region or geopolitical entity/country.

In one study,[24] exploratory factor analysis (EFA) of a thirteen item scale showed there were two particularly important levels of needs in the US during the peacetime of 1993 to 1994: survival (physiological and safety) and psychological (love, self-esteem, and self-actualization). In 1991, a retrospective peacetime measure was established and collected during the Persian Gulf War and US citizens were asked to recall the importance of needs from the previous year. Once again, only two levels of needs were identified; therefore, people have the ability and competence to recall and estimate the importance of needs. For citizens in the Middle East (Egypt and Saudi Arabia), three levels of needs regarding importance and satisfaction surfaced during the 1990 retrospective peacetime. These three levels were completely different from those of the US citizens.

Changes regarding the importance and satisfaction of needs from the retrospective peacetime to the wartime due to stress varied significantly across cultures (the US vs. the Middle East). For the US citizens, there was only one level of needs since all needs were considered equally important. With regards to satisfaction of needs during the war, in the US there were three levels: physiological needs, safety needs, and psychological needs (social, self-esteem, and self-actualization). During the war, the satisfaction of physiological needs and safety needs were separated into two independent needs while during peacetime, they were combined as one. For the people of the Middle East, the satisfaction of needs changed from three levels to two during wartime.[25][26]

A 1981 study looked at how Maslow's hierarchy might vary across age groups.[27] A survey asked participants of varying ages to rate a set number of statements from most important to least important. The researchers found that children had higher physical need scores than the other groups, the love need emerged from childhood to young adulthood,

the esteem need was highest among the adolescent group, young adults had the highest self-actualization level, and old age had the highest level of security, it was needed across all levels comparably. The authors argued that this suggested Maslow's hierarchy may be limited as a theory for developmental sequence since the sequence of the love need and the self-esteem need should be reversed according to age.

31.4.2 Definition of terms

Self-actualization

The term "self-actualization" may not universally convey Maslow's observations; this motivation refers to focusing on becoming the best person that one can possibly strive for in the service of both the self and others.[4] Maslow's term of self-actualization might not properly portray the full extent of this level; quite often, when a person is at the level of self-actualization, much of what they accomplish in general may benefit others, or "the greater good".

31.5 See also

- Energy hierarchy
- Engel's law, an economic model for how well basic needs are met
- ERG theory, which further expands and explains Maslow's theory
- Fundamental human needs, Manfred Max-Neef's model
- John Curtis Gowan
- Juan Antonio Pérez López, spontaneous and rational motivation
- Management innovation
- Murray's psychogenic needs
- Need theory
- Positive disintegration

31.6 References

[1] Maslow's Hierarchy of Needs

[2] Maslow, A.H. (1943). "A theory of human motivation". *Psychological Review*. **50** (4): 370–96. doi:10.1037/h0054346 – via psychclassics.yorku.ca.

[3] M.,, Wills, Evelyn. *Theoretical basis for nursing*. ISBN 9781451190311. OCLC 857664345.

[4] Maslow, A (1954). *Motivation and personality*. New York, NY: Harper. ISBN 0-06-041987-3.

[5] Mittelman, W. (1991). "Maslow's study of self-actualization: A reinterpretation". *Journal of Humanistic Psychology*. **31** (1): 114–135. doi:10.1177/0022167891311010.

[6] Kremer, William Kremer; Hammond, Claudia (31 August 2013). "Abraham Maslow and the pyramid that beguiled business". *BBC news magazine*. Retrieved 1 September 2013.

[7] Steere, B. F. (1988). *Becoming an effective classroom manager: A resource for teachers*. Albany, NY: SUNY Press. ISBN 0-88706-620-8.

[8] Goble, F. (1970). *The third force: The psychology of Abraham Maslow*. Richmond, CA: Maurice Bassett Publishing. pp. 62.

[9] A.H. Maslow, "Critique of self-actualization theory", in: E. Hoffman (Ed.), *Future visions: The unpublished papers of Abraham Maslow* (Thousand Oaks, CA: Sage, 1996), pp. 26–32

[10] Cfr. A.H. Maslow, "The farther reaches of human nature", in: *Journal of Transpersonal Psychology* 1(1969)1, pp. 1-9; A. Maslow, *The farther reaches of human nature* (New York: The Viking Press, 1971); Mark E. Koltko-Rivera, "Rediscovering the Later Version of Maslow's Hierarchy of Needs: Self-Transcendence and Opportunities for Theory, Research, and Unification", in: *Review of General Psychology* 10(2006)4, pp. 302-317 (PDF); Albert Garcia-Romeu, "Self-transcendence as a measurable transpersonal construct", in: *Journal of Transpersonal Psychology*, 42(2010)1, p. 26-47 (PDF)

[11] Taylor, Carol. Fundamentals of Nursing: The Art and Science of Person-centered Nursing Care. 8th ed. Philadelphia: Wolters Kluwer, 2015. Print.

[12] Villarica, H. (August 17, 2011). "Maslow 2.0: A new and improved recipe for happiness". *theatlantic.com*.

[13] Tay, L.; Diener, E. (2011). "Needs and subjective well-being around the world". *Journal of Personality and Social Psychology*. **101** (2): 354–365. doi:10.1037/a0023779.

[14] Bretherton, I. (1992). "The Origins of Attachment Theory: John Bowlby and Mary Ainsworth". *Developmental Psychology*. **28** (5): 759–775. doi:10.1037/0012-1649.28.5.759.

[15] Bugental DB (2000). "Acquisition of the Algorithms of Social Life: A Domain-Based Approach". *Psychological Bulletin*. **126** (2): 178–219. PMID 10748640. doi:10.1037/0033-2909.126.2.187.

[16] Wahba, M. A.; Bridwell, L. G. (1976). "Maslow reconsidered: A review of research on the need hierarchy theory". *Organizational Behavior and Human Performance*. **15** (2): 212–240. doi:10.1016/0030-5073(76)90038-6.

[17] Hofstede, G. (1984). "The cultural relativity of the quality of life concept" (PDF). *Academy of Management Review*. **9** (3): 389–398. doi:10.5465/amr.1984.4279653.

[18] Cianci, R.; Gambrel, P. A. (2003). "Maslow's hierarchy of needs: Does it apply in a collectivist culture". *Journal of Applied Management and Entrepreneurship*. **8** (2): 143–161.

[19] Kenrick, D. (May 19, 2010). "Rebuilding Maslow's pyramid on an evolutionary foundation". *psychologytoday.com/*.

[20] Kenrick, D. T.; Griskevicius, V.; Neuberg, S. L.; Schaller, M. (2010). "Renovating the pyramid of needs: Contemporary extensions built upon ancient foundations". *Perspectives on Psychological Science*. **5**: 292. doi:10.1177/1745691610369469.

[21] Bogaert, Anthony F (2006). "Toward a conceptual understanding of asexuality". *Review of General Psychology*. **10** (3): 241–250. doi:10.1037/1089-2680.10.3.241.

[22] Kelly, Gary F. (2004). "Chapter 12". *Sexuality Today: The Human Perspective* (7 ed.). McGraw-Hill. p. 401. ISBN 978-0-07-255835-7 Asexuality is a condition characterized by a low interest in sex.

[23] Prause, Nicole; Cynthia A. Graham (August 2004). "Asexuality: Classification and Characterization" (PDF). *Archives of Sexual Behavior*. **36** (3): 341–356. PMID 17345167. doi:10.1007/s10508-006-9142-3. Archived from the original (PDF) on 27 September 2007. Retrieved 31 August 2007.

[24] Tang, T. L.; West, W. B. (1997). "The importance of human needs during peacetime, retrospective peacetime, and the Persian Gulf War". *International Journal of Stress Management*. **4** (1): 47–62.

[25] Tang, T. L.; Ibrahim, A. H. (1998). "Importance of human needs during retrospective peacetime and the Persian Gulf War: Mid-eastern employees". *International Journal of Stress Management*. **5** (1): 25–37.

[26] Tang, T. L.; Ibrahim, A. H.; West, W. B. (2002). "Effects of war-related stress on the satisfaction of human needs: The United States and the Middle East". *International Journal of Management Theory and Practices*. **3** (1): 35–53.

[27] Goebel, B. L.; Brown, D. R. (1981). "Age differences in motivation related to Maslow's need hierarchy". *Developmental Psychology*. **17**: 809–815. doi:10.1037/0012-1649.17.6.809.

31.7 Further reading

- Heylighen, Francis (1992). "A cognitive-systemic reconstruction of maslow's theory of self-actualization" (PDF). *Behavioral Science.* **37** (1): 39–58. doi:10.1002/bs.3830370105.

- Kress, Oliver (1993). "A new approach to cognitive development: ontogenesis and the process of initiation". *Evolution and Cognition.* **2** (4): 319–332.

31.8 External links

- A Theory of Human Motivation, original 1943 article by Maslow.

Chapter 32

Positive psychology

For the positive mental attitude, see Optimism. For the belief that positive thoughts result in positive life events, see Law of attraction (New Thought).

Positive psychology is "the scientific study of what makes life most worth living",[1] or "the scientific study of positive human functioning and flourishing on multiple levels that include the biological, personal, relational, institutional, cultural, and global dimensions of life".[2] Positive psychology is concerned with eudaimonia, "the good life", reflection about what holds the greatest value in life – the factors that contribute the most to a well-lived and fulfilling life.

Positive psychology began as a new domain of psychology in 1998 when Martin Seligman chose it as the theme for his term as president of the American Psychological Association.[3][4] Mihaly Czikszentmihalyi and Christopher Peterson are regarded as co-initiators of this development.[5] It is a reaction against psycho-analysis and behaviorism, which have focused on "mental illness", meanwhile emphasising maladaptive behavior and negative thinking. It builds further on the humanistic movement, which encouraged an emphasis on happiness, well-being, and positivity, thus creating the foundation for what is now known as positive psychology.[4]

Guiding theories are Seligman's *P.E.R.M.A.*, and Csikszentmihalyi's theory of flow, while Seligman and Peterson's *Character Strengths and Virtues* was a major contribution to the methodological study of positive psychology.

Positive psychologists have suggested a number of ways in which individual happiness may be fostered. Social ties with a spouse, family, friends and wider networks through work, clubs or social organisations are of particular importance, while physical exercise and the practice of meditation may also contribute to happiness. Happiness may rise with increasing financial income, though it may plateau or even fall when no further gains are made.[6]

32.1 Definition and basic assumptions

32.1.1 Definition

Seligman and Csikszentmihalyi define positive psychology as

> ... the scientific study of positive human functioning and flourishing on multiple levels that include the biological, personal, relational, institutional, cultural, and global dimensions of life.[2]

Christopher Peterson defines positive psychology as

> ...the scientific study of what makes life most worth living,"[1]

32.1.2 Basic concepts

Positive psychology is concerned with eudaimonia, "the good life" or flourishing, living according to what holds the greatest value in life – the factors that contribute the most to a well-lived and fulfilling life. While not attempting a strict definition of the good life, positive psychologists agree that one must live a happy, engaged, and meaningful life in order to experience "the good life". Martin Seligman referred to "the good life" as "using your signature strengths every day to produce authentic happiness and abundant gratification".[7] According to Christopher Peterson, "eudaimonia trumps hedonism".[1]

Related concepts are happiness, well-being, quality of life, contentment,[8] and meaningful life.

32.1.3 Research topics

According to Seligman and Peterson, positive psychology is concerned with three issues: positive emotions, positive individual traits, and positive institutions. Positive emotions are concerned with being content with one's past, being happy in the present and having hope for the future. Positive individual traits focus on one's strengths and virtues. Finally, positive institutions are based on strengths to better a community of people.[9]

According to Peterson, positive psychologists are concerned with four topics: (1) positive experiences, (2) enduring psychological traits, (3) positive relationships, and (4) positive institutions.[10] According to Peterson, topics of interest to researchers in the field are: states of pleasure or flow, values, strengths, virtues, talents, as well as the ways that these can be promoted by social systems and institutions.[11]

32.1.4 Basic assumptions

Positive psychology complements, without intending to replace or ignore, the traditional areas of psychology. By emphasizing the study of positive human development this field helps to balance other approaches that focus on disorder, and which may produce only limited understanding.[12]

The basic premise of positive psychology is that human beings are often drawn by the future more than they are driven by the past. A change in our orientation to time can dramatically affect how we think about the nature of happiness. Seligman identified other possible goals: families and schools that allow children to grow, workplaces that aim for satisfaction and high productivity, and teaching others about positive psychology.[9]

Those who practice positive psychology attempt psychological interventions that foster positive attitudes toward one's subjective experiences, individual traits, and life events.[13] The goal is to minimize pathological thoughts that may arise in a hopeless mindset, and to, instead, develop a sense of optimism toward life.[13]

32.2 Origins and development

32.2.1 Origin

Positive psychology began as a new area of psychology in 1998 when Martin Seligman chose it as the theme for his term as president of the American Psychological Association.[3] In the first sentence of his book *Authentic Happiness*, Seligman claimed: "for the last half century psychology has been consumed with a single topic only – mental illness",[14]:xi expanding on Maslow's comments.[15] He urged psychologists to continue the earlier missions of psychology of nurturing talent and improving normal life.[16]

To Martin Seligman, psychology (particularly its positive branch) can investigate and promote realistic ways of fostering more wellbeing in individuals and communities.

The term originates with Maslow, in his 1954 book *Motivation and Personality*,[17] and there have been indications that psychologists since the 1950s have been increasingly focused on the promotion of mental health rather than merely treating illness.[18][19]

32.2.2 Development

The first positive psychology summit took place in 1999. The First International Conference on Positive Psychology took place in 2002.[16] More attention was given by the general public in 2006 when, using the same framework, a course at Harvard University became particularly popular.[20] In June 2009, the First World Congress on Positive Psychology took place at the University of Pennsylvania.[21]

The International Positive Psychology Association (IPPA) is a recently established association that has expanded to thousands of members from 80 different countries. The IPPA's missions include: (1) "further the science of positive psychology across the globe and to ensure that the field continues to rest on this science" (2) "work for the effective and responsible application of positive psychology in diverse areas such as organizational psychology, counselling and clinical psychology, business, health, education, and coaching", (3) "foster education and training in the field".[22]

The field of positive psychology today is most advanced in the United States and Western Europe. Even though positive psychology offers a new approach to the study of positive emotions and behavior, the ideas, theories, research, and motivation to study the positive side of human behavior is as old as humanity.[23]

32.2.3 Historical antecedents

Positive psychology is the latest effort by human beings to understand the nature of happiness and well-being, but it is by no means the first attempt to solve that particular puzzle.

Religion and philosophy

5 steps to a happier life. Infographic based on Haidt, Jonathan (2005). The Happiness Hypothesis: Finding Modern Truth in Ancient Wisdom. New York: Basic Books

Judaism has a 3,000-year tradition of wisdom regarding happiness.[16][24][25] It also promotes a Divine command theory of happiness: happiness and rewards follow from following the commands of the divine. The Early Hebrews believed in the divine command theory which finds happiness by living according to the commands or rules set down by a Supreme Being.[26]

The Greeks thought that happiness could be discovered through logic and intellectual contemplation. The ancient Greeks had many schools of thought. Socrates advocated self-knowledge as the path to happiness. Plato's allegory of the cave influenced western thinkers who believed that happiness is found by finding deeper meaning. Aristotle believed happiness, or eudaimonia is constituted by rational activity in accordance with virtue over a complete life. The Epicureans believed in reaching happiness through the enjoyment of simple pleasures. The Stoics believed they could remain happy by being objective and reasonable, and described many "spiritual exercises" comparable to the psychological exercises employed in cognitive behavioral therapy and positive psychology.[16][27]

Christianity continued to follow the Divine command theory of happiness. In the Middle Ages, Christianity taught that true happiness would not be found until the afterlife.

The seven deadly sins are about earthly self-indulgence and narcissism. On the other hand, the Four Cardinal Virtues and Three Theological Virtues were supposed to keep one from sin.[16]

Modern Age

During the Renaissance and Age of Enlightenment, individualism was valued. Simultaneously, creative individuals gained prestige, as they were now considered artists, not just craftsmen. Utilitarian philosophers such as John Stuart Mill believed moral actions were actions that maximized happiness for the most number of people; they suggested an empirical science of happiness should be used to determine which actions are moral (a science of morality). Thomas Jefferson and other proponents of democracy believed "Life, liberty and the pursuit of happiness" are inalienable rights, and violation of these justifies the overthrow of the government.[16]

The Romantics valued individual emotional expression and sought their emotional "true selves," which were unhindered by social norms. At the same time, love and intimacy became main motivations for marriage.[16]

Psychology

See also: Well-being

Several humanistic psychologists, most notably Abraham Maslow, Carl Rogers, and Erich Fromm, developed theories and practices pertaining to human happiness and flourishing. More recently, *positive* psychologists have found empirical support for the humanistic theories of flourishing. In addition, positive psychology has moved ahead in a variety of new directions.

In 1984, Diener published his tripartite model of subjective well-being, positing "three distinct but often related components of wellbeing: frequent positive affect, infrequent negative affect, and cognitive evaluations such as life satisfaction".[28] In this model, cognitive, affective and contextual factors contribute to subjective well-being.[29] According to Diener and Suh, subjective well-being is "...based on the idea that how each person thinks and feels about his or her life is important".[30]

Carol Ryff's Six-factor Model of Psychological Well-being was initially published in 1989, and additional testing of its factors was published in 1995. It postulates six factors which are key for well-being, namely self-acceptance, personal growth, purpose in life, environmental mastery, autonomy, and positive relations with others.[31]

According to Corey Keyes, who collaborated with Carol

Ryff and uses the term flourishing as a central concept, mental well-being has three components, namely hedonic (c.q. subjective or emotional[32]), psychological, and social well-being.[33] Hedonic well-being concerns emotional aspects of well-being, whereas psychological and social well-being, c.q eudaimonic well-being, concerns skills, abilities, and optimal functioning.[34] This tripartite model of mental well-being has received extensive empirical support across cultures.[34][32][35][36]

32.3 Theory and methods

32.3.1 Flow

Main article: Flow (psychology)

In the 1970s Csikszentmihalyi's started to study *flow*, a state of absorption where one's abilities are well-matched to the demands at-hand. Flow is characterized by intense concentration, loss of self-awareness, a feeling of being perfectly challenged (neither bored nor overwhelmed), and a sense "time is flying". Flow is intrinsically rewarding; it can also assist in the achievement of goals (e.g., winning a game) or improving skills (e.g., becoming a better chess player).[37] Anyone can experience flow, in different domains, such as play, creativity, and work.

Flow is achieved when the challenge of the situation meets one's personal abilities. A mismatch of challenge for someone of low skills results in a state of anxiety; insufficient challenge for someone highly skilled results in boredom.[37] The effect of challenging situations means that flow is often temporarily exciting and variously stressful, but this is considered Eustress, which is also known as "good" stress. Eustress is arguably less harmful than chronic stress, although the pathways of stress-related systems are similar. Both can create a "wear and tear" effect; however, the differing physiological elements and added psychological benefits of eustress might well balance any wear and tear experienced.

Csikszentmihalyi identified nine indicator elements of flow: 1. Clear goals exist every step of the way, 2. Immediate feedback guides one's action, 3. There is a balance between challenges and abilities, 4. Action and awareness are merged, 5. Distractions are excluded from consciousness, 6. Failure is not worrisome, 7. Self-consciousness disappears, 8. Sense of time is distorted, and 9. The activity becomes "autotelic" (an end in itself, done for its own sake)[38] His studies also show that flow is greater during work while happiness is greater during leisure activities.[39]

32.3.2 PERMA

Initial theory: three paths to happiness

In *Authentic Happiness* (2002) Seligman proposed three kinds of a happy life which can be investigated:[40][41]

1. *Pleasant life*: research into the Pleasant Life, or the "life of enjoyment", examines how people optimally experience, forecast, and savor the positive feelings and emotions that are part of normal and healthy living (e.g., relationships, hobbies, interests, entertainment, etc.). Despite the attention given, Martin Seligman says this most transient element of happiness may be the least important.[42]

2. *Good Life*: investigation of the beneficial effects of immersion, absorption, and *flow*, felt by individuals when optimally engaged with their primary activities, is the study of the Good Life, or the "life of engagement". Flow is experienced when there is a positive match between a person's strength and their current task, i.e., when one feels confident of accomplishing a chosen or assigned task.[note 1]

3. *Meaningful Life*: inquiry into the Meaningful Life, or "life of affiliation", questions how individuals derive a positive sense of well-being, belonging, meaning, and purpose from being part of and contributing back to something larger and more permanent than themselves (e.g., nature, social groups, organizations, movements, traditions, belief systems).

These categories appear neither widely disputed nor adopted by researchers across the 12 years that this academic area has been in existence.

Development into PERMA-theory

Simple exercise, such as running, is cited as key to feeling happy.[43]

In *Flourish* (2011) Seligman argued that the last category, "meaningful life", can be considered as 3 different categories. The resulting acronym is PERMA: Positive Emotions, Engagement, Relationships, Meaning and purpose, and Accomplishments. It is a mnemonic for the five elements of Martin Seligman's well-being theory:[41][44]

- *Positive emotions* include a wide range of feelings, not just happiness and joy.[45] Included are emotions like excitement, satisfaction, pride and awe, amongst others. These emotions are frequently seen as connected to positive outcomes, such as longer life and healthier social relationships.[46]

- *Engagement* refers to involvement in activities that draws and builds upon one's interests. Mihaly Csikszentmihalyi explains true engagement as flow, a feeling of intensity that leads to a sense of ecstasy and clarity.[47] The task being done needs to call upon higher skill and be a bit difficult and challenging yet still possible. Engagement involves passion for and concentration on the task at hand and is assessed subjectively as to whether the person engaged was completely absorbed, losing self-consciousness.[45]

- *Relationships* are all important in fueling positive emotions, whether they are work-related, familial, romantic, or platonic. As Dr. Christopher Peterson puts it simply, "Other people matter."[48] Humans receive, share, and spread positivity to others through relationships. They are important not only in bad times, but good times as well. In fact, relationships can be strengthened by reacting to one another positively. It is typical that most positive things take place in the presence of other people.[49]

- *Meaning* is also known as purpose, and prompts the question of "why". Discovering and figuring out a clear "why" puts everything into context from work to relationships to other parts of life.[50][51] Finding meaning is learning that there is something greater than one's self. Despite potential challenges, working with meaning drives people to continue striving for a desirable goal.

- *Accomplishments* are the pursuit of success and mastery.[45] Unlike the other parts of PERMA, they are sometimes pursued even when accomplishments do not result in positive emotions, meaning, or relationships. That being noted, accomplishments can activate the other elements of PERMA, such as pride, under positive emotion.[52] Accomplishments can be individual or community-based, fun- or work-based.

Selection-criteria

The five PERMA elements were selected according to three criteria:

1. It contributes to well-being.

2. It is pursued for its own sake.

3. It is defined and measured independently of the other elements.

Alternate theories on well-being[53]

Main article: Well-being

Although intended as comprehensive theory on well-being, Seligmann's model competes with older models of well-being, such as Carol Ryff's Six-factor Model of Psychological Well-being and Diener's tripartite model of subjective well-being. Due to the existing investments in theory and research into these models, Seligmann's model has not become the "golden standard" of well-being research.[41]

32.3.3 Character Strengths and Virtues

The development of the *Character Strengths and Virtues* (CSV) handbook (2004) represented the first attempt by Seligman and Peterson to identify and classify positive psychological traits of human beings. Much like the *Diagnostic and Statistical Manual of Mental Disorders* (DSM) of general psychology, the CSV provided a theoretical framework to assist in understanding strengths and virtues and for developing practical applications for positive psychology. This manual identified 6 classes of virtues (i.e., "core virtues"), underlying 24 measurable character strengths.[54]

The CSV suggested these 6 virtues have a historical basis in the vast majority of cultures; in addition, these virtues and strengths can lead to increased happiness when built upon. Notwithstanding numerous cautions and caveats, this suggestion of universality hints threefold: 1. The study of positive human qualities broadens the scope of psychological research to include mental wellness, 2. the leaders of the positive psychology movement are challenging moral relativism, suggesting people are "evolutionarily predisposed" toward certain virtues, and 3. virtue has a biological basis.[54]:51

The organization of the 6 virtues and 24 strengths is as follows:

1. **Wisdom and Knowledge:** creativity, curiosity, open-mindedness, love of learning, perspective, innovation

2. **Courage:** bravery, persistence, integrity, vitality, zest

3. **Humanity:** love, kindness, social intelligence

4. **Justice:** citizenship, fairness, leadership

5. **Temperance:** forgiveness and mercy, humility, prudence, self control

6. **Transcendence:** appreciation of beauty and excellence, gratitude, hope, humor, spirituality

Recent research challenged the need for 6 virtues. Instead, researchers suggested the 24 strengths are more accurately grouped into just 3 or 4 categories: Intellectual Strengths, Interpersonal Strengths, and Temperance Strengths[55] or alternatively, Interpersonal Strengths, Fortitude, Vitality, and Cautiousness[56] These strengths, and their classifications, have emerged independently elsewhere in literature on values. Paul Thagard described examples; these included Jeff Shrager's workshops to discover the habits of highly creative people.[57] Some research indicates that well-being effects that appear to be due to spirituality are actually better described as due to virtue.[58]

32.4 Applications and research findings

Main article: Well-being contributing factors

Research on positive psychology, well-being, eudaimonia and happiness, and the theories of Diener, Ryff, Keyes and Seligmann covers a broad range of levels and topics, including "the biological, personal, relational, institutional, cultural, and global dimensions of life".[2]

Recently some websites and apps such as Happier, Uplifter, and Boonoob have tried to incorporate positive psychology interventions to increase happiness.[59] [60] Maybe the most notorious goal have been set by Boonoob to increase the world happiness by 10% by the end of year 2022.[61]

32.5 Criticism

According to Kirk Schneider, positive psychology fails to explain past heinous behaviors such as those perpetrated by the Nazi party, Stalinist marches and Klan gatherings, to identify but a few. Furthermore, Schneider pointed to a body of research showing high positivity correlates with positive illusion, which effectively distorts reality.[62] The extent of the downfall of high positivity (also known as flourishing) is one could become incapable of psychological growth, unable to self-reflect, and tend to hold racial biases. By contrast, negativity, sometimes evidenced in mild to moderate depression, is correlated with less distortion of reality. Therefore, negativity might play an important role within the dynamics of human flourishing. To illustrate, conflict engagement and acknowledgement of appropriate negativity, including certain negative emotions like guilt, might better promote flourishing.[63] Overall, Schneider provided perspective: "perhaps genuine happiness is not something you aim at, but is a by-product of a life well lived, and a life well lived does not settle on the programmed or neatly calibrated".[64] Seligman has acknowledged in his work the point about positive illusion,[65] and is also a critic of merely feeling good about oneself apart from reality and recognises the importance of negativity / dysphoria.[66]

Ian Sample, writing for *The Guardian*, noted that, "Positive psychologists also stand accused of burying their heads in the sand and ignoring that depressed, even merely unhappy people, have real problems that need dealing with." Sample also quoted Steven Wolin, a clinical psychiatrist at George Washington University, as saying that the study of positive psychology is just a reiteration of older ways of thinking, and that there is not much scientific research to support the efficacy of this method.[67] Gable responds to criticism on their pollyanna view on the world by saying that they are just bringing a balance to a side of psychology that is glaringly understudied.[68]

Barbara Held argued that while positive psychology makes contributions to the field of psychology, it has its faults. She offered insight into topics including the negative side effects of positive psychology, negativity within the positive psychology movement, and the current division in the field of psychology caused by differing opinions of psychologists on positive psychology. In addition, she noted the movement's lack of consistency regarding the role of negativity. She also raised issues with the simplistic approach taken by some psychologists in the application of positive psychology. A "one size fits all" approach is arguably not beneficial to the advancement of the field of positive psychology; she suggested a need for individual differences to be incorporated into its application.[69]

32.6 See also

Precursors

- New Thought

- Humanistic psychology

- Maslow's hierarchy of needs

- Needs and Motives (Henry Murray)
- Logotherapy (Viktor Frankl)

Various

- *Anatomy of an Epidemic*
- Aversion to happiness
- Cool To Be Kind
- Culture and positive psychology
- Happiness economics
- Meaning of life
- Meaningful life
- Outline of psychology
- Positive education
- Positive Youth Development
- Pragmatism
- Psychological resilience
- Random Acts of Kindness Day
- Rational ignorance
- Second wave positive psychology
- Sex-positive movement
- Theory of humor
- World Kindness Day
- World Kindness Movement

32.7 Notes

[1] See related concepts: Self-efficacy and play.

32.8 References

[1] Christopher Peterson (2008), *What Is Positive Psychology, and What Is It Not?*

[2] Seligman & Csikszentmihalyi 2000.

[3] "Time Magazine's cover story in the special issue on "The Science of Happiness", 2005" (PDF). Retrieved 2011-02-07.

[4] Srinivasan, T. S. (2015, February 12). The 5 Founding Fathers and A History of Positive Psychology. Retrieved February 4, 2017, from https://positivepsychologyprogram.com/founding-fathers/

[5] *The 5 Founding Fathers and A History of Positive Psychology*

[6] Positive Psychology Progress: Empirical Validation of Interventions. Seligman, Martin E. P.; Steen, Tracy A.; Park, Nansook; Peterson, Christopher American Psychologist, Vol 60(5), Jul-Aug 2005, 410-421. http://dx.doi.org/10.1037/0003-066X.60.5.410

[7] Seligman, M.E.P. (2009). Authentic Happiness. New York: Free Press.

[8] Graham, Michael C. (2014). *Facts of Life: ten issues of contentment*. Outskirts Press. pp. 6–10. ISBN 978-1-4787-2259-5.

[9] Seligman, Martin E.P. "Positive Psychology Center." Positive Psychology Center. University of Pennsylvania, 2007. Web. 12 Mar. 2013.

[10] Peterson, C. (2009). *Positive Psychology*. Reclaiming Children and Youth. Vol.18, Issue 2, pp. 3–7.

[11] Peterson, Christopher (27 July 2006). *A Primer in Positive Psychology*. Oxford University Press. ISBN 978-0-19-518833-2.

[12] Peterson, C (2009). "Positive psychology". *Reclaiming Children and Youth*. **18** (2): 3–7.

[13] Seligman, M. E., & Csikszentmihalyi, M. (2014). Positive psychology: An introduction (pp. 279-298). Springer Netherlands.

[14] Seligman, Martin E. P. (2002). *Authentic Happiness: Using the New Positive Psychology to Realize Your Potential for Lasting Fulfillment*. New York: Free Press. ISBN 0-7432-2297-0.

[15] "The science of psychology has been far more successful on the negative than on the positive side. It has revealed to us much about man's shortcomings, his illness, his sins, but little about his potentialities, his virtues, his achievable aspirations, or his full psychological height. It is as if psychology has voluntarily restricted itself to only half its rightful jurisdiction, the darker, meaner half" (Maslow, Motivation and Psychology, p. 354).

[16] Compton, William C (2005). "1". *An Introduction to Positive Psychology*. Wadsworth Publishing. pp. 1–22. ISBN 0-534-64453-8.

[17] Note: the last chapter is entitled "Toward a Positive Psychology".

[18] Secker J (1998). "Current conceptualizations of mental health and mental health promotion" (PDF). **13** (1). Health Education Research. p. 58. Retrieved 2010-05-18. ... Amongst psychologists ... the importance of promoting health rather than simply preventing ill-health date back to the 1950s (Jahoda, 1958)

32.8. REFERENCES

[19] Dianne Hales (2010). "An Invitation to Health, Brief: Psychological Well-Being" (2010–2011 ed.). Wadsworth Cengage Learning. p. 26. Retrieved 2010-05-18.

[20] Ben-Shahar, Ben (2007) "Happier -Learn the Secrets to Daily Joy and Lasting Fulfillment", First Edition, McGraw-Hill Co.

[21] Reuters, Jun 18, 2009: First World Congress on Positive Psychology Kicks Off Today With Talks by Two of the World's Most Renowned Psychologists

[22] International Positive Psychology Association (IPPA) (2011). international positive psychology association. Retrieved from http://www.ippanetwork.org/about_ippa/

[23] Compton, William C., and Edward Hoffman. *Positive Psychology: The Science of Happiness and Flourishing.* 2nd ed. Belmont, CA: Wadsworth Cengage Learning, 2013. Print.

[24] Wilkinson, Phaedra (October 21, 2014). "From the community: Exciting Class on Jewish Positive Psychology to be Presented in Northbrook". Chicago: Chicago Tribune. Retrieved 16 November 2014. The Jewish Learning Institute's (JLI) Newest Class Looks at Positive Psychology through the 3,000-year-old lens of Jewish thought. Northbrook, IL – When Israeli-born psychologist Tal Ben-Shahar began teaching a class called Positive Psychology at Harvard in 2006, a record 855 undergraduate students signed up for his class. Droves of students at the academically-intense university came to learn, as the course description puts it, about "psychological aspects of a fulfilling and flourishing life."

[25] "Chabad Jewish Center to present 'How Happiness Thinks: Jewish Perspectives on Positive Psychology'". Cape Coral Daily Breeze. October 31, 2014. Retrieved 3 November 2014. Rabbi Zalman Abraham of JLI's headquarters in New York says that being happy can depend on one's perspective, explaining, "How Happiness Thinks is based on the premise that to be happy, you can either change the world, or you can change your thinking". While drawing on 3,000 years of Jewish wisdom on happiness, the course, which was prepared in partnership between JLI and the Washington School of Psychiatry, builds on the latest observations and discoveries in the field of positive psychology.

[26] "Positive psychology began as a new area of psychology in 1998 - PSYC - 101". *www.coursehero.com*. Retrieved 2017-02-08.

[27] Robertson, D (2010). *The Philosophy of Cognitive-Behavioural Therapy: Stoicism as Rational and Cognitive Psychotherapy*. London: Karnac. ISBN 978-1-85575-756-1.

[28] Tov & Diener (2013), *Subjective Well-Being*. Research Collection School of Social Sciences. Paper 1395. http://ink.library.smu.edu.sg/soss_research/1395

[29] Costa Galinha, Iolanda; Pais-Ribeiro, José Luís (2011). "Cognitive, affective and contextual predictors of subjective wellbeing". *International Journal of Wellbeing*. **2** (1): 34–53. doi:10.5502/ijw.v2i1.3.

[30] Diener, Suh, Ed, Eunkook (2000). *Culture and Subjective Well-being*. A Bradford Book. p. 4.

[31] *Carol Ryff's Model of Psychological Well-being. The Six Criteria of Well-Being*

[32] Robitschek, Christine; Keyes, Corey L. M. "Keyes's model of mental health with personal growth initiative as a parsimonious predictor.". *Journal of Counseling Psychology*. **56** (2): 321–329. doi:10.1037/a0013954.

[33] Keyes, Corey L. M. (2002-01-01). "The Mental Health Continuum: From Languishing to Flourishing in Life". *Journal of Health and Social Behavior*. **43** (2): 207–222. JSTOR 3090197. doi:10.2307/3090197.

[34] Joshanloo, Mohsen (2015-10-23). "Revisiting the Empirical Distinction Between Hedonic and Eudaimonic Aspects of Well-Being Using Exploratory Structural Equation Modeling". *Journal of Happiness Studies*: 1–14. ISSN 1389-4978. doi:10.1007/s10902-015-9683-z.

[35] Joshanloo, Mohsen; Lamers, Sanne M. A. (2016-07-01). "Reinvestigation of the factor structure of the MHC-SF in the Netherlands: Contributions of exploratory structural equation modeling". *Personality and Individual Differences*. **97**: 8–12. doi:10.1016/j.paid.2016.02.089.

[36] Gallagher, Matthew W.; Lopez, Shane J.; Preacher, Kristopher J. (2009-08-01). "The Hierarchical Structure of Well-Being". *Journal of Personality*. **77** (4): 1025–1050. ISSN 1467-6494. PMC 3865980. PMID 19558444. doi:10.1111/j.1467-6494.2009.00573.x.

[37] Csikszentmihalyi, Mihaly (1990). *Flow: The Psychology of Optimal Experience*. New York: Harper & Row. ISBN 0-06-016253-8.

[38] ""In the zone": enjoyment, creativity, and the nine elements of "flow"". MeaningandHappiness.com. Retrieved 2010-11-11.

[39] Lopez, S. J., Snyder, C. R. *The Oxford Handbook of Positive Psychology*. Oxford University Press, 2009. p. 200.

[40] Seligman 2002, p. 275.

[41] David Sze (2015), *The Father of Positive Psychology and His Two Theories of Happiness*

[42] Wallis, Claudia (2005-01-09). "Science of Happiness: New Research on Mood, Satisfaction". TIME. Retrieved 2011-02-07.

[43] Best Benefit of Exercise? Happiness, Robin Loyd, Fox News, May 30, 2006.

[44] "The World Question Center 2011— Page 2". Edge.org. Retrieved 2011-02-07.

[45] Seligman, M. E. P. (2011). Flourish: A Visionary New Understanding of Happiness and Well-being. New York: Free Press. Ch 1

[46] "The Pursuit of Happiness".

[47] "Mihaly Csikszentmihalyi TED talk".

[48] "Other People Matter".

[49] "Using Positive Psychology in Your Relationships".

[50] "Start with Why".

[51] "Why do You do What You Do?".

[52] "The Science of a Happy Startup".

[53] "Uplifter".

[54] Peterson, Christopher; Seligman, Martin E.P. (2004). *Character strengths and virtues: A handbook and classification*. Oxford: Oxford University Press. ISBN 0-19-516701-5.

[55] Shryack, J.; Steger, M. F.; Krueger, R. F.; Kallie, C. S. (2010). "The structure of virtue: An empirical investigation of the dimensionality of the virtues in action inventory of strengths". *Personality and Individual Differences*. **48** (6): 714–719. doi:10.1016/j.paid.2010.01.007.

[56] Brdr, I.; Kashdan, T.B. (2010). "Character strengths and well-being in Croatia: An empirical investigation of structure and correlates". *Journal of Research in Personality*. **44**: 151–154. doi:10.1016/j.jrp.2009.12.001.

[57] Thagard, P. (2005). How to be a successful scientist. In M. E. Gorman, R. D. Tweney, D. C. Gooding & A. P. Kincannon (Eds.), Scientific and technological thinking (pp. 159-171). Mawah, NJ: Lawrence Erlbaum Associates.

[58] Schuurmans-Stekhoven, James (2011). "Is it God or just the data that moves in mysterious ways? How wellbeing researchers may be mistaking faith for virtue". *Social Indicators Research*. **100** (2): 313–330. doi:10.1007/s11205-010-9630-7.

[59] "Happier Website".

[60] "Uplifter Website".

[61] "Boonoob Website".

[62] Schneider, K. (2011). "Toward a Humanistic Positive Psychology". *Existential Analysis: Journal of the Society for Existential Analysis*. **22** (1): 32–38.

[63] Fredrickson, B. L.; Losada, M. F. (2005). "Positive Affect and the Complex Dynamics of Human Flourishing". *American Psychologist*. **60** (7): 678–686. PMC 3126111. PMID 16221001. doi:10.1037/0003-066X.60.7.678.

[64] Schneider 2011, p. 35

[65] Seligman, Martin E. (1995). *The Optimistic Child*. Houghton Mifflin Company. pp. 295–299.

[66] Seligman 1995, pp. 41–42

[67] Sample, I. (19 November 2003). "How to be happy". *The Guardian*.

[68] Gable, S. L., & Haidt, J. (2005). What (and why) is positive psychology?. Review of general psychology, 9(2), 103.

[69] Held, Barbara S. (January 2004). "The Negative Side of Positive Psychology". *Journal of Humanistic Psychology*. **44** (1): 9–41. doi:10.1177/0022167803259645.

32.9 Sources

- Argyle, Michael (2001). *The Psychology of Happiness*. Routledge.

- Benard, Bonnie (2004). *Resiliency: What We Have Learned*. San Francisco: WestEd.

- Biswas-Diener, Robert, & Diener, Ed. (2004). "The psychology of subjective well-being". *Daedalus*. **133** (2): 18–25. doi:10.1162/001152604323049352.

- Fromm, Eric (1973). *The Anatomy of Human Destructiveness*. New York: New York, Holt, Rinehart and Winston. ISBN 0-03-007596-3.

- Held, BS (January 2004). "The Negative Side of Positive Psychology". *Journal of Humanistic Psychology*. **44** (1): 9–41. doi:10.1177/0022167803259645.

- Kahneman, Daniel; Diener, Ed; Schwarz, Norbert (2003). *Well-Being: The Foundations of Hedonic Psychology*. Russell Sage Foundation Publications.

- Kashdan, T.B. (2009). *Curious? Discover the Missing Ingredient to a Fulfilling Life*. New York, NY: HarperCollins.

- Keyes & J. Haidt (eds.). *Flourishing: Positive Psychology and the Life Well-lived*. Washington DC: American Psychological Association. pp. 275–289.

- McMahon, Darrin M. (2006). *Happiness: A History*. Atlantic Monthly Press.

- Peterson, Christopher (2009). "Positive Psychology". *Reclaiming Children and Youth*. **18** (2): 3–7.

- Robbins B.D (2008). "What is the good life? Positive psychology and the renaissance of humanistic psychology" (PDF). *The Humanistic Psychologist*. **36** (2): 96–112. doi:10.1080/08873260802110988.

- Seligman, Martin (1990). *Learned Optimism: How to Change Your Mind and Your Life*. Free Press.

- Seligman, M.E.P.; Csikszentmihalyi, M. (2000), "Positive Psychology: An introduction", *American Psychologist*, **55** (1): 5–14, PMID 11392865, doi:10.1037/0003-066x.55.1.5

- Seligman, M.E.P. (Spring 2004). "Can Happiness be Taught?". *Daedalus*. **133** (2): 80–87. doi:10.1162/001152604323049424.

- Seligman, M.E.P. (2011), *Flourish. A Visionary New Understanding of Happiness and Well-being*, Simon & Schuster

- Snyder, C.R., and Lopez, Shane J. (2001). *Handbook of Positive Psychology*. Oxford University Press.

- Stebbins, R.A. (2015). *Leisure and Positive Psychology: Linking Activities with Positiveness.* Houndmills, UK: Palgrave Macmillan.

- Zagano, Phyllis; Gillespie, C. Kevin (2006). "Ignatian Spirituality and Positive Psychology". *The Way*. **45** (4): 41–58. (Tr. to Italian: "La Spiritualita Ignaziana e la psicologia positiva" La relazione d'aiuto: il counseling tra psicologia e fede, Ed. Andrea Toniolo, Padua, (November 2008) 29–44)

32.10 Further reading

- Seligman & Csikszentmihalyi (2000), *Positive Psychology. An Introduction.*

- Seligman, M.E.P. (2011), *Flourish. A Visionary New Understanding of Happiness and Well-being*, Simon & Schuster

- Howard Cutler and the Dalai Lama, *The Art of Happiness*

32.11 External links

Origins

- Christopher Peterson, *What Is Positive Psychology, and What Is It Not?*

- *The 5 Founding Fathers and A History of Positive Psychology*

- *The Father of Positive Psychology and His Two Theories of Happiness*

Resources

- University of Pennsylvania, Authentic Happiness, website of Martin Seligman

Various

- Martin Seligman presentation on positive psychology (Video) at TED conference

- The Karma of Happiness: A Buddhist Monk Looks at Positive Psychology by Thanissaro Bhikkhu

- The positive words dictionary: An online resource of positive words for use in Positive Psychology https://positivewordsdictionary.com

Chapter 33

Posthumanism

This article is about a critique of humanism. For the futurist ideology and movement, see transhumanism.

Posthumanism or **post-humanism** (meaning "after humanism" or "beyond humanism") is a term with at least seven definitions according to philosopher Francesca Ferrando:[1]

1. **Antihumanism**: any theory that is critical of traditional humanism and traditional ideas about humanity and the human condition.[2]

2. **Cultural posthumanism**: a branch of cultural theory critical of the foundational assumptions of humanism and its legacy[3] that examines and questions the historical notions of "human" and "human nature", often challenging typical notions of human subjectivity and embodiment[4] and strives to move beyond archaic concepts of "human nature" to develop ones which constantly adapt to contemporary technoscientific knowledge.[5]

3. **Philosophical posthumanism**: a philosophical direction which draws on cultural posthumanism, the philosophical strand examines the ethical implications of expanding the circle of moral concern and extending subjectivities beyond the human species[4]

4. **Posthuman condition**: the deconstruction of the human condition by critical theorists.[6]

5. **Transhumanism**: an ideology and movement which seeks to develop and make available technologies that eliminate aging and greatly enhance human intellectual, physical, and psychological capacities, in order to achieve a "posthuman future".[7]

6. **AI takeover**: A more pessimistic alternative to transhumanism in which humans will not be enhanced, but rather eventually *replaced* by artificial intelligences. Some philosophers, including Nick Land, promote the view that humans should embrace and accept their eventual demise.[8] This is related to the view of "cosmism" which supports the building of strong artificial intelligence even if it may entail the end of humanity as in their view it "would be a cosmic tragedy if humanity freezes evolution at the puny human level".[9][10][11]

7. **Voluntary Human Extinction**, which seeks a "posthuman future" that in this case is a future *without humans*.

33.1 Philosophical posthumanism

Philosopher Ted Schatzki suggests there are two varieties of posthumanism of the philosophical kind:[12]

One, which he calls 'objectivism', tries to counter the overemphasis of the subjective or intersubjective that pervades humanism, and emphasises the role of the nonhuman agents, whether they be animals and plants, or computers or other things.[12]

A second prioritizes practices, especially social practices, over individuals (or individual subjects) which, they say, constitute the individual.[12]

There may be a third kind of posthumanism, propounded by the philosopher Herman Dooyeweerd. Though he did not label it as 'posthumanism', he made an extensive and penetrating immanent critique of Humanism, and then constructed a philosophy that presupposed neither Humanist, nor Scholastic, nor Greek thought but started with a different religious ground motive.[13] Dooyeweerd prioritized law and meaningfulness as that which enables humanity and all else to exist, behave, live, occur, etc. "*Meaning* is the *being* of all that has been *created*," Dooyeweerd wrote, "and the nature even of our selfhood."[14] Both human and nonhuman alike function subject to a common 'law-side', which is diverse, composed of a number of distinct law-spheres or *aspects*.[15] The temporal being of both human and nonhuman is multi-aspectual; for example, both plants and humans are bodies, functioning in the biotic aspect, and both computers and humans function in the formative and lingual aspect, but humans function in the aesthetic, juridical, ethical and faith aspects too. The Dooyeweerdian version is able to incorporate and integrate both the objectivist version and the practices version, because it allows nonhuman agents their own subject-functioning in various aspects and places emphasis on aspectual functioning.[16]

33.2 Emergence of philosophical posthumanism

Ihab Hassan, theorist in the academic study of literature, once stated:

> Humanism may be coming to an end as humanism transforms itself into something one must helplessly call posthumanism.[17]

This view predates most currents of posthumanism which have developed over the late 20th century in somewhat diverse, but complementary, domains of thought and practice. For example, Hassan is a known scholar whose theoretical writings expressly address postmodernity in society. Beyond postmodernist studies, posthumanism has been developed and deployed by various cultural theorists, often in reaction to problematic inherent assumptions within humanistic and enlightenment thought.[4]

Theorists who both complement and contrast Hassan include Michel Foucault, Judith Butler, cyberneticists such as Gregory Bateson, Warren McCullouch, Norbert Wiener, Bruno Latour, Cary Wolfe, Elaine Graham, N. Katherine Hayles, Donna Haraway, Peter Sloterdijk, Stefan Lorenz Sorgner, Evan Thompson, Francisco Varela, Humberto Maturana and Douglas Kellner. Among the theorists are philosophers, such as Robert Pepperell, who have written about a "posthuman condition", which is often substituted for the term "posthumanism".[5][6]

Posthumanism differs from classical humanism by relegating humanity back to one of many natural species, thereby rejecting any claims founded on anthropocentric dominance.[18] According to this claim, humans have no inherent rights to destroy nature or set themselves above it in ethical considerations *a priori*. Human knowledge is also reduced to a less controlling position, previously seen as the defining aspect of the world. Human rights exist on a spectrum with animal rights and posthuman rights.[19] The limitations and fallibility of human intelligence are confessed, even though it does not imply abandoning the rational tradition of humanism.

Proponents of a posthuman discourse, suggest that innovative advancements and emerging technologies have transcended the traditional model of the human, as proposed by Descartes among others associated with philosophy of the Enlightenment period.[20] In contrast to humanism, the discourse of posthumanism seeks to redefine the boundaries surrounding modern philosophical understanding of the human. Posthumanism represents an evolution of thought beyond that of the contemporary social boundaries and is predicated on the seeking of truth within a postmodern context. In so doing, it rejects previous attempts to establish 'anthropological universals' that are imbued with anthropocentric assumptions.[18]

The philosopher Michel Foucault placed posthumanism within a context that differentiated humanism from enlightenment thought. According to Foucault, the two existed in a state of tension: as humanism sought to establish norms while Enlightenment thought attempted to transcend all that is material, including the boundaries that are constructed by humanistic thought.[18] Drawing on the Enlightenment's challenges to the boundaries of humanism, posthumanism rejects the various assumptions of human dogmas (anthropological, political, scientific) and takes the next step by attempting to change the nature of thought about what it means to be human. This requires not only decentering the human in multiple discourses (evolutionary, ecological, technological) but also examining those discourses to uncover inherent humanistic, anthropocentric, normative notions of humanness and the concept of the human.[4]

33.3 Contemporary posthuman discourse

Posthumanistic discourse aims to open up spaces to examine what it means to be human and critically question the concept of "the human" in light of current cultural and historical contexts[4] In her book *How We Became Posthuman*, N. Katherine Hayles, writes about the struggle between different versions of the posthuman as it continually co-evolves alongside intelligent machines.[21] Such coevolution, according to some strands of the posthuman discourse, allows one to extend their subjective understandings of real experiences beyond the boundaries of embodied existence. According to Hayles's view of posthuman, often referred to as technological posthumanism, visual perception and digital representations thus paradoxically become ever more salient. Even as one seeks to extend knowledge by deconstructing perceived boundaries, it is these same boundaries that make knowledge acquisition possible. The use of technology in a contemporary society is thought to complicate this relationship.

Hayles discusses the translation of human bodies into information (as suggested by Hans Moravec) in order to illuminate how the boundaries of our embodied reality have been compromised in the current age and how narrow definitions of humanness no longer apply. Because of this, according to Hayles, posthumanism is characterized by a loss of subjectivity based on bodily boundaries.[4] This strand of posthumanism, including the changing notion of subjectivity and the disruption of ideas concerning what it means to be human, is often associated with Donna Haraway's concept of the cyborg.[4] However, Haraway has distanced herself from posthumanistic discourse due to other theorists' use of the term to promote utopian views of technological innovation to extend the human biological capacity[22] (even though these notions would more correctly fall into the realm of transhumanism[4]).

While posthumanism is a broad and complex ideology, it has relevant implications today and for the future. It attempts to redefine social structures without inherently humanly or even biological origins, but rather in terms of social and psychological systems where consciousness and communication could potentially exist as unique disembodied entities. Questions subsequently emerge with respect to the current use and the future of technology in shaping human existence,[18] as do new concerns with regards to language, symbolism, subjectivity, phenomenology, ethics, justice and creativity.[23]

33.4 Relationship with transhumanism

Posthumanism is sometimes used as a synonym for a cultural and philosophical movement known as "transhumanism" because it proposes a transition to a "posthuman future", achieved through the application of technology to expand human capacities.

James Hughes comments that there is considerable confusion between the two terms.[24][25]

33.5 Criticism

Some critics have argued that all forms of posthumanism, including transhumanism, have more in common than their respective proponents realize.[26] Linking these different approaches, Paul James suggests that 'the key political problem is that, in effect, the position allows the human as a category of being to flow down the plughole of history':

However, some posthumanists in the humanities and the arts are critical of transhumanism (the brunt of Paul James's criticism), in part, because they argue that it incorporates and extends many of the values of Enlightenment humanism and classical liberalism, namely scientism, according to performance philosopher Shannon Bell:[28]

While many modern leaders of thought are accepting of nature of ideologies described by posthumanism, some are more skeptical of the term. Donna Haraway, the author of *A Cyborg Manifesto*, has outspokenly rejected the term, though acknowledges a philosophical alignment with posthumanism. Haraway opts instead for the term of companion species, referring to nonhuman entities with which humans coexist.[22]

Questions of race, some argue, are suspiciously elided within the "turn" to posthumanism. Noting that the terms "post" and "human" are already loaded with racial meaning, critical theorist Zakiyyah Iman Jackson argues that the impulse to move "beyond" the human within posthumanism too often ignores "praxes of humanity and critiques produced by black people,"[29] including Frantz Fanon and Aime Cesaire to Hortense Spillers and Fred Moten.[29] Interrogating the conceptual grounds in which such a mode of "beyond" is rendered legible and viable, Jackson argues that it is important to observe that *"blackness conditions and constitutes the very nonhuman disruption and/or disruption"* which posthumanists invite.[29] In other words, given that race in general and blackness in particular constitutes the very terms through which human/nonhuman distinctions are made, for example in enduring legacies of scientific racism, a gesture toward a "beyond" actually "returns us to a Eurocentric transcendentalism long challenged".[30]

33.6 See also

- Posthuman
- Metahuman

33.7 References

[1] Ferrando, Francesca (2013). "Posthumanism, Transhumanism, Antihumanism, Metahumanism, and New Materialisms: Differences and Relations" (PDF). ISSN 1932-1066. Retrieved 2014-03-14.

[2] J. Childers/G. Hentzi eds., *The Columbia Dictionary of Modern Literary and Cultural Criticism* (1995) p. 140-1

[3] Esposito, Roberto (2011). "Politics and human nature" (PDF). doi:10.1080/0969725X.2011.621222. Retrieved 2013-06-06.

[4] Miah, A. (2008) A Critical History of Posthumanism. In Gordijn, B. & Chadwick R. (2008) Medical Enhancement and Posthumanity. Springer, pp.71-94.

[5] Badmington, Neil (2000). *Posthumanism (Readers in Cultural Criticism)*. Palgrave Macmillan. ISBN 0-333-76538-9.

[6] Hayles, N. Katherine (1999). *How We Became Posthuman: Virtual Bodies in Cybernetics, Literature, and Informatics*. University Of Chicago Press. ISBN 0-226-32146-0.

[7] Bostrom, Nick (2005). "A history of transhumanist thought" (PDF). Retrieved 2006-02-21.

[8] "The Darkness Before the Right".

[9] Hugo de Garis (2002). "First shot in Artilect war fired". Archived from the original on 17 October 2007.

[10] "Machines Like Us interviews: Hugo de Garis". 3 September 2007. gigadeath – the characteristic number of people that would be killed in any major late 21st century war, if one extrapolates up the graph of the number of people killed in major wars over the past 2 centuries

[11] Garis, Hugo de. "The Artilect War - Cosmists vs. Terrans" (PDF). *agi-conf.org*. Retrieved 14 June 2015.

[12] Schatzki, T.R. 2001. Introduction: Practice theory, in *The Practice Turn in Contemporary Theory* eds. Theodore R.Schatzki, Karin Knorr Cetina & Eike Von Savigny.

[13] http://www.dooy.info/ground.motives.html

[14] Dooyeweerd, H. (1955/1984). A new critique of theoretical thought (Vol. 1). Jordan Station, Ontario, Canada: Paideia Press. P. 4

[15] 'law-side'

[16] his radical notion of subject-object relations

[17] Hassan, Ihab (1977). "Prometheus as Performer: Toward a Postmodern Culture?". In Michel Benamou, Charles Caramello. *Performance in Postmodern Culture. Performance in Postmodern Culture*. Madison, Wisconsin: Coda Press. ISBN 0-930956-00-1.

[18] Wolfe, C. (2009). **'What is Posthumanism?'** University of Minnesota Press. Minneapolis, Minnesota.

[19] Evans, Woody (2015). "Posthuman Rights: Dimensions of Transhuman Worlds". Madrid: Teknokultura.

[20] Badmington, Neil. "Posthumanism". Blackwell Reference Online. Retrieved 22 September 2015.

[21] Cecchetto, David (2013). *Humanesis: Sound and Technological Posthumanism*. Minneapolis, MN: University of Minnesota Press.

[22] Gane, Nicholas (2006). "When We Have Never Been Human, What Is to Be Done?: Interview with Donna Haraway". *Theory, Culture & Society*. **23** (7-8): 135–158.

[23] Roudavski, Stanislav; McCormack, Jon (2016). "Post-Anthropocentric Creativity". *Digital Creativity*. **27** (1): 3–6. doi:10.1080/14626268.2016.1151442.

[24] Ranisch, Robert (January 2014). "Post- and Transhumanism: An Introduction". Retrieved 25 August 2016.

[25] MacFarlane, James. "Boundary Work: Post- and Transhumanism, Part I, James Michael MacFarlane". Retrieved 25 August 2016.

[26] Winner, Langdon. "Resistance is Futile: The Posthuman Condition and Its Advocates". In Harold Bailie, Timothy Casey. *Is Human Nature Obsolete?*. Massachusetts Institute of Technology, October 2004: M.I.T. Press. pp. 385–411. ISBN 0262524287.

[27] James, Paul (2017). "Alternative Paradigms for Sustainability: Decentring the Human without Becoming Posthuman". In Karen Malone, Son Truong, and Tonia Gray. *Reimagining Sustainability in Precarious Times*. Ashgate. p. 21.

[28] Zaretsky, Adam (2005). "Bioart in Question. Interview.". Retrieved 2007-01-28.

[29] Jackson 2015, p. 216.

[30] Jackson 2015, p. 217.

33.7.1 Works cited

- Jackson, Zakiyyah Iman (June 2015). "Outer Worlds: The Persistence of Race in Movement 'Beyond the Human'". *Gay and Lesbian Quarterly (GLQ)*.

33.8 Text and image sources, contributors, and licenses

33.8.1 Text

- **Secular humanism** *Source:* https://en.wikipedia.org/wiki/Secular_humanism?oldid=793305737 *Contributors:* Tobias Hoevekamp, Magnus Manske, Eloquence, Wesley, Bryan Derksen, Koyaanis Qatsi, Ed Poor, Roadrunner, Ant, Heron, Stevertigo, Nealmcb, Michael Hardy, Gabbe, Cyde, Hermeneus, JWSchmidt, Ijon, Salsa Shark, Cyan, Palfrey, LordK, Maximus Rex, Saltine, VeryVerily, Buridan, Nomaed, Topbanana, Scott Sanchez, Wetman, William Trevor Blake, Bearcat, Robbot, Moriori, Fredrik, Altenmann, Sam Spade, TimothyPilgrim, Postdlf, Tualha, Blainster, Hoot, Rrjanbiah, Jeroen, Witbrock, Alan Liefting, Gwalla, Nadavspi, Luis Dantas, Derobert, JeffBobFrank, PseudoThinker, Popefauvexxiii, Antandrus, Benw, PDH, JimWae, Rattlesnake, Sohcahtoa, Icairns, GeoGreg, Nickptar, Freakofnurture, Wikkrockiana, Wfaulk, Rich Farmbrough, Rhobite, Alby, Vsmith, Empedocles~enwiki, Euthydemos, Cladist, Bender235, Curufinwe, Ben Standeven, El C, Lycurgus, Zenohockey, Kross, Ricardo E. Trelles, Mikeh, Stesmo, Smalljim, Adraeus, Viriditas, Jez, Alansohn, Mcduarte2000, Mr Adequate, AzaToth, !melquiades, Batmanand, Rohirok, Papau, Ravenhull, TaintedMustard, Starman 1976, Ndteegarden, Najja, Dismas, Stemonitis, FrancisTyers, Boothy443, Mindmatrix, FeanorStar7, WadeSimMiser, MGTom, Bkwillwm, Rchamberlain, Karbinski, Mandarax, Graham87, BD2412, Rjwilmsi, Nightscream, Mikepjones, Cadfile, Harro5, Fish and karate, Dionyseus, FayssalF, Osprey39, Inspirewithhope, Danielsp, Alexjohnc3, Str1977, Andriesb, 2ct7, Franko2nd, Bgwhite, Roboto de Ajvol, Wavelength, GMT, Misterwindupbird, Mark Ironie, Gaius Cornelius, Leutha, Rjensen, Daanschr, THB, Dannyno, Vastu, Nick C, Ackie00, EEMIV, Zythe, Wknight94, JLaTondre, Spliffy, Rhwentworth, RG2, David Hockey, Yahnatan, Victor falk, TechBear, A bit iffy, SmackBot, Selfworm, Brammers, Mira, Cavenba, Xblkx, ProveIt, Yamaguchi??, Anarkisto, Ohnoitsjamie, Andy M. Wang, Durova, Chris the speller, SuMadre, Autarch, Pietaster, SvGeloven, WikiPedant, JonHarder, Wikipedia brown, Leep4life, Savidan, George, Dacoutts, Schnarr, Metamagician3000, Starghost, Ace ETP, Clayto, Ohconfucius, Byelf2007, Rory096, Zahid Abdassabur, Chodorkovskiy, Count Caspian, Osbus, Fig wright, Jaywubba1887, Stjamie, The Bread, Tasc, Avedomni, Wwagner, Dl2000, Mindbodyfitness, Twas Now, Adambiswanger1, Jjoplin, George100, FatalError, Peter1c, Thomasmeeks, Thetasigmapi, Gregbard, Cydebot, Daniel J. Leivick, Studerby, Doug Weller, Diarmada, Garik, ChrisTW, Coelacan, Nakomis, Nowimnthing, Carole Jean, Raeven0, RFerreira, Rotundo, Matthew Proctor, Heroeswithmetaphors, SomeHuman, RDT2, Clan-destine, Numerator42, ARTEST4ECHO, Gdo01, Rubensni, Odinbolt, Storkk, Ghmyrtle, Somerset219, Gökhan, JAnDbot, Husond, KevinCLovesU, Giler, TAnthony, Windinmysails, Magioladitis, Tctopcat, Cooper24, VoABot II, Jhw57, Silentaria, Seansinc, Baristarim, Jasonid, FisherQueen, MartinBot, Anaxial, Earthdenizen, Sheep2000, Alanbkahn, MistyMorn, R613vlu, Master shepherd, KylieTastic, Darkfrog24, Varunpramanik, Idioma-bot, VolkovBot, Jfspine, Ryan032, Philip Trueman, Zamphuor, Cahill1, Tomsega, Wikidemon, Steven J. Anderson, IllaZilla, Gekritzl, Don4of4, Vgranucci, Goatghost, Mauriceivy, Marijuanarchy, Falcon8765, Xasthuresque, MCTales, Paradoxos, Ziphilt, Theoneintraining, Dewine, SieBot, Malcolmxl5, Dawn Bard, Bugg1979, Jojalozzo, ThAtSo, OKBot, Gorrrillla5, Bowei Huang 2, Emsantiago, JL-Bot, Faithlessthewonderboy, Twinsday, ClueBot, IceUnshattered, Hoejamma, Saddhiyama, Goochylittlepig, LizardJr8, Masterpiece2000, MollyMac13, Perjef, Rangestand, Stevenrasnick, Editor2020, Ban Bridges, DumZiBoT, Bridies, AgnosticPreachersKid, Nathan Johnson, Jonxwood, SilvonenBot, Frood, Duckwing, Aunt Entropy, Tayste, Tcravens, Addbot, Laurinavicius, Socerizard, HandThatFeeds, Yobot, TaBOT-zerem, Azcolvin429, Dmarquard, Bbb23, AnomieBOT, Jim1138, Shambalala, Mann jess, Materialscientist, Acmartin, Xqbot, Killugh, Inferno, Lord of Penguins, Srich32977, AV3000, Sekaniscott, GrouchoBot, Cos Oro, Amperman, Neil Clancy, FrescoBot, Ibleedgreencda, AriTotle, Haeinous, Buddhaamaatya, FriedrickMILBarbarossa, Hamtechperson, Brad Polard, Thinking of England, Serols, Night Jaguar, Jandalhandler, Thrissel, Danishblue, Athene cheval, Lgyure, TjBot, Noommos, Tesseract2, Rayman60, Animan360, Appu001, WikitanvirBot, Dewritech, Tommy2010, Dcirovic, Moonlight8888, Tb20000, Elektrik Shoos, LikeableLefty, Wikignome0530, Nanib, Jesanj, Noodleki, Jess, FelixG1995, Wisdomtenacityfocus, AxiomOfFaith, Pokbot, Jimbonano, AUN4, ClueBot NG, Sofa Cylon, Somedifferentstuff, Heni123, Catlemur, ForgottenHistory, Newfoundland25, DonaldRichardSands, Lahill94, Snorrymouth, Widr, PtC61891, Helpful Pixie Bot, Sinestar, Curb Chain, TheBarchettaMan, Theinfamousm, Headlearn55, A.K.Khalifeh, Snow Rise, MrBill3, Anonymous02138, Humayun360, ChrisGualtieri, Saedon, Mjsteiner, Rhlozier, Cerabot~enwiki, Lugia2453, JustAMuggle, Me, Myself, and I are Here, Tentinator, Lboniello, JoanReed, Gary Berg-Cross, Wimmsk, Atho Phink, Hclasalle, Humanismws, CAPTAIN RAJU, Babymissfortune, Castellanepurdy, Bar Sargis, Marvellous Spider-Man, Bender the Bot, Differentnotbetter, EpicMan, Steve button the first, The Spaghetti Kid, Mcbrarian, Neik Krane and Anonymous: 471

- **Humanism and Its Aspirations** *Source:* https://en.wikipedia.org/wiki/Humanism_and_Its_Aspirations?oldid=797411383 *Contributors:* Delirium, VeryVerily, Jeffq, Bearcat, GreatWhiteNortherner, Duncharris, Hydriotaphia, Skollur, Toussaint, WhyBeNormal, RussBot, SmackBot, Jim62sch, Zyxw, Eav, Eudaemonic3, Dacoutts, CPAScott, Cydebot, Bellerophon5685, Ghmyrtle, David Eppstein, Grantsky, DadaNeem, Jevansen, BertSen, VVVBot, LeadSongDog, Reyka, Addbot, Tassedethe, OlEnglish, Happyhuman, Faolin42, Spicemix, KLBot2, BigJim707, MusikAnimal, Allecher, Jfhutson, Ninmacer20, AMMeier, Darwin machine, Wcrea6 and Anonymous: 10

- **Humanist Manifesto** *Source:* https://en.wikipedia.org/wiki/Humanist_Manifesto?oldid=787389436 *Contributors:* Ed Poor, VeryVerily, Duncharris, Brockert, Vsmith, Stephan Leeds, Vcelloho, Derktar, Toussaint, Tpkunesh, 2ct7, WhyBeNormal, Mercury McKinnon, SmackBot, Swed Simon, ProveIt, JonHarder, Austinfidel, Kpearson99, Cydebot, Bellerophon5685, Ghmyrtle, VoABot II, Murraylove, Ted.strauss, HarZim, Tdadamemd, VVVBot, Fra59e, Vvevo, Anarchocelt, Masterpiece2000, Addbot, CL, Tide rolls, Luckas-bot, Yobot, AnomieBOT, Happyhuman, TechBot, FrescoBot, Airborne84, Yudvirsidhu, Tesseract2, BillyPreset, ZéroBot, Marcocapelle, Jfhutson, Aisteco, Ivangarcia44, Medecine Wheel, Sondra.kinsey, Da rascal, Zana Abdulkareem and Anonymous: 38

- **Reason** *Source:* https://en.wikipedia.org/wiki/Reason?oldid=797688283 *Contributors:* Mav, Toby Bartels, Enchanter, Ortolan88, Zoe, Juwiley, Heron, Ryguasu, Stevertigo, Michael Hardy, Pit~enwiki, Lousyd, Zeno Gantner, Dcljr, Glenn, Andres, TonyClarke, JamesReyes, Hashar, Peter Damian (original account), Timwi, Vancouverguy, Hyacinth, RedWolf, Wereon, Alan Liefting, Giftlite, Ausir, Mboverload, Chowbok, Andycjp, MisfitToys, Karol Langner, Oneiros, Bodnotbod, TiMike, Herschelkrustofsky, Picapica, GreedyCapitalist, Poccil, CALR, Discospinster, ElTyrant, Rich Farmbrough, Guanabot, Florian Blaschke, Bender235, Chalst, Wareh, Dustinasby, Bobo192, Cmdrjameson, Aquillion, Martinjoh, Pearle, Mdd, Szczels, Andrewpmk, SlimVirgin, Hu, Snowolf, Wtmitchell, Staeiou, Jesvane, SteinbDJ, Capecodeph, Ceyockey, AlexTiefling, Mel Etitis, Woohookitty, Mindmatrix, Missdipsy, Djilk, WadeSimMiser, Firien, Matilda, BD2412, Kbdank71, Rjwilmsi, Mayumashu, Vary, Anrwlias, Marax, Nihiltres, RexNL, Alexjohnc3, Jrtayloriv, Spencerk, Chobot, DVdm, YurikBot, Wavelength, RussBot, Pi Delport, Ferrydun, Odysses, Rick Norwood, Nirvana2013, Jaxl, Apokryltaros, Zwobot, ZhaoHong, Denis C., Tomisti, Wknight94, Andrew Lancaster, Mike Dillon, E Wing, Thelb4, Magic.dominic, Fastifex, Sardanaphalus, Anton n, SmackBot, Lestrade, Unyoyega, Vald, Jagged 85, Zaqarbal, Gilliam, Saros136, Trebor, Octahedron80, DHN-bot~enwiki, Oatmeal batman, Zachorious, George Ho, E946, Johndemers, Yidisheryid, Computerman45, COMPFUNK2, Flyguy649, Radagast83, Whoistheroach, Richard001, Jon Awbrey, Bdiscoe, Salamurai, SashatoBot, ArglebargleIV,

RTejedor, OcarinaOfTime, Žiga, NeveVsMackie, 16@r, Robert Bond, Meco, Waggers, SandyGeorgia, E-Kartoffel, Ginkgo100, Kissedsmiley, Dreftymac, JoeBot, Catherineyronwode, Mrdthree, Lenoxus, Cyrusc, Ezadarque, N2e, Penbat, MaxEnt, Gregbard, Julian Mendez, Niculaegeorgepion~enwiki, Click23, Letranova, Thijs!bot, Dogaroon, Helgus, Esowteric, Pelz2k, Escarbot, Porqin, Widefox, Emeraldcityserendipity, Dylan Lake, Gh5046, Alphachimpbot, JAnDbot, Skomorokh, The Transhumanist, BenB4, Gaeddal, TAnthony, Zorro CX, Magioladitis, Xenogyst, Tito-, The Enlightened, Ensign beedrill, Cathalwoods, Edward321, WLU, Silqworm, Anarchia, R'n'B, Ooga Booga, J.delanoy, Tikiwont, Idontthinkso, Lonjers, Tyducv, Fredeaker, Belovedfreak, Ajfweb, Pchackal, CardinalDan, Idioma-bot, Signalhead, Paulherrick, TXiKiBoT, Jomasecu, Jazzwick, Hensa, Jorgamun, The Tetrast, Apollojai, Saturn star, FunkDemon, Profvsprasad, Yk Yk Yk, Synthebot, Sapphic, HiDr-Nick, Aednichols, Prom2008, Newbyguesses, KyZan, SieBot, Euryalus, DarknessEnthroned, Bentogoa, Mimihitam, Ellusion, Sunrise, Iahklu, SlackerMom, Loren.wilton, ClueBot, IceUnshattered, Rrsullivan, Shinpah1, Jwyg, Dmyersturnbull, SchreiberBike, Thingg, JDPhD, ClanCC, XLinkBot, Poli08, WikHead, The Rationalist, JohnyGoodman, Addbot, Cxz111, Non-dropframe, Queenmomcat, Danie Tei, Reas0nn, MrOllie, Redheylin, Avazelda13, AnnaFrance, Tassedethe, Abiyoyo, Tide rolls, Lightbot, Liang9993, Legobot, Ptmohr, Luckas-bot, Yobot, Reindra, Ningauble, AnomieBOT, Jim1138, Jeff Muscato, Carturo222, MauritsBot, ⁇, Unscented, AV3000, Petropoxy (Lithoderm Proxy), Omnipaedista, Lelouch.Angelo., Aaron Kauppi, 11cookeaw1, Alika13, Metalindustrien, Zara9, FrescoBot, Tobby72, Parth24, Machine Elf 1735, Skylarity, Pinethicket, Momergil, Σ, AmesJussellR, Jandalhandler, Wotnow, CircularReason, Lotje, LilyKitty, Edinwiki, Prmwp, Stryder29, Stroppolo, Editor99999, Mean as custard, Walkinxyz, WildBot, Tesseract2, EmausBot, Themastertree, Faolin42, Finn Bjørklid, Solarra, Alfredo ougaowen, Mburdis, Shmilyshy, Staszek Lem, Ratpow, Crown Prince, Шиманський Василь, DASHBotAV, Mjbmrbot, ClueBot NG, CocuBot, Loew Galitz, Widr, Sarahmworthy, MerIIwBot, Helpful Pixie Bot, Nightenbelle, Uknowwhatsup, Strike Eagle, Night-changer, Vilij, PhnomPencil, Langchri, ElphiBot, Interchangeable, Dzforman, JacobTrue, Davidiad, Marcocapelle, Mariraja2007, FiveColourMap, Houn29, Altaïr, G0ldenphoenyx, BLACK COMMANDO, Philosus~enwiki, Hardjl4, BattyBot, Pratyya Ghosh, ChrisGualtieri, Sfgiants1995, Me, Myself, and I are Here, ElaineF423, Biogeographist, Mumtaz muhammed, Thennicke, Claireney, Loraof, Lancerness, Blahomg, KingChainsaw13, KasparBot, Mikekuang108, Philosophynut1952, Jhonnygzu, VirtueofKnowledge, Doulph88, InternetArchiveBot, Motivação, FluteANINJA123, Sensisparis, Stoicjoe, VeritasSpiritus, Wmdly, Magic links bot, 18hagak, ReasoningBrain and Anonymous: 267

- **Ethics** *Source:* https://en.wikipedia.org/wiki/Ethics?oldid=797049921 *Contributors:* AxelBoldt, General Wesc, Bryan Derksen, The Anome, Malcolm Farmer, Ed Poor, RK, Christian List, Anthere, Camembert, KF, Mkmcconn, Stevertigo, Michael Hardy, EddEdmondson, Wshun, Owl, Nixdorf, Ixfd64, Lquilter, Sannse, Mdebets, Copsewood, Ronz, EntmootsOfTrolls, Notheruser, Angela, Andrewa, Glenn, Ciphergoth, LouI, Poor Yorick, Andres, Jeandré du Toit, Evercat, Xgkkp, TonyClarke, Faré~enwiki, Qwert, Adam Conover, Emperorbma, Timwi, Bemoeial, Andrewman327, Rednblu, Wik, Selket, Furrykef, Saltine, Lsolum, Shizhao, Banno, Carlossuarez46, Jni, Phil Boswell, Robbot, Astronautics~enwiki, Yelyos, Mayooranathan, Kesuari, Academic Challenger, Rursus, Ivan~enwiki, Hadal, Michael Snow, Dhodges, Pengo, DocWatson42, ShaunMacPherson, Bogdanb, Meursault2004, Cool Hand Luke, Ajgorhoe, Guanaco, Sundar, Siroxo, JRR Trollkien, Maclyn611, Thijs!, Andycjp, Alexf, Mike R, Sparticus, Antandrus, Beland, Augusta~enwiki, Thorn969, Elembis, Savant1984, Kaldari, Jeroboambramblejam, Jossi, Karol Langner, Rdsmith4, JimWae, DanielDemaret, EuroTom, Karl-Henner, Jh51681, Davidstrauss, Canterbury Tail, Flex, Lucidish, Rfl, Mindspillage, Yuren~enwiki, Discospinster, Rich Farmbrough, FranksValli, Fsvallare, Vsmith, Ardonik, Calion, Bishonen, Notinasnaid, Antaeus Feldspar, SocratesJedi, Dbachmann, Bender235, JoeSmack, Brian0918, CanisRufus, El C, Mwanner, Sietse Snel, RoyBoy, Tbannist, Causa sui, Bobo192, Icut4you, NetBot, Viriditas, Skywalker, L33tminion, SpeedyGonsales, Kaganer, Sasquatch, TheProject, Hagerman, HasharBot~enwiki, Danski14, Alansohn, Ungtss, Ryanmcdaniel, Walter Görlitz, Andrewpmk, Kalle~enwiki, Logologist, Lightdarkness, TeresaTeng, Snowolf, Velella, Simplebrain, Cburnett, RJII, RainbowOfLight, Pauli133, Paraphelion, Axeman89, HenryLi, Red dwarf, Ultramarine, Sylvain Mielot, Velho, Woohookitty, Twobitsprite, Yansa, Serche, Kzollman, Rimmeraj, Jeff3000, Joshdick, Gimboid13, PeregrineAY, Dysepsion, Graham87, Marskell, BD2412, FreplySpang, Grammarbot, Edison, Josh Parris, Jorunn, Rjwilmsi, Joe Decker, Саша Стефановић, Vary, Tawker, Ligulem, DouglasGreen~enwiki, Afterwriting, Bhadani, TheGWO, Aapo Laitinen, Sango123, Lotu, Vuong Ngan Ha, Daegred, Titoxd, FlaBot, Nivix, Rbonvall, RexNL, Gurch, Jrtayloriv, Nick81, EronMain, Common Man, JonathanFreed, ...adam..., Chobot, Turidoth, DVdm, Korg, Bgwhite, Uriah923, YurikBot, Wavelength, Sceptre, Paxik~enwiki, Phantomsteve, Justpetehere, Arex, SpuriousQ, Stephenb, Flo98, Gaius Cornelius, Eleassar, Vincej, Pseudomonas, KSchutte, Wimt, Cunado19, Ugur Basak, Anomalocaris, NawlinWiki, Nirvana2013, Welsh, SAE1962, 24ip, Amcfreely, Kwh, Adhall, Stevenwmccrary58, Sebleblanc, Elkman, CLW, Gilemon, Phaedrus86, Richardxthripp, CQ, BenBildstein, Ninly, RDF, Closedmouth, Endomion, KGasso, Dspradau, Anclation~enwiki, Willtron, ArielGold, Rhwentworth, Zip3~enwiki, Error slow, Allens, Ephilei, Mebden, Voiceimitator, Tim314, Infinity0, DVD R W, Tom Morris, Luk, Sardanaphalus, Veinor, SmackBot, Unschool, Cubs Fan, KnowledgeOfSelf, DCDuring, Unyoyega, Pgk, KocjoBot~enwiki, Pkirlin, Delldot, HalfShadow, Kikuchiyo, Vassyana, Gilliam, Donama, Ohnoitsjamie, JAn Dudík, Kurykh, NCurse, Tito4000, JDCMAN, Thumperward, Oli Filth, R dizzle, JONJONAUG, Go for it!, DHN-bot~enwiki, Gyrobo, Ig0774, Can't sleep, clown will eat me, Jahiegel, Onorem, John C PI, Yidisheryid, Kennovak, Xyzzyplugh, Addshore, Edivorce, Dramageek, AndySimpson, Rubbaducky42, Cybercobra, Nibuod, Legaleagle86, AnaisSatin, Cubbi, SnappingTurtle, Whoistheroach, Dreadstar, Dacoutts, Richard001, Clean Copy, KI, Kleuske, Fuzzypeg, DMacks, Acdx, Where, Curly Turkey, Horiavulpe, BrianH123, Byelf2007, Cast, Visium, Caravaca, Zearin, AmiDaniel, Dialectic~enwiki, Aragorn23, Heimstern, Nomirror, Gobonobo, Disavian, Bo99, NongBot~enwiki, Physis, Ckatz, A. Parrot, Stwalkerster, SQGibbon, Mr Stephen, Lifeartist, TastyPoutine, Condem, Hu12, Vanished user, Fan-1967, Iridescent, Dekaels~enwiki, Seanmason, NativeForeigner, Sam Clark, Aeternus, IvanLanin, Tony Fox, GoDawgs, Gil Gamesh, Clay, Tawkerbot2, Chris55, Dan1679, Jim Wilkinson, Wolfdog, CRGreathouse, Postmodern Beatnik, High Elf, Wafulz, Aherunar, KyraVixen, Dgw, Penbat, Gregbard, LCP, Shanoman, Islington warrior, Cydebot, Jasperdoomen, Hudec, Mato, Gogo Dodo, Carl Turner, GRBerry, Pascal.Tesson, Crudnick, Julian Mendez, Dancter, Dr.enh, HitroMilanese, DumbBOT, Chrislk02, JoshHolloway, Pdemecz, Janoside, UberScienceNerd, Thijs!bot, Epbr123, Bot-maru, Gojiro0, Redhotone, Ramendra Nath, Jeffrey.Rodriguez, Marek69, Carole Jean, NorwegianBlue, James086, X201, Apierrot~enwiki, Leon7, Nick Number, Crazy head, Natalie Erin, Escarbot, KrakatoaKatie, AntiVandalBot, Sheilrod, Luna Santin, Wiki4fun, Jj137, Cinnamon42, Sonofecthelion, Modernist, MECU, Myanw, Sluzzelin, JAnDbot, Husond, Davewho2, Barek, MER-C, Skomorokh, The Transhumanist, Andonic, OckRaz, Hut 8.5, TAnthony, Meeples, Magioladitis, Bongwarrior, VoABot II, Sdcrym, Ling.Nut, Y0kai, Pushnell, Sharon habel, Prokaryote, Snowded, Theroadislong, Nposs, Shim'on, Loonymonkey, AsgardBot, LookingGlass, Allstarecho, Ludvikus, Glen, DerHexer, JaGa, Grunge6910, Edward321, Johnbrownsbody, Juliangarside, Steevven1, Posidonious, Adriaan, Jackson Peebles, MartinBot, Robert Daoust, Grandia01, Arjun01, Anarchia, Axlq, Anaxial, Danjyates, Dabringer, Cadorj, EverSince, Creol, Cyrus Andiron, J.delanoy, Pharaoh of the Wizards, UBeR, Numbo3, Uncle Dick, Drsuzyb, Cymbalta, Maurice Carbonaro, Randomword, Ginsengbomb, Maproom, Katalaveno, McSly, Gthb, Mikael Häggström, Kelvin Knight, HiLo48, NewEnglandYankee, SJP, Doomsday28, Madhava 1947, Juliancolton, Evb-wiki, DorganBot, Pastordavid, Grizzlegritz, Ja 62, S (usurped also), Useight, Poulton~enwiki, The enemies of god, CardinalDan, Idioma-bot, Funandtrvl, Lights, 28bytes, Macedonian, VasilievVV, Gurpreet gupy, Jimmaths, Aesopos, Philip Trueman, Mkcmkc, TXiKiBoT, Oshwah, Semper Ama, Ann Stouter, Markus Dragonblood, Charlesdrakew, Bbulkow, Oxfordwang, John Carter, COLBERTRULZ,

Abdullais4u, Jackfork, LeaveSleaves, Zsalamander, Seb az86556, JonathanDLehman, Witchzilla, CO, Yk Yk Yk, Poltair, @pple, Metadat, Pete Hartree, Doc James, AlleborgoBot, Symane, Quantpole, Ljay2two, Danmo99, SieBot, Williamlindgren, Tosun, Fixer1234, Zebardyn, Jauerback, JoeMauer, Washdivad, Dawn Bard, Caltas, Mackenzie kenzy, G0dsweed, Gravitan, Arda Xi, Bentogoa, Tiptoety, Jamesbdunn, Pm master, Faradayplank, Ks0stm, Mkeranat, Fatallight, Kudret abi, Denisarona, Kanonkas, Erik Daniel P, XDanielx, Lethesl, ClueBot, Jbening, FoxDiamond, Kai-Hendrik, Justin W Smith, Fyyer, The Thing That Should Not Be, Jagun, Gawaxay, Meekywiki, Razimantv, Pete unseth, Shaded0, Niceguyedc, LizardJr8, Neverquick, ChandlerMapBot, Jimlyttle, DragonBot, Deselliers, Excirial, Anonymous101, Jusdafax, Robbie098, Benedictadolson, Coralmizu, Eeekster, Lartoven, Rhododendrites, JamieS93, ShowToddSomeLove, Dekisugi, Editor510, Liaun, Thingg, Aitias, 7, Carlroddam, Editor2020, YouRang?, XLinkBot, Jytdog, Pfhorrest, Poli08, Boyd Reimer, David Delony, Moralsandethics, Stitchill, Tee2008, Cmr08, WikiDao, Chaos4.6692, Adrian Scholl, Dysart09, Addbot, Proofreader77, Grayfell, Some jerk on the Internet, Friginator, Ronhjones, Elmondo21st, Mhkhng, Jpoelma13, Stephendupont69, LaaknorBot, Hi jagdish, AndersBot, Favonian, LemmeyBOT, LinkFA-Bot, West.andrew.g, Joshdenomey76, Justian, Phdarts, Z6mu28OO, Numbo3-bot, Tide rolls, Bfigura's puppy, OlEnglish, Verazzano, Willondon, Gail, Jarble, Ben Ben, Legobot, Luckas-bot, ZX81, Yobot, Themfromspace, TaBOT-zerem, Edoe, THEN WHO WAS PHONE?, The Philosophical Penguin, Eric-Wester, Synchronism, Wiki Roxor, AnomieBOT, DemocraticLuntz, Demonhunter698, 1exec1, Ignacio500, Killiondude, Jim1138, Galoubet, JackieBot, BlazerKnight, Piano non troppo, AdjustShift, Shambalala, Flewis, HRV, Materialscientist, Citation bot, Acidrainyday, Xqbot, Sketchmoose, Capricorn42, 4twenty42o, Timmyshin, J1440, Tomwsulcer, Coretheapple, RadiX, Bioethica Americana, Omnipaedista, WhitePlazma, Frankie0607, Wiki emma johnson, Hadesplague, Eisfbnore, Rogozub, Sesu Prime, ⁇, Nixón, Nagualdesign, Dreampsy, Magic.Wiki, Grucio, FrescoBot, UrghartSatisPeter, T of Locri, Tobby72, Buziatov, Razor820, Nonoffensive1234, Sky Attacker, R9obert, Calixtekabore, Lightningfox, Galorr, D'ohBot, Zero Thrust, Zackmiri, HJ Mitchell, Markalanfoster, Yoichi123, Airborne84, Pathwrote, Larissa19662, Biker Biker, Pekayer11, Pinethicket, I dream of horses, Haaqfun, LittleWink, Serols, Σ, Vardabooks18, Abethecop, Meaghan, Jauhienij, Du dilsta, R12556, FoxBot, TobeBot, Prototypebeta, Pollinosisss, DixonDBot, ItsZippy, Dr. Rubinsaw, Metalhead103, LilyKitty, Mishae, Aoidh, Reaper Eternal, Kid kiru, Diannaa, Hitnrun321, Sirkablaam, Tbhotch, Reach Out to the Truth, Literateur, Kamiel79, Deanmullen09, DARTH SIDIOUS 2, Onel5969, Mean as custard, TjBot, Ripchip Bot, Gleaman, George Richard Leeming, EmausBot, Orphan Wiki, Acather96, WikitanvirBot, Ken95, Super48paul, Gildedtiger, Taymaz.azimy, Faolin42, Razor2988, GoingBatty, RA0808, Minimac's Clone, NotAnonymous0, Tommy2010, Wikipelli, Dcirovic, K6ka, Zero939, PBS-AWB, Franthor, Ida Shaw, Fæ, Somebodhi, The Nut, Anaselshamy, Rails, SporkBot, AManWithNoPlan, Freakshownerd, Mostafa.Hassan, Wayne Slam, Homama~enwiki, Fishbefish, Erianna, Giuseppe Fusco, Hiernonymous, IGeMiNix, L Kensington, Barrrower, Donner60, Wikiloop, Beautiful.wave, Phronetic, Loot23, Pierpietro, Terra Novus, 28bot, ClueBot NG, Egbarker, Manubot, This lousy T-shirt, Chester Markel, Frietjes, O.Koslowski, Dream of Nyx, M. Ahsanul Haq, Widr, WikiPuppies, Lawsonstu, Helpful Pixie Bot, Ashirwad Gogoi, HMSSolent, BG19bot, Deborahallen13, MPSchneiderLC, Langchri, Graham11, Wiki13, MusikAnimal, Frze, Hurricane2u, Marcocapelle, Squig2510, Mark Arsten, Rm1271, Nflaxington, Hero777, Idealisticnihilist, Ethical89, InstLocalGovt, MrBill3, Theconsequentialist, Brad7777, Polmandc, Glacialfox, Gavinparr2, Klilidiplomus, Group Cirrcunciser, Felixthehamster, BattyBot, User name 12, ~riley, Rickybrewhaha, Cyberbot II, Mediran, TheJJJunk, Khazar2, Tbeasley0504, IjonTichyIjonTichy, Dohaschmoha, Dexbot, FoCuSandLeArN, Webclient101, Mogism, Saehry, TwoTwoHello, Kc avatar, Lugia2453, ShawnTang, Jamesx12345, Eredner, Jdc1197, Adam2828, Kevin12xd, GeoffHoeber, Me, Myself, and I are Here, Bot8880090, Reatlas, Paulie2212, Iapgeoethics, CsDix, Spinozone, Pdecalculus, Michipedian, Ketxus, Syntaxerrorz, Brotherxandepuss, Ginsuloft, Little katrina, I.am.a.qwerty, Edouarddp, Patty.rich, Kind Tennis Fan, Secular1, Liz, Alishayankhan0, Gubino, Senorjefe420, JKMan0123, Boy238128, Kathryn Erskine, Cammavin85, Yabut01, Fridlida, Sbchristianlowe, 7Sidz, Campsite55, Akakane2, Starfishprime1989, Monkbot, Cybersecurity101, Itamarkalimi, Mike2085, Jakeiscool1223, Btlastic, J8089709, Bstone22, Janepharper, Ethics!fred, Almajestic, Jiten Dhandha, Adrianna4405, Ghosthux, BrendonYuri, Gypsy Danger Dynamite, Giuseppe.dicapua, Sfg69, Mypowerpuff, Hulash Barupal, Zortwort, Aguilarv, MurderByDeadcopy, Dumfish15, Jnbeta, Jerodlycett, Andrewmeehan17, KasparBot, Sweepy, Honoriodeh, Feminist, Amherstpsychometrics, You better look out below!, Mohammad1985k, Moonlightx75, Rast goftar, Rjp1909, Hatripet, JoelGreen, Mssoydan, Opgroove, InternetArchiveBot, Cathrotterdam, Neha Nand, Marianna251, Gulumeemee, Grammar Man Jack, MMC's, FutureOfEthicallyEffectiveLeadership, SarahDavies, Imran Sarwar, Quasar G., Justeditingtoday, Here2help, DrStrauss, Medrobbins, Julius Kay, Wmdly, Mikeb686, Daniel0Wellby sch, Magic links bot and Anonymous: 1261

- **Social justice** *Source:* https://en.wikipedia.org/wiki/Social_justice?oldid=795404841 *Contributors:* The Cunctator, Derek Ross, -- April, William Avery, Bomfog, AdamRetchless, Heron, Lir, Jrcrin001, Vera Cruz, Kku, Paul A, Kaihsu, Selket, Hyacinth, Greglocock, Populus, Pollinator, UninvitedCompany, Francs2000, Chrism, Altenmann, Hadal, Ancheta Wis, Nikodemos, Tom harrison, Lupin, Orangemike, Everyking, FeloniousMonk, DO'Neil, SWAdair, Gyrofrog, MSTCrow, Neilc, Isidore, Teak (usurped), Utcursch, Andycjp, Antandrus, Tal642, Ice Czar, Rdsmith4, Sam Hocevar, Neutrality, Ukexpat, Montanean, Squash, Grunt, Esperant, Quill, DanielCD, Discospinster, Rich Farmbrough, Florian Blaschke, Paul August, Bender235, JoeSmack, Lycurgus, RoyBoy, Bobo192, Janna Isabot, Chapium, Viriditas, Elipongo, Aquillion, Shanen, MPerel, Knucmo2, Alansohn, Diego Moya, Primalchaos, Wikidea, Markjlyon, Kanodin, Cgmusselman, Shmooth, RainbowOfLight, Kaibabsquirrel, Jdege, Nightstallion, Ceyockey, Hijiri88, Stemonitis, Bobrayner, Mel Etitis, Woohookitty, Manjeetchaturvedi, I64s, Randy2063, Isnow, Macaddct1984, Tetraminoe, Laurel Bush, Toussaint, KHM03, BD2412, FreplySpang, TobyJ, Sjö, Nightscream, Janosabel, TJive, Red King, TexasDawg, The wub, Falphin, Eubot, WWC, Who, RexNL, Ewlyahoocom, Str1977, Fephisto, Jfraatz, Common Man, King of Hearts, Frappyjohn, David91, Bgwhite, Gwernol, RogerK, Wavelength, Hawaiian717, Stan2525, RussBot, Sillybilly, Bhny, Mark Ironie, Yamara, Rsrikanth05, Wimt, Thane, NawlinWiki, Bruxism, Mike18xx, Yahya Abdal-Aziz, Rbarreira, Raven4x4x, Number 57, Danlaycock, Izuko, Pierpontpaul2351, Jpeob, SgtPepper, Theda, Closedmouth, Mais oui!, Spliffy, Caballero1967, Allens, ModernGeek, DVD R W, CIreland, Tom Morris, NickelShoe, Luk, C mon, Rcronk, Resolute, Crystallina, SmackBot, MattieTK, Nkrupans, Demonweed, Marcusscotus1, Nihonjoe, Reedy, InverseHypercube, McGeddon, C.Fred, Ramdrake, Kahuzi, Bill3000, Randy Schutt, Delldot, Cla68, Brossow, Jwestbrook, Gilliam, Hmains, Master Jay, Sabriel~enwiki, Rajeevmass~enwiki, Baa, Firetrap9254, Verrai, Can't sleep, clown will eat me, Lantrix, Sidious1701, Jmlk17, Flyguy649, Dreadstar, Zdravko mk, BullRangifer, Jklin, Kukini, SashatoBot, ArglebargleIV, Euchiasmus, UberCryxic, JohannaHypatia, Heimstern, Gobonobo, Disavian, Benesch, Bilby, Northmeister, Liberty4u, Sxeptomaniac, LaMenta3, Lasifcac, Xionbox, Dl2000, Hu12, BranStark, Boxa~enwiki, Andrew Davidson, Solificus, PapayaSF, Tamino, Joseph Solis in Australia, Onefinalstep, Sam Clark, Tony Fox, Amakuru, TheOCflash, Tawkerbot2, Confutarus, DKqwerty, Rhcameron, Fvasconcellos, CmdrObot, Asteriks, Nunquam Dormio, Jsrhem, Thomasmeeks, Lentower, Leujohn, Marosha, Memills, Bill Sayre, Spoxox, Yaris678, Atomaton, Ntsimp, Goldfritha, Destructor2006, Wildnox, Tawkerbot4, Doug Weller, EqualRights, Salvor Hardin, Pustelnik, Golffan0112, PKT, Thijs!bot, Epbr123, Jdm64, Anupam, Headbomb, Simeon H, John254, Frank, Stratvic, Mezlo, KrakatoaKatie, AntiVandalBot, Luna Santin, Seaphoto, Bharshaw, Emeraldcityserendipity, Baiettis1, Zero g, Datavortex, David Shankbone, 1Rabid Monkey, JenLouise, Koncorde, MER-C, Acroterion, Magioladitis, VoABot II, Jackbirdsong, Swpb, Animum, Fang 23, Glen, DerHexer, Edward321, Syntacticus, Seba5618, MartinBot, R'n'B, J.delanoy, Pharaoh of the Wizards, Numbo3, Groove01, Discott, Dbiel, Cocoaguy, Dbeckingham, Kataleveno, Hakufu Sonsaku, Redundance, HiLo48, Zackisnot, NewEnglandYankee, Drake

Dun, DadaNeem, J Readings, Juliancolton, Foofighter20x, Abdars, Scott Illini, Barak181, Malik Shabazz, VolkovBot, Jimmaths, Childhoodsend, TXiKiBoT, Oshwah, Wikidemon, Spoisp, John Carter, Martin451, LeaveSleaves, Room429, Snowbot, B1ack1ma63rs, Jevergreen, Gen. von Klinkerhoffen, NinjaRobotPirate, Wavehunter, Kehrbykid, SecretaryNotSure, Gessa~enwiki, Scotts33, SieBot, StAnselm, Euryalus, Bachcell, Winchelsea, Caltas, Rtdem, Flyer22 Reborn, Ursasapien, Rodney Shakespeare, Jruderman, Macy, Egoldstein84, MrsKrishan, Vice regent, Capitalismojo, Ascidian, MarcWmA, Faithlessthewonderboy, Martarius, Elassint, ClueBot, Tdlewis77, The Thing That Should Not Be, Lwalkingwoman, Der Golem, Mezigue, Shaded0, CounterVandalismBot, Cirt, Passargea, Solar-Wind, Athenav00, Excirial, Goodone121, Hughsont, Rhododendrites, NuclearWarfare, Utopial, Dryir Lent, TheRedPenOfDoom, JayWhitney, SchreiberBike, Xyz1000, Aitias, Iota 9, Doriftu, Zookiss, DumZiBoT, XLinkBot, Jytdog, Jovianeye, SilvonenBot, Frood, Addbot, Tcncv, Kongr43gpen, Jncraton, Linderdog, CanadianLinuxUser, Fluffernutter, Cst17, MrOllie, DFS454, Liassic, Clearlight418, Chzz, Xxdriftwoodxx, Numbo3-bot, Lerpiniere, Lightbot, Urpunkt, Jarble, Rhodospirillum, Luckas-bot, Yobot, Jasongreene463, Wikivol, Kman1111000, KamikazeBot, SwisterTwister, There'sNoTime, South Bay, AnomieBOT, Jim1138, IRP, Cptnono, OpenFuture, Kingpin13, Britans, RobertEves92, Citation bot, ArthurBot, Rdr66, MauritsBot, Xqbot, Zad68, Cskinns89, Capricorn42, CrushSoda, Hammersbach, MakeBelieveMonster, Jmundo, Locos epraix, Livrocaneca, Srich32977, Andywozhere, S0790601, RibotBOT, Learner001, Slava1234, FrescoBot, LucienBOT, Pepper, Vishnu2011, Cannolis, HamburgerRadio, Seanrife, Pinethicket, Edderso, LinDrug, Sketch66, Thinking of England, Serols, Fixer88, Hand creamer, Motorizer, Foobarnix, Artefactme, Cullen328, White Shadows, Karnawalski, Cooperate23, Issacomm, Lotje, Fox Wilson, Vrenator, LilyKitty, Denzl480, Nothing but words, JoeCarter888, Diannaa, Philos008488, Tinpac, Rafafederer, Alph Bot, MShabazz, Greenmint, Saunan, Esoglou, Orphan Wiki, Pjzed, Socialjustice23, Britannic124, GoingBatty, RA0808, Jadeslair, Shamhat456, Tommy2010, Winner 42, Wikipelli, Dcirovic, Djembayz, Moxtheox, JWestX, ZéroBot, John Cline, Chuck Baggett, Scutter7282, Akerans, A930913, MbcoopMD, Unreal7, EWikist, Gaarmyvet, LastDodo, Erianna, Tperla, Donner60, Yasbhar, Pochsad, Tricee, Mcc1789, ClamDip, Rocketrod1960, Sonicyouth86, ClueBot NG, Entienne01, J.korwin-mikke, Snotbot, Delusion23, Djodjo666, Mesoderm, Hrsiddique, Widr, Theopolisme, Jk2q3jrklse, Helpful Pixie Bot, Astute neophyte, Strike Eagle, Gob Lofa, WNYY98, Jeraphine Gryphon, BG19bot, Emikoala, Keepingleft, Markushudson, JohnChrysostom, User1961914, North911, Meclee, Polmandc, News Historian, Wannabemodel, RGloucester, BattyBot, Solntsa90, Lukas[23], Cyberbot II, Noodles Addiction, Tech77, Isarcontario, Eoxenford, IjonTichyIjonTichy, Iman servatjoo, AutomaticStrikeout, Mogism, VictorD7, Lugia2453, Frosty, Ell96, The Triple M, Edgeofdesire0, Ueutyi, Epicgenius, CsDix, Seralinda, Quartalist245, Sooosw, 3abos, NYBrook098, DavidLeighEllis, Yelysavet, BartStewart1, Jianhui67, Drchriswilliams, Erinflynn, Akalank, Liz, Hisashiyarouin, Aidansheerin, Stamptrader, Csutric, DuchessofGuyenne, Taxi502, Sabdoo, Lagoset, Werhdnt, Zumoarirodoka, Riceissa, TerryAlex, 3primetime3, DissidentAggressor, A falderal, Bammie73, Davisonio, Karmanatory, Parsonderperson, TrainerRedDer, Mr. Magoo and McBarker, Theeditorofpages, FourViolas, JQTriple7, Sangdeboeuf, Godsy, PeterTheFourth, 1989, Vivekkush1983, GYaneli789, User000name, Feministwhiteknight666, MurderByDeadcopy, Povertydave, Nøkkenbuer, CyanoTex, KasparBot, Willshetterly, MusikBot, Equinox, Tumblrayeaye, Theeditor55, Highlime, Peter Damian Quill, Hbs~dewiki, Dabrams13, SwordofStorms, ImHere2015, HenryGarden1000, Sturgeontransformer, AshyPhi, Parsley Man, InternetArchiveBot, OurTy2, Ghost of hugh glass, Got55666, Matthew James Lewis, Ih8sjws1, Jennapope, GreenC bot, Motivação, Bender the Bot, CheeseMasterX, Stoicjoe, Thegame318, Roupe1sm, BDarsow, Weegeeweeg, Dlp34, USstateofmind, Magic links bot, Jole66, Zaenon, Angela Gh, OrthoDemoExpert, Wikieditor2764 and Anonymous: 788

- **Naturalism (philosophy)** *Source:* https://en.wikipedia.org/wiki/Naturalism_(philosophy)?oldid=795023860 *Contributors:* Bryan Derksen, Ed Poor, FvdP, GrahamN, Michael Hardy, Gabbe, Peter Damian (original account), Markhurd, E23~enwiki, Lord Kelvin, Chealer, Daelin, Goethean, Netizen, Nurg, Modulatum, Sunray, Giftlite, FeloniousMonk, Guanaco, JRR Trollkien, Andycjp, Piotrus, PSzalapski, Robin Hood~enwiki, Lucidish, Dave souza, Bender235, Rick MILLER~enwiki, Livajo, Bookofjude, Circeus, Nectarflowed, Unquietwiki, 9SGjOSfyHJaQVsEmy9NS, Aquillion, Thialfi, Pearle, Mdd, Alansohn, Ungtss, Batmanand, Mel Etitis, Kzollman, Jok2000, Tabletop, Ivar Y, GregorB, Isnow, Toussaint, Graham87, Alienus, BD2412, Rjwilmsi, MarSch, Tangotango, TheIncredibleEdibleOompaLoompa, FayssalF, FuelWagon, Ian Pitchford, Doc glasgow, Andystreich, Str1977, Lmatt, Common Man, BradBeattie, YurikBot, Kafziel, Mark Ironie, Markus Schmaus, Srinivasasha, Stijn Calle, Bucketsofg, Tomisti, TheMadBaron, Terfgiu, Chriswaterguy, Fram, Kansaikiwi, Katieh5584, Jade Knight, Veinor, SmackBot, Rtc, InverseHypercube, Swmeyer, ProveIt, Jbyers2, Nethency, Srnec, Yamaguchi??, Portillo, Folajimi, Cowman109, JONJONAUG, Ctbolt, Pasado, Zachorious, Leinad-Z, Vanished User 0001, JonHarder, EvelinaB, Rrburke, BullRangifer, Pickelbarrel, IGod, Omelianchuk, Giovanni33, Atkinson 291, Saek, Hu12, K, Chovain, Jafet, Chris55, Ken McRitchie, CmdrObot, Agathman, GHe, Thomasmeeks, Gregbard, Mattj2, Cydebot, Gogo Dodo, Frosty0814snowman, Ttiotsw, Rajkiran g, Doug Weller, Underpants, RichardCarrier, Peter morrell, Callmarcus, Neil916, Plantago, GordonRoss, Bmorton3, AntiVandalBot, Amcguinn, Cstreet, Danny lost, North Shoreman, Mercury543210, Somerset219, Edwardtbabinski, JAnDbot, Skomorokh, The Transhumanist, PhilKnight, Wasell, General Nolledge, Daniel Cordoba-Bahle, Professor marginalia, Snowded, Animum, Gabrielthursday, Memotype, Dan Pelleg, Rickard Vogelberg, NMaia, MartinBot, Anarchia, E.Shubee, J.delanoy, Nigholith, Dbiel, SharkD, Chiswick Chap, Uhai, Tiggerjay, Geekdiva, Zazizoma, Idioma-bot, VolkovBot, Magnvss, OakMt, Philip Trueman, Myscience, Synthebot, Zarcoen, @pple, W4chris, Jreagan~enwiki, Vitor pk, SieBot, Darrenhendrie, Jrun, Nummer29, Keilana, Jvs, Maynard-Clark, Mimihitam, General Synopsis, RyanParis, Sunrise, DancingPhilosopher, Rumostra, Amanafu, JustinBlank, Dlrohrer2003, Revirvlkodlaku, ClueBot, Binksternet, The Thing That Should Not Be, Jamesmatthiascowpersmithdow, Niceguyedc, Kalem, Sustainablefutures2015, Masterpiece2000, Christian Skeptic, Excirial, CohesionBot, Jusdafax, Brews ohare, Mikaey, MickCallaghan, Chrisscrewball, Versus22, Johnuniq, SoxBot III, BarretB, Aunt Entropy, Jmkim dot com, Good Olfactory, Tayste, Addbot, Some jerk on the Internet, Jojhutton, Danbur, Kenneth Cooke, Preston Wescott Sr., AdrianW.Elder, Fieldday-sunday, Ironholds, Megster0202, Bassbonerocks, Dolphinrider16, LinkFA-Bot, Vasiľ, Xenobot, Quantumobserver, Aletheon, Yobot, Tohd8BohaithuGh1, Theology10101, AnomieBOT, Jim1138, UnitarianUniversalism, Trabucogold, Materialscientist, Xqbot, Freebirth Toad, DermottBanana, AbigailAbernathy, Omnipaedista, Tgreach, Melelaswe, Rickproser, Metaphysicalnaturalist, FreeKnowledgeCreator, FrescoBot, Paine Ellsworth, Adam9389, D'ohBot, Machine Elf 1735, Gdje je nestala duša svijeta, Harlequin Phoenix, AstaBOTh15, Winterst, Pinethicket, Adlerbot, Dude1818, Pollinosisss, Philocentric, Standardfact, WebEdHC, Popovvk, Livingrm, LamboMan7, Diannaa, OlderIgor, Noraft, RjwilmsiBot, Tesseract2, Logictj, John of Reading, The Mysterious El Willstro, Dcirovic, K6ka, PBS-AWB, CanonLawJunkie, Phelsume, AshforkAZ, Wikignome0530, TyA, L Kensington, Emptyviewers, Mcc1789, Scottnmatt, ClueBot NG, Gilderien, YumaTuba, RVscholar, Luxorlover, AlmondRocaFanatic, FleeTheCaptor, Abracadabra777, Helpful Pixie Bot, Nashhinton, Arvardo, ParkerGibbons, Mthoodhood, Yerevantsi, CedricElijahHenry, Harizotoh9, SmittysmithIII, Frontseatdog, Davidcpearce, Nathanielfirst, Dexbot, Sacramentosam, Dfpolis, Jukmnl, Srorourke, Tpopper, Ugog Nizdast, Aubreybardo, Aristotle2013, Ms Sarah Welch, Sergio7z, Welcome1To1The1Jungle, Gouncbeatduke, Eurodyne, Atho Phink, OtisDixon, Engineer1559, CorrectionPleaseXL, BiosocialPolymath, Manguemure, PrimeBOT, DanTheDaniel, Zaenon and Anonymous: 267

- **Utilitarianism** *Source:* https://en.wikipedia.org/wiki/Utilitarianism?oldid=792973183 *Contributors:* Paul Drye, Mav, Bryan Derksen, RK, Enchanter, Detritus, William Avery, Heron, Steverapaport, Michael Hardy, Earth, Dante Alighieri, MartinHarper, Gabbe, Tregoweth, Sir

Paul, Ciphergoth, Rossami, Susurrus, Jeandré du Toit, Evercat, Conti, Charles Matthews, Radgeek, Dtgm, Nv8200pa, Aqualung, Jni, Robbot, Chealer, ChrisG, Altenmann, Nurg, Hadal, Wereon, Aknxy, Nikodemos, Aphaia, Wilfried Derksen, Lussmu~enwiki, Michael Devore, Henry Flower, Gzornenplatz, Jackol, SWAdair, Antandrus, Elembis, PhDP, Oneiros, JimWae, Elz dad, M.e, FrozenUmbrella, Pmanderson, Vanguard, EuroTom, Karl-Henner, Ezekiel Cheever, Joyous!, MakeRocketGoNow, Frikle, Absinf, JasticE, Grunt, Gazpacho, Lucidish, Discospinster, Twinxor, Vsmith, Dave souza, Raistlinjones, User2004, Bender235, Kbh3rd, Sgeo, RJHall, Borofkin, Mjk2357, Sietse Snel, Causa sui, Bobo192, Icut4you, WCityMike, Ruszewski, SpeedyGonsales, Li3crmp, Amerindianarts, ADM, Storm Rider, Abolitionist, Alansohn, Gary, Pinar, SlimVirgin, Lightdarkness, Pion, Wtmitchell, Velella, Leoadec, Omphaloscope, RJII, Pauli133, Versageek, NPswimdude500, Redvers, Ultramarine, Sars~enwiki, Velho, Kelly Martin, Simetrical, Mel Etitis, Timo Laine, Justinlebar, Swamp Ig, Cmillspaige366, JCY2K, Ruud Koot, Tabletop, Bkwillwm, Ch'marr, Bluemoose, X127, Prashanthns, BryanKaplan, Nema Fakei, TrentonLipscomb, Eluchil, MrSomeone, Sin-man, Magister Mathematicae, BD2412, Lev Lafayette, Electionworld, Rjwilmsi, Lars T., KYPark, Bob A, XP1, SMC, Nneonneo, ElKevbo, CalPaterson, Durin, Yamamoto Ichiro, Exeunt, FlaBot, Ian Pitchford, Margosbot~enwiki, Nihiltres, Alex is awake, Pathoschild, Gurch, Hansamurai, Jeremygbyrne, Davidbrake, Common Man, King of Hearts, Frappyjohn, Turidoth, Volunteer Marek, Bgwhite, HoCkEy PUCK, Adoniscik, Gwernol, Vmenkov, YurikBot, Wavelength, Dannycas, Mclayto, Mark Ironie, Gaius Cornelius, Rsrikanth05, KSchutte, Skotte, NawlinWiki, Chick Bowen, Darker Dreams, Crasshopper, Zwobot, Epa101, Zythe, DryaUnda, M3taphysical, Showem, Haon, Roy Lee's Junior, Gregzeng, JWH-Pryor, Tsunaminoai, Canley, Nae'blis, Tolle, Anclation~enwiki, Ben golub, Cassandraleo, Katieh5584, Kungfuadam, Axfangli, Voiceimitator, The Way, Infinity0, That Guy, From That Show!, C mon, Sardanaphalus, Amalthea, Ivolucien, SmackBot, Reedy, InverseHypercube, McGeddon, Unyoyega, Lawrencekhoo, Ryanlintelman, Yamaguchi☒☒, Gilliam, Hraefen, Vincent Vecera, Chris the speller, J.L.Main, Bartimaeus, NCurse, Master of Puppets, Asasa64, Liamdaly620, Miquonranger03, Moshe Constantine Hassan Al-Silverburg, WeniWidiWiki, Skyrocket, RomaC, Whispering, Solidusspriggan, Darth Panda, Dragice, Mladifilozof, Ig0774, Kotra, DéRahier, Chesaguy, Jennica, Mhaeberli, Rrburke, Addshore, Stevenmitchell, Benz240, Khoikhoi, Fuhghettaboutit, Breadandroses, Alexander VII, Whoistheroach, Richard001, Jinksy, Twiffy, Derek R Bullamore, Badgerpatrol, Clean Copy, Sammy1339, Kukini, BrianH123, Byelf2007, The undertow, JackpotDen, Giovanni33, JzG, CPMcE, Shadowlynk, JorisvS, Alex Stacey, Minna Sora no Shita, RomanSpa, Kirbytime, Santa Sangre, Zelfirelli, Catquas, Dl2000, Gredma, Iridescent, WGee, JHP, Sam Clark, Mugwumpman, Dartelaar, FairuseBot, Tawkerbot2, Daniel5127, Firewall62, George100, Kurtan~enwiki, Lahiru k, Orangutan, Megatronium, Devourer09, JForget, Mdsandul, Postmodern Beatnik, High Elf, ScriptBlue, Eewild, Awb49, Bdubois, Thomasmeeks, Yellowtailshark, Sdorrance, Nnp, Gregbard, Ttiotsw, ST47, AbyssWyrm, NaLalina, Pdemecz, Sweikart, Elmarand, Ncjones, Epbr123, Livedevilslivedevil, Litbr, Headbomb, Marek69, Woody, Keelm, Tlp, Matthew Proctor, Mph99, Escarbot, Dantheman531, AntiVandalBot, Voyaging, ChurchOfReason, Mackan79, Siege b, Modernist, Dylan Lake, Shlomi Hillel, Mutt Lunker, Myanw, Knotwork, Ingolfson, Obeattie, JAnDbot, Narssarssuaq, NBeale, Gatemansgc, Barek, Skomorokh, LinkinPark, OhanaUnited, .anacondabot, Michytoo, Magioladitis, Unused0029, Bongwarrior, VoABot II, Burninate 58, WikieWikieWikie, KConWiki, Catgut, ☒☒☒, Atlemk, Phileosophian, DerHexer, Edward321, Coffeepusher, Charitwo, MartinBot, Robert Daoust, Poeloq, Shafeeq882005, Sarah Bishop Merrill, Kostisl, CommonsDelinker, Smokizzy, Pomte, Valaggar, Thirdright, Shellwood, AlphaEta, J.delanoy, Pharaoh of the Wizards, Bogey97, FriendlyRiverOtter, Gzkn, It Is Me Here, Katalaveno, McSly, Notreallydavid, Mikael Häggström, Invidus, Davidmorrow, Robertson-Glasgow, Arms & Hearts, Another Philosopher, Al B. Free, Juliancolton, Nkb15uk, WinterSpw, Djr13, Idioma-bot, James Kidd, VolkovBot, DDSaeger, MonsterfUnC, Jeff G., Ecaepekam, LeilaniLad, Philip Trueman, TXiKiBoT, Tomsega, IllaZilla, DennyColt, BabaDraconis, Centipedian, Kenshin, Bearian, Latulla, January2007, Larklight, Billinghurst, Enigmaman, Bporopat, Michael Frind, Logan, Stringman5, Deconstructhis, Thomas94~enwiki, Strombollii, SieBot, WereSpielChequers, Phoenix2007, DeathByNukes, Yintan, Mothmolevna, Stratman07, Toddst1, Curuxz, RawEgg1, JSpung, Thirdeyeopen33, Miguel.mateo, Sunrise, Philosophy470, Bowei Huang 2, Ascidian, Pinkadelica, Sitush, Escape Orbit, Operation Spooner, JustinBlank, Loren.wilton, Martarius, ClueBot, The Thing That Should Not Be, Bthomson100, Plastikspork, EoGuy, Mx3, Ndenison, Mild Bill Hiccup, Neverquick, Callum Martin, Mbcudmore, Diagramma Della Verita, Excirial, Alexbot, PixelBot, Jopo11, Kaiba, Rusty505, Dekisugi, Pharwood, Llaezyn, Tired time, Acabashi, Celebreth, Aitias, Fledgeaaron, Smarkflea, Berean Hunter, DumZiBoT, Finalnight, Jax 0677, XLinkBot, Eliran Levi, Rreagan007, SilvonenBot, Philosophyclass HSOG, Addbot, Tcncv, KorinoChikara, Proxima Centauri, Jurj, FCSundae, Tomhandley111, LinkFA-Bot, AgadaUrbanit, Tide rolls, ForesticPig, OlEnglish, Jarble, JEN9841, Ettrig, Wikifan12345, Luckas-bot, Yobot, Timeroot, Hairy poker monster, Sumail, Vidur10, Eduen, Synchronism, AnomieBOT, Erel Segal, Jim1138, Ddoomdoom, Materialscientist, Citation bot, Jtshelton, MauritsBot, Xqbot, Sirspamalotiii, JimVC3, Felisophia, Capricorn42, Mandez01, RamziNahawi, Lewishal25, Shizuka Kamishima, GetLinkPrimitiveParams, Srich32977, GrouchoBot, Omnipaedista, Designalife, Johnson175, PineScented, Sunils2cool, Kcauley, Yelnod, FrescoBot, Nicolas Perrault III, ChikeJ, Zero Thrust, Macaneave, Airborne84, Pinethicket, I dream of horses, Sargeantneo, Venomous-Concept, December21st2012Freak, EphemeralKnowledge, Caspian Rehbinder, Lotje, Dasha14, Grifftob, Lucobrat, ABarnes94, Dyrankor, Mandolinface, Tesseract2, Shabidoo, The soul is unknown, EmausBot, John of Reading, WikitanvirBot, Marco Roy, Star7827, Unimpeder, Anthonypearce54, GoingBatty, CoincidentalBystander, Solarra, Gransta, Slawekb, Grantpant, ZéroBot, PBS-AWB, Fæ, PRABHAT PINGREJA, GZ-Bot, GrindtXX, Sailsbystars, Pochsad, Orange Suede Sofa, ChuispastonBot, Peter Karlsen, Wakebrdkid, Freddy eduardo, ClueBot NG, Gareth Griffith-Jones, FourLights, MelbourneStar, Intoronto1125, Fioravante Patrone, Snotbot, Gbsnlspl, Frietjes, Chriscook54321, Marcus Pivato, Widr, Squareanimal, Helpful Pixie Bot, Poo-goo, BG19bot, ISTB351, KevinEnders, Langchri, Graham11, Hallows AG, Nathan59, Stelpa, Carlstak, Soerfm, CitationCleanerBot, NotWith, Theconsequentialist, AdamJazzVt, Chomsky, Felixthehamster, Davidcpearce, BattyBot, NGC 2736, Teammm, Alfasst, ChrisGualtieri, Citizen Gardens, Neils51, SD5bot, Deathlasersonline, Vanished user sdij4rtltkjasdk3, Futurist110, InformationvsInjustice, Dexbot, Polsky215, Mr. Guye, Kbog, Nasmith1234, The Vintage Feminist, Lugia2453, Cupco, WolfgangAzureus, Me, Myself, and I are Here, Zviroth, BurritoBazooka, Dschslava, Epicgenius, CsDix, I am One of Many, BreakfastJr, Jodosma, VoiceOfTheCommons, PhantomTech, Rolf h nelson, Sophiahounslow, Ugog Nizdast, The Herald, Aubreybardo, Philofiler, Lest We Forget, Anrnusna, Monkbot, Mike2085, Kinetic37, Cassandra3001, Annamarmus, Janepharper, House of Mogh, Sigehelmus, Secretkeeper12, 00090R, Econtruthseeker, 65HCA7, Wikispring, Cusku'i, Puraki, Bconte, SquidHomme, KenTancwell, Pandresen, KasparBot, Johngot, Toughdan, Thinkofanumber, MaxWillyo, MartinZ, Kellydhru, Thefinncarter, InternetArchiveBot, Istandwiththesilent, See-3 Pee-Oh, Acopyeditor, Gihs1999, PrimeBOT, Wynton.lam, Edthat2, Djkahled and Anonymous: 965

- **Ethical movement** *Source:* https://en.wikipedia.org/wiki/Ethical_movement?oldid=787948879 *Contributors:* Leandrod, Admiralh, Vik-Thor, Kalki, Nikai, Hyacinth, Wjhonson, GreatWhiteNortherner, RoyBoy, Defrosted, Carbon Caryatid, Jheald, Nightstallion, Woohookitty, FeanorStar7, Toussaint, BD2412, Rjwilmsi, Koavf, Jweiss11, JLM~enwiki, SchuminWeb, Wavelength, RussBot, Zythe, Closedmouth, Rhwentworth, Tom Morris, SmackBot, Pkirlin, The Rhymesmith, Jreedy21, JonHarder, C0h3n, Kennovak, Huon, Couchand, Dacoutts, UVnet, JzG, LadyofShalott, BeenAroundAWhile, Fordmadoxfraud, Cydebot, Damifb, Ghmyrtle, Caper13, Greg Comlish, Kacela, Magioladitis, VoABot II, Xb2u7Zjzc32, Troystew, Emil Volcheck, Jim.henderson, Ethicalsusan, CommonsDelinker, RJBurkhart3, Hotaru in meditation, DadaNeem, Master shepherd, John Carter, Supertask, Joanna London, Bearian, Anarchangel, Milowent, Cirdan747, SE7, Barttsky, JL-Bot, Escape Or-

33.8. TEXT AND IMAGE SOURCES, CONTRIBUTORS, AND LICENSES

bit, Jobas, ClueBot, M J Mason, Chitowner57, Chessophile, Parkwells, Leadwind, Masterpiece2000, Bergendude, Editor2020, DumZiBoT, Addbot, Doctorspecial, Eaglesmile, Mentisock, Lightbot, Luckas-bot, Yobot, Legobot II, Matanya, AnomieBOT, Mike Hayes, Amovieman, Maniadis, Crzer07, Ethicalrysec, Grswld, AnotherOnymous, Tnaovi, Geraldmpiper, Moonraker, Lotje, SuperStan100, Lopifalko, Griswaldo, Haroldk 2001, Jbbiker, SporkBot, Noodleki, RayneVanDunem, Deestr, ClueBot NG, David C Bailey, EricWR, Cyberbot II, Nathanielfirst, Ethicalhuman, Me, Myself, and I are Here, Alan.trevethan, InternetArchiveBot, GreenC bot, Bender the Bot, SVES and Anonymous: 43

- **Ethical naturalism** *Source:* https://en.wikipedia.org/wiki/Ethical_naturalism?oldid=793634635 *Contributors:* Larry Sanger, Zanimum, Chealer, Alan Liefting, Eequor, Edcolins, JRR Trollkien, Elembis, Rich Farmbrough, Guanabot, Pearle, SteinBDJ, Velho, Graham87, BD2412, Common Man, RussBot, RL0919, Real World, Tevildo, Canadianism, Ivolucien, SmackBot, InverseHypercube, Gilliam, Chris the speller, Bluebot, Kurykh, Darth Panda, Whoistheroach, Byelf2007, Dr. Sunglasses, Sidmow, Kirbytime, TheOtherStephan, Seqsea, Andkore, Gregbard, Cydebot, Skomorokh, TAnthony, Anarchia, Eubulides, MARKELLOS, Pfhorrest, J. Johnson, Legobot, Yobot, Eric-Wester, AnomieBOT, F.morett, Materialscientist, Prari, FrescoBot, Adam9389, Machine Elf 1735, Wikididact, Bluszczokrzew, Tesseract2, PBS-AWB, ClueBot NG, Hendrikgommer, Vicente810, 3primetime3, MRD2014, BrianPansky, IWillBuildTheRoads, PVPWINNER and Anonymous: 31

- **Evolutionary ethics** *Source:* https://en.wikipedia.org/wiki/Evolutionary_ethics?oldid=797457465 *Contributors:* Bueller 007, TonyClarke, Rursus, Elembis, Jackhynes, Marskell, Rjwilmsi, Mitsukai, Muntuwandi, Aldux, Epipelagic, 2over0, SmackBot, Hardyplants, Fuzzform, EPM, Richard001, Byelf2007, Lapaz, Kripkenstein, Dl2000, Ken Gallager, Thijs!bot, Marek69, Dekkanar, Julia Rossi, PelleSmith, EverSince, S.dedalus, Charlesdrakew, Macdonald-ross, Dan Polansky, Wessmaniac, PixelBot, Lingo pen, Rankiri, Addbot, Chzz, Zorrobot, Yobot, Speedy la cucaracha, Mr. Muntuwandi, Azcolvin429, AnomieBOT, In case you're wondering, Citation bot 1, I dream of horses, Wikididact, CircularReason, Tesseract2, EmausBot, PBS-AWB, Alrik, Miradre, Helpful Pixie Bot, BG19bot, Harizotoh9, Krimin killr21, Academica Orientalis, Krzys kurd, Arronne, Stamptrader, Annn07, Kproff, Zokusai, J.A.W.Abrams, Magic links bot, KolbertBot and Anonymous: 27

- **Secular ethics** *Source:* https://en.wikipedia.org/wiki/Secular_ethics?oldid=791902617 *Contributors:* Bryan Derksen, Edward, Dino, Lockeownzj00, Elembis, Silence, Dave souza, Euthydemos, Maqsarian, Ungtss, Mindmatrix, Naraht, Nihiltres, No Swan So Fine, BMF81, Wavelength, Trovatore, RL0919, Newagelink, Aryah, Canadianism, SmackBot, Thomas Ash, McGeddon, ProveIt, Sorceress Jade, Gilliam, Kleuske, Starghost, Yogesh Khandke, Capt Jack Doicy, Chrisrivers, Neelix, Cydebot, Woody, SomeHuman, JamesBWatson, Deus911, PelleSmith, Lenschulwitz, CommonsDelinker, KTo288, Acalamari, Lbeaumont, Evb-wiki, Charlesdrakew, Jojalozzo, Anchor Link Bot, Bowei Huang 2, Darobian, Imperatoromnium, Heironymous Rowe, NellieBly, Addbot, Proxima Centauri, Bnordlund, Lightbot, Luckas-bot, Yobot, Eazyskankin, AnomieBOT, Darolew, LilHelpa, Xqbot, Jeffrey Mall, Rasnaboy, GrouchoBot, Aldo samulo, FrescoBot, Airborne84, Mjsaz4018, Wikididact, Samuel Salzman, ItsZippy, Jnewton37, Noommos, Tesseract2, Zujine, Cousin Kevin, GoingBatty, Wensei78, Wikignome0530, Alrik, Donner60, ChuispastonBot, AxiomOfFaith, ClueBot NG, Gpaul1, Helpful Pixie Bot, Wbm1058, Atari25, Headlearn55, Pogogyneabramsii, Etothepi, ChrisGualtieri, Dexbot, Mogism, Lugia2453, Ira Leviton, Mateoski06, Cristian.nt and Anonymous: 32

- **Law of three stages** *Source:* https://en.wikipedia.org/wiki/Law_of_three_stages?oldid=793524258 *Contributors:* Bearcat, Piotrus, DVdm, Whoisjohngalt, Alynna Kasmira, Jyoshimi, Onorem, Iridescent, Nethac DIU, Chris55, Cydebot, Richhoncho, Epbr123, Riverscene, Dentren, Hbent, Tomsega, SieBot, ShiftFn, Faradayplank, Firefly322, ClueBot, SummerWithMorons, RafaAzevedo, Alejandrocaro35, Muro Bot, Vegetator, Addbot, Ptbotgourou, Eumolpo, ArthurBot, DrRom, Jay-Sebastos, ClueBot NG, Cwmhiraeth, Satellizer, Marcocapelle, Meclee, Brad7777, Kimballer760, Maryam816, BoltonSM3, Puthoni, KH-1, 89sec, Hellomynameiscarlandiloveyou, Apollo The Logician and Anonymous: 51

- **Science of morality** *Source:* https://en.wikipedia.org/wiki/Science_of_morality?oldid=769678177 *Contributors:* Edward, William M. Connolley, Strebe, Pjedicke, JimWae, Thorwald, Maclean25, Firsfron, Woohookitty, Waldir, Sjakkalle, Nightscream, War, Wragge, Diliff, Gaius Cornelius, RL0919, Tony1, Naasking, SmackBot, InverseHypercube, Cazort, Portillo, Hmains, Pure~enwiki, DoctorW, Matt2h, Neutronium, Clean Copy, Byelf2007, Ckatz, Grapplequip, Lathrop1885, Aeternus, CmdrObot, Gregbard, Calvinballing, Nick Number, BenMcLean, Rothorpe, Gert7, Oicumayberight, Robert Daoust, EverSince, Fconaway, Laurusnobilis, Nfoglesbee, Belovedfreak, Landroving Linguist, Shadowlapis, Robert1947, StAnselm, Wing gundam, Jojalozzo, SmallRepair, IdreamofJeanie, SpareSimian, Cloonmore, Unbuttered Parsnip, Mild Bill Hiccup, Trivialist, SchreiberBike, Alextanium, Jytdog, Pfhorrest, Eightbitlegend, Prefetch, Annielogue, J. Johnson, Yobot, Azcolvin429, AnomieBOT, Jim1138, Materialscientist, Eumolpo, LilHelpa, Tomwsulcer, Kfrancist, Pereant antiburchius, FrescoBot, Machine Elf 1735, Airborne84, Blubro, Wikididact, RjwilmsiBot, Tesseract2, GoingBatty, Moswento, H3llBot, MisterDub, DM4242, Mcc1789, ClueBot NG, Hendrikgommer, ZarlanTheGreen, Snotbot, IvoryMeerkat, Keitaro202, ScienceDawns, Helpful Pixie Bot, Langchri, AdventurousSquirrel, Snow Blizzard, Felixthehamster, Iank125, Academica Orientalis, Khazar2, Khimaris, Hmainsbot1, Critical-interval, Reatlas, Skepticinfo1234, Negative24, Cassandra3001, Jackbollda, BrianPansky, Milbog, Billyswong, Bender the Bot and Anonymous: 70

- **Secular morality** *Source:* https://en.wikipedia.org/wiki/Secular_morality?oldid=793720042 *Contributors:* Shd~enwiki, JASpencer, Markhurd, JackofOz, Alan Liefting, Bfinn, TaintedMustard, Rjwilmsi, Cassowary, SmackBot, Portillo, Chris the speller, Whoistheroach, BullRangifer, Memills, Headbomb, Missvain, LeedsKing, JamesBWatson, R'n'B, Adavidb, Vanished user g454XxNpUVWvxzlr, Belovedfreak, Colinpearse, Charlesdrakew, Andrewaskew, MCTales, Ignacio Bibcraft, StAnselm, Eberber, Jruderman, Yesuto15, SchreiberBike, Editor2020, Addbot, MrOllie, Luckas-bot, Yobot, WikiDan61, Eazyskankin, Chiubaca, AnomieBOT, Shambalala, Dhidalgo, Rasnaboy, FrescoBot, Cdw1952, Airborne84, Wikididact, Yunshui, TheoloJ, Jnewton37, Nederlandse Leeuw, RjwilmsiBot, Tesseract2, Cousin Kevin, John of Reading, IHMAPK, Shuipzv3, Scrutineer1, Wensei78, AUN4, Tehuacana, ClueBot NG, Thaimoon, Bahasaindo, Helpful Pixie Bot, Epicurus B., Wheeke, BattyBot, MadameXX, Ava333, OmarPinaPena, Saltybone, Monkbot, Ethvoyager, Knowledgebattle, Bender the Bot, Texasrattlesnake31698, Magic links bot and Anonymous: 41

- **Metaphysical naturalism** *Source:* https://en.wikipedia.org/wiki/Metaphysical_naturalism?oldid=790651153 *Contributors:* Ed Poor, Gabbe, Julesd, Big iron, Banno, Chealer, Nurg, AceMyth, Michael2, Barbara Shack, Castaa, Wolfkeeper, Anville, Andycjp, Dave souza, Bender235, ESkog, John Vandenberg, Rbj, ADM, Jumbuck, Monado, Toussaint, BD2412, Qwertyus, Rjwilmsi, Koavf, Quiddity, Alexjohnc3, Common Man, Bgwhite, YurikBot, Mark Ironie, Sylvanius75, Gaius Cornelius, Erielhonan, Veinor, Laurence Boyce, SmackBot, ProveIt, Srnec, Gilliam, Portillo, Chris the speller, DT Strain, Tigerhawkvok, Standardrobot, Cybercobra, Dreadstar, Richard001, Andrew c, Metamagician3000, Byelf2007, Joshua Scott, SolarBreeze, Noleander, Iridescent, K, Shoeofdeath, Jlrobertson, Agathman, Editorius, LarianLeQuella, Gregbard, Cydebot, Doug Weller, Void main, RichardCarrier, Mbell, Masonroy, Bobblehead, Voyaging, Ioeth, Skomorokh, Leolaursen, Wasell, Coffee2theorems, Jetstreamer, Banderbe, Swpb, Snowded, Marvol, Davidpapineau, Upholder, PraetorDrew, Lyonscc, LoneFox, Gabrielthursday, Talon Artaine, Raiph, Zalmaki, Chiswick Chap, Endlessmike 888, Erwin.lengauer, Rucha58, TXiKiBoT, Jacqueslacansan, Steven J. Anderson, Redrocker, Dpleibovitz, Hrafn, AdRock, Moonriddengirl, ClueBot, Philosophy.dude, Christian Skeptic, Johnuniq, XLinkBot, Duncan, SilvonenBot, Aunt

Entropy, Eightbitlegend, Elmo iscariot, Addbot, DFS454, AnnaFrance, Soupforone, Lightbot, Yobot, Jcsam70, We66er, Carleas, Enki hits, Mnation2, AnomieBOT, Archon 2488, Richardbrucebaxter, IRP, UnitarianUniversalism, Trabucogold, Mann jess, Dhidalgo, Materialscientist, ArthurBot, Xqbot, Cureden, Spotfixer, Omnipaedista, Spamcatcher, Shadowjams, Metaphysicalnaturalist, FrescoBot, Paine Ellsworth, MathHisSci, Adam9389, Tormine, TruthIIPower, Machine Elf 1735, 8teenfourT4, MTDinoHunter, RedBot, CircularReason, Zonafan39, WebEdHC, Livingrm, Noraft, RjwilmsiBot, Deftil, Tesseract2, Mr. Thrasymachus, John of Reading, Faceless Enemy, Dcirovic, ZéroBot, AshforkAZ, Tolly4bolly, MisterDub, Mcc1789, ClueBot NG, Gilderien, RVscholar, FleeTheCaptor, Irlboy, Helpful Pixie Bot, Nashhinton, BG19bot, Morganhillbilly, Henryallenedison, Wallawallaonion, Mthoodhood, Sparkie82, SmittysmithIII, Cyberbot II, Elizah379, Saedon, Nathanielfirst, NaturaNaturans, Dfpolis, PhilosophyOfScience, HerbertHuey, Klak of Klak, Apidium23, BreakfastJr, Aristotle2013, Monkbot, BrianPansky, GoodNews970, Rowland938, Nøkkenbuer, OtisDixon, JemimaJameson, GreenC bot, Apollo The Logician, Quasar G., Magic links bot and Anonymous: 148

- **Religion of Humanity** *Source:* https://en.wikipedia.org/wiki/Religion_of_Humanity?oldid=795776723 *Contributors:* Paul Barlow, Dan Koehl, Menchi, Vanieter, Rursus, DanielDemaret, Olaf Simons, BD2412, Zythe, T. Anthony, Lifebaka, Benzocaine, George100, AtticusX, KConWiki, Themoodyblue, Tomsega, Lechatjaune, Shans2001pk, Malcolmxl5, NicolasFox, Editor2020, XLinkBot, Addbot, Ajverink, Denispir, AnomieBOT, Materialscientist, ArthurBot, AnotherOnymous, FrescoBot, Bhagat.bb, DrilBot, GoingBatty, The Blade of the Northern Lights, H. Klaus M. Hoffmann, Historyinc, ClueBot NG, Graham11, Marcocapelle, Brad7777, Zumwalte, LahmacunKebab, Monkbot, User000name, ChouxMonster, Bender the Bot, Magic links bot and Anonymous: 25

- **International Humanist and Ethical Union** *Source:* https://en.wikipedia.org/wiki/International_Humanist_and_Ethical_Union?oldid=796125448 *Contributors:* Kku, Chinju, Lquilter, Jeandré du Toit, Dino, Filemon, Tom harrison, Duncharris, Icairns, Mike Rosoft, Vsmith, John Vandenberg, Rohirok, Derktar, MGTom, Koavf, 2ct7, VolatileChemical, Bgwhite, Adonisick, RussBot, Welsh, Trovatore, Zythe, John-Lloyd, Rhwentworth, Bob.Churchill, SmackBot, Dwanyewest, ProveIt, Mre5765, Chris the speller, Dacoutts, Zdravko mk, Bejnar, Dl2000, George100, Ingersoll~enwiki, Davidhume, Gregbard, Cydebot, Thijs!bot, SomeHuman, Zeitlupe, Mmyotis, Ghmyrtle, MikeLynch, Magioladitis, AtticusX, Wku2m5rr, StewE17, TXiKiBoT, Nazgul02, Malcolmxl5, Twinsday, Paul K., Masterpiece2000, Pmpdurrant, XLinkBot, Albambot, Addbot, NjardarBot, Download, LaaknorBot, Zorrobot, Yobot, Amirobot, AnomieBOT, Open 2, Frankol77, Jtamad, Obersachsebot, GrouchoBot, Berserkur, Mark Schierbecker, IHEU, MastiBot, Orenburg1, Kristofferst, Lotje, LilyKitty, Nederlandse Leeuw, Nihola, EmausBot, John of Reading, Arneberg, Griswaldo, ZéroBot, H3llBot, Erianna, ChiZeroOne, Catlemur, Md.altaf.rahman, Harsimaja, Wbm1058, Marcocapelle, Cyberbot II, Dobie80, Dexbot, Makecat-bot, XandeBox, RaphaelQS, Rhanabanana, Finnusertop, DTNewman, Vieque, Will Talk2, Timdevries1003, Vanisheduser00348374562342, Zoppeting, Awesome user123, GreenC bot and Anonymous: 50

- **Renaissance humanism** *Source:* https://en.wikipedia.org/wiki/Renaissance_humanism?oldid=790849379 *Contributors:* Shii, Michael Hardy, Nixdorf, Charles Matthews, Adam Bishop, Tpbradbury, Saltine, Wetman, Kev, Dimadick, Bearcat, Goethean, Rursus, DocWatson42, Avaragado, James Crippen, Zeimusu, Antandrus, The Singing Badger, JoJan, Neutrality, Aioth, Discospinster, Quiensabe, Dbachmann, Bender235, Orlady, Bobo192, Elipongo, Thewayforward, Pearle, ADM, Jumbuck, Aeolien, Alansohn, Gary, Miranche, Fwb44, Jaw959, Bart133, Vcelloho, Kazvorpal, Jeff3000, Tabletop, Gimboid13, Rjwilmsi, Jake Wartenberg, Vary, FlaBot, Kmorozov, Hackloon, EronMain, Gurubrahma, TheSun, Mark Yen, Uriah923, RussBot, Stephenb, Eleassar, Fnorp, Bachrach44, Journalist, DeadEyeArrow, Private Butcher, Deville, Zzuuzz, Palthrow, Samuel Blanning, DVD R W, Nick Michael, That Guy, From That Show!, SmackBot, Elonka, InverseHypercube, Tonyr68uk, Rbreen, Srnec, Gilliam, Carl.bunderson, Wilson Delgado, Chris the speller, Persian Poet Gal, Rex Germanus, Jprg1966, Afasmit, Dustimagic, Mulder416, Rrburke, RedHillian, Vault~enwiki, Savidan, Geoffr, Kleuske, Jbergquist, Filpaul, Ck lostsword, Kukini, Jontveit, Bloomradio, Kuru, Hawjam, Bilby, Ckatz, RandomCritic, Bless sins, Rizome~enwiki, Armadel, Igoldste, RekishiEJ, Beno1000, Courcelles, Tawkerbot2, Dlohcierekim, Igni, Edward Vielmetti, NessBird, Richaraj, Longshot.222, Gregbard, FastLizard4, SpK, Epbr123, Bot-maru, CSvBibra, John254, CharlotteWebb, SomeHuman, Mentifisto, AntiVandalBot, Yonatan, Mendedcloak, Zappernapper, Danny lost, Farosdaughter, Cgram@adelphia.net, Ghmyrtle, JAnDbot, MER-C, Henkk78, Cynwolfe, Boleslaw, Moni3, Connormah, VoABot II, Vordabois, Awwiki, Invisible Flying Mangoes, Schumi555, Rettetast, DBlomgren, Shellwood, J.delanoy, Maurice Carbonaro, -jmac-, It Is Me Here, Johnbod, Bailo26, AntiSpamBot, Girlfawkes, NewEnglandYankee, Kjetilb, GreatJT, Jamesofur, Ken g6, Mike V, Diana Kostyrko, Boombaard, Sam Blacketer, DarkArcher, Mcewan, Oshwah, Gen. Quon, Xresonance, Martin451, Rooke42, Mnfiero, Viator slovenicus, Eric9876, Madhero88, Koenraad Cl, Xenovatis, SaltyBoatr, Tpb, StAnselm, Mikemoral, Nihil novi, WereSpielChequers, SE7, Flyer22 Reborn, Tiptoety, Aruton, Loren.wilton, ClueBot, SummerWithMorons, Fyyer, The Thing That Should Not Be, Shark96z, Arakunem, Steelwool, Masterpiece2000, Excirial, CohesionBot, Jusdafax, Abrech, Lartoven, Maser Fletcher, Tyler, Tired time, Thingg, Aitias, Party, XLinkBot, Count1, Skarebo, WikHead, ZooFari, HexaChord, Addbot, Master michael 90, Grayfell, Heavenlyblue, Bushcutter, Ronhjones, Marx01, Bobfronk, Download, LaaknorBot, West.andrew.g, 5 albert square, Tide rolls, Lightbot, Pietrow, Blah28948, Yobot, 2D, Fraggle81, Mballen, AnomieBOT, AdjustShift, Citation bot, Dewan357, Gsmgm, Xqbot, Haputdas, Cureden, Pontificalibus, HaleyZZ, Omnipaedista, RibotBOT, Mattis, Astatine-210, Ecclesiasticalhistory, Sky Attacker, Citation bot 1, AstaBOTh15, Rushbugled13, Skyerise, Traversari~enwiki, MastiBot, Serols, 夢の魚, Im not afraid, Steve2011, FoxBot, XxX wonderwoman XxX, Yong, Cowlibob, Shouri226, Purple333, NottheAverageJoe, Diannaa, Nascar1996, DARTH SIDIOUS 2, RjwilmsiBot, Madelyn Garcia, Saltydog101, Djordje-dz, EmausBot, Super48paul, Boundarylayer, Tommy2010, Wikipelli, Dcirovic, ZéroBot, Susfele, Josve05a, Eyadhamid, EWikist, Wayne Slam, Crochet, Dagko, L Kensington, Donner60, Damirgraffiti, CouchPeanut, ClueBot NG, Satellizer, अमर्, Rattakorn c, O.Koslowski, Drjoshario, Mannanan51, Widr, Helpful Pixie Bot, Calabe1992, Fuzzybunny911, Northamerica1000, AvocatoBot, Mark Arsten, Gorthian, Hamish59, Lgoldsten, Polmandc, Merlaysamuel, Dexbot, Chicbyaccident, Табалдыев Ысламбек, Lugia2453, GoThere2000, Errorranger69, Iantio12, Doodadoo, AndyHarwell, Speahlman, DavidLeighEllis, Transphasic, JWNoctis, MrScorch6200, TWC6, Matiia, Tetra quark, OilandTempura, IlCorso, GeneralizationsAreBad, CAPTAIN RAJU, Babymissfortune, Dilidor, Niceboy97, Vikings101, Chickenlatte, Chickentaco123, BambiMajumdar, PackMecEng, Teconomus, Clemby, Wmdly, Magic links bot, Dammitkevin and Anonymous: 435

- **Renaissance humanism in Northern Europe** *Source:* https://en.wikipedia.org/wiki/Renaissance_humanism_in_Northern_Europe?oldid=770651988 *Contributors:* Edward, Charles Matthews, Qertis, Grstain, Bellhalla, Kmorozov, RussBot, C mon, SmackBot, TimBentley, Afasmit, Colonies Chris, Iridescent, RCS, Barticus88, CSvBibra, Geniac, Balloonguy, VirtualDelight, Johnbod, Alnokta, Pastordavid, StromBer, SPECVLVMSINCERVS, Auntof6, Iohannes Animosus, Addbot, Download, Lightbot, Fatepur, GrouchoBot, I dream of horses, Purple333, Dewritech, GoingBatty, Dcirovic, Midas02, LWG, ClueBot NG, BG19bot, BattyBot, GoShow, Thewikiguru1, Viamor, Duma, CAPTAIN RAJU and Anonymous: 19

- **Humanism in France** *Source:* https://en.wikipedia.org/wiki/Humanism_in_France?oldid=631499787 *Contributors:* Frecklefoot, Charles Matthews, Ardfern, Kmorozov, RussBot, Johndarrington, Fram, Colonies Chris, Tazmaniacs, Lazulilasher, Ken Gallager, Hemlock Marti-

33.8. TEXT AND IMAGE SOURCES, CONTRIBUTORS, AND LICENSES

nis, Travelbird, Johnbod, Pastordavid, Lightbot, Eduen, Ulric1313, Tongbongschong, Muleiolenimi, AsceticRose, Jacobisq, Chillllls, GoShow, Sunday NPR Lover and Anonymous: 7

- **Center for Inquiry** *Source:* https://en.wikipedia.org/wiki/Center_for_Inquiry?oldid=797163129 *Contributors:* Edward, Lquilter, Dcljr, Nina, Nine Tail Fox, Big iron, Dino, Rednblu, Carlossuarez46, Bearcat, RcktScientistX, Vsmith, Martpol, Bender235, Kurieeto, Wtmitchell, Jheald, RJFJR, Woohookitty, Havermayer, Tabletop, Magister Mathematicae, BD2412, Rjwilmsi, Nightscream, Jweiss11, Alaney2k, Bubba73, Askolnick, Mindme, Wavelength, Backburner001, Malcolma, THB, Jcurious, Rathfelder, That Guy, From That Show!, Skeptic sid, SmackBot, Vassyana, Betacommand, Jeff5102, Austinfidel, Tdudkowski, Arbustoo, BillFlis, Yuk Yuk Yec, George100, CmdrObot, TheOtter, Ken Gallager, Gregbard, Cydebot, Jonathan Tweet, Crash.car.star, Richhoncho, LetsGoAngels, Smeazel, Nick Number, Alphachimpbot, Gwynarina, Ghmyrtle, Nickieee, Cpl Syx, DerHexer, DGG, Keith D, Johnpacklambert, PrestonH, Nigholith, Drkoepsell, The King of Spain's beard, Ontarioboy, FGT2, DebGod, Steven J. Anderson, Gekritzl, Torontonian1, Laval, Hmwith, Cirdan747, Behind The Wall Of Sleep, Lightmouse, Kjtobo, Randy Kryn, Twinsday, Plastikspork, Niceguyedc, Solar-Wind, Skepticpsychic, Qwfp, DumZiBoT, Mitch Ames, Addbot, Zarcadia, Tassedethe, Luckas-bot, Yobot, AmericanHumanist, AnomieBOT, Krelnik, Galoubet, Ipatrol, Funcrunch, Srich32977, Peter Damian, Mark Schierbecker, FrescoBot, Paine Ellsworth, Skeppyrron, Full-date unlinking bot, ThinkCritically, Lotje, Nederlandse Leeuw, Nmillerche, GoingBatty, ZéroBot, Sgerbic, Aisaak, H3llBot, Frietjes, BG19bot, Neptune's Trident, Cap020570, Marcocapelle, Allecher, Harizotoh9, MrBill3, Nopath, BattyBot, FuryAndMire, Mjsteiner, Dobie80, Grandmentor, Dexbot, Epicgenius, Wuerzele, Coladar, RaphaelQS, Freethoughtlibrarian, Noyster, Ocdctx, Marchjuly, Melcous, Monkbot, Epicurus78, BethNaught, Vanisheduser00348374562342, RaffVitali, K.e.coffman, Rubbish computer, Jardouin, Dalek Supreme X, Orange500, InternetArchiveBot, Jmcgnh, Bender the Bot, Cynulliad3, KolbertBot and Anonymous: 34

- **A Secular Humanist Declaration** *Source:* https://en.wikipedia.org/wiki/A_Secular_Humanist_Declaration?oldid=787389498 *Contributors:* SimonP, Rsabbatini, Charles Matthews, Steinsky, Itai, Premeditated Chaos, Duncharris, Alan Nicoll, Icairns, Kate, Vsmith, DcoetzeeBot~enwiki, Susvolans, Woohookitty, Toussaint, Quuxplusone, NawlinWiki, Adicarlo, Tom Morris, SmackBot, ProveIt, Dacoutts, Austinfidel, George100, Cydebot, Bellerophon5685, AlexanderLevian, JustAGal, Calaka, AntiVandalBot, OinkOink~enwiki, Jevansen, Fra59e, Gold1618, Trivialist, Addbot, Tassedethe, Lightbot, Yobot, Unara, Ulric1313, Obersachsebot, Mark Schierbecker, Rastko Pocesta, Faolin42, ZéroBot, Cymru.lass, Jfhutson, Rschneider29, MPFishSticks and Anonymous: 12

- **Amsterdam Declaration** *Source:* https://en.wikipedia.org/wiki/Amsterdam_Declaration?oldid=787389528 *Contributors:* Delirium, Jeffq, Mboverload, STHayden, Batmanand, Rohirok, Toussaint, Rjwilmsi, Tim!, ProveIt, Dacoutts, Zymurgy, Cydebot, Bellerophon5685, Thijs!bot, SomeHuman, Ghmyrtle, AtticusX, R'n'B, Mirithing, Supertask, Malcolmxl5, Jrun, Tameamseo, Addbot, Lightbot, Lennim, BenzolBot, EmausBot, ZéroBot, Illegitimate Barrister, Stevfan, Syrinx666pan, Marcocapelle, Jfhutson, Shirudo, Hazelares and Anonymous: 6

- **American Humanist Association** *Source:* https://en.wikipedia.org/wiki/American_Humanist_Association?oldid=788822450 *Contributors:* Nealmcb, JakeVortex, Water Seven, Lquilter, Sannse, Delirium, Bdonlan, Jiang, Dysprosia, VeryVerily, Alan Liefting, Brockert, CryptoDerk, Icairns, Neutrality, Lacrimosus, Bcat, Func, Dr Gangrene, Woohookitty, Lsmithlas, Marudubshinki, BD2412, Rjwilmsi, Naraht, Ground Zero, 2ct7, Severa, Gaius Cornelius, Welsh, EntChickie, Dtemp, SmackBot, Iopq, Niro5, Fiziker, Rkelly74, New World Man, Dacoutts, Statsone, Killua, Drinibot, ShelfSkewed, Myasuda, Jeffseaver, Biruitorul, David wallace croft, SomeHuman, Whugotit22, Ghmyrtle, Cgingold, DBWikis, Cpl Syx, Britishmafia, James.E.Watson, DadaNeem, Master shepherd, Jevansen, GyRo567, Djflem, F.chiodo, Wbforbes, Weelijimmy, VVVBot, Anchor Link Bot, WurmWoode, Solar-Wind, Masterpiece2000, Rhododendrites, Fdsbgdsf, Muro Bot, BOTarate, NixManes, Mitch Ames, Tameamseo, Airplaneman, Odin 85th gen, Addbot, Supermandan, Locationx3, Yobot, Hadrian89, Ulric1313, Dhidalgo, Materialscientist, Happyhuman, Psychiatrick, Chrismiceli, Mark Schierbecker, FreeKnowledgeCreator, FrescoBot, Wikihumanist, Full-date unlinking bot, TRBP, Standardfact, Nederlandse Leeuw, RjwilmsiBot, Lebrouillard, Gcastellanos, Wikipelli, Itchesavvy, The Nut, Filmfluff, Senator2029, ClueBot NG, MelbourneStar, Catlemur, PedR, Mannanan51, BG19bot, Marcocapelle, Allecher, BattyBot, Justincheng12345-bot, Cyberbot II, CarrieVS, Khazar2, Uploaderaha, Kvonnegut, Mogism, PhillipCK1, RaphaelQS, MFleischhacker, Filedelinkerbot, Thedailyclark, InternetArchiveBot, GreenC bot, Bender the Bot, Magic links bot and Anonymous: 71

- **Humanists UK** *Source:* https://en.wikipedia.org/wiki/Humanists_UK?oldid=797347777 *Contributors:* Ant, Gabbe, Lquilter, Leigh, Charles Matthews, IceKarma, Proteus, Tim Ivorson, David Edgar, Guy Peters, Duncharris, Chowbok, Bodnotbod, Icairns, Grunners, Thorwald, Kate, Vsmith, Grutter, Famousdog, Pearle, Rohirok, H2g2bob, Crosbiesmith, Woohookitty, BlaiseFEgan, Toussaint, Tim!, Nigosh, Gavinatkinson, CalJW, JdforresterBot, GünniX, 2ct7, Billpg, Gaius Cornelius, CambridgeBayWeather, Nirvana2013, Escheffel, Doggadogdog, Gabrielbodard, Dannyno, Zythe, Mr-Thomas, Chanheigeorge, JRawle, Mais oui!, Rathfelder, Tom Morris, SmackBot, ProveIt, Autarch, Dacoutts, Derek R Bullamore, BrownHairedGirl, John, -ramz-, John Cumbers, JDAWiseman, ChazYork, Cydebot, EssentialParadox, SomeHuman, Cstreet, Ghmyrtle, Hut 8.5, Rothorpe, VoABot II, AtticusX, Johnnyfunktastic, BHA, StewE17, MarkMarek, Itemirus, AdRock, GirasoleDE, Malcolmxl5, Lightmouse, Jruderman, OKBot, Sean.hoyland, Jjjimg, Dabomb87, Martarius, Solar-Wind, Muro Bot, Addbot, Lightbot, Bermicourt, Yobot, Captainclegg, AnomieBOT, Materialscientist, Bchurchill, Skinnysock, Ohso, Mark Schierbecker, Anna Roy, Bontshet, DrilBot, TRBP, Nayintheoaks, Diannaa, Nederlandse Leeuw, Signyred, RjwilmsiBot, Northern Arrow, AvicBot, ZéroBot, OnePt618, Erianna, Amencorner1, Philafrenzy, Wolfgar Kodiak, ClueBot NG, Spanishdave, Hazhk, Basitasif, BG19bot, Frze, MisterMorton, Felixthehamster, Samwalton9, BattyBot, Th4n3r, ChrisGualtieri, Khazar2, Nathanielfirst, Redrhoc, Hmainsbot1, Mogism, Xwoodsterchinx, Michipedian, Jodosma, RaphaelQS, Alan Cossey, DTNewman, CWM 93, Graihagh, Narky Blert, Mohanbhan, JSutton93, Helper201, Singaporesling1508, Bender the Bot, Mramoeba and Anonymous: 97

- **Humanistischer Verband Deutschlands** *Source:* https://en.wikipedia.org/wiki/Humanistischer_Verband_Deutschlands?oldid=758612796 *Contributors:* Rathfelder, Cydebot, Yobot, Mark Schierbecker, Nederlandse Leeuw, HeywoodFloyd, Aisteco, KasparBot and Anonymous: 3

- **Humanist Society Scotland** *Source:* https://en.wikipedia.org/wiki/Humanist_Society_Scotland?oldid=786562498 *Contributors:* Lquilter, Icairns, Alai, Ceyockey, BlaiseFEgan, Franko2nd, Zythe, Mais oui!, SmackBot, Colonies Chris, JonHarder, Dacoutts, Charivari, John, Youngdegsy, Statsone, Dl2000, MuttGirl, Cydebot, SomeHuman, AtticusX, Cpl Syx, CommonsDelinker, Malcolmxl5, Humanisto~enwiki, MystBot, Addbot, Yobot, Omnipaedista, FrescoBot, Full-date unlinking bot, Thrissel, Ariadne 13, Nederlandse Leeuw, OnePt618, Scotshumanist, MrBill3, Drchriswilliams, UglowT, CFindlay12, Atlantic306 and Anonymous: 11

- **Norwegian Humanist Association** *Source:* https://en.wikipedia.org/wiki/Norwegian_Humanist_Association?oldid=756759167 *Contributors:* Lquilter, Jni, Rj, Icairns, Picapica, Geschichte, Toussaint, Ground Zero, 2ct7, Sus scrofa, RussBot, Richardcavell, Mais oui!, Rathfelder, Hathaldir~enwiki, SmackBot, InverseHypercube, Dacoutts, Statsone, Meco, Dea-Renate, Cydebot, NorwegianBlue, SomeHuman, Ghmyrtle, AtticusX, Terjen, CommonsDelinker, Strappado~enwiki, Malcolmxl5, Hordaland, Solar-Wind, Mlaffs, Addbot, Zorrobot, AnomieBOT,

Zad68, Mark Schierbecker, Kim-Zhang-Hong, HRoestBot, Kristofferst, Thomasmh, Nederlandse Leeuw, Bento00, Plommespiser, Arneberg, Dewritech, Tommy2010, AvicAWB, Neil P. Quinn, ClueBot NG, Tusnus, Cyberbot II, Indiasummer95, NoNotes, GreenC bot, 69Joker21 and Anonymous: 25

- **Alternatives to the Ten Commandments** *Source:* https://en.wikipedia.org/wiki/Alternatives_to_the_Ten_Commandments?oldid=796954323 *Contributors:* Dcljr, Tpbradbury, Dehumanizer, Mmj, Viriditas, Rjwilmsi, Koavf, Rsrikanth05, Alarob, Thumperward, Sadads, MovGP0, Mordantkitten, Cybercobra, Yogidoo, Noleander, Iridescent, Egoadvocate, Cydebot, Alaibot, Sirmylesnagopaleentheda, Marek69, Carole Jean, LinkinPark, Eurobas, Jessicapierce, Adrian J. Hunter, Edward321, Pharaoh of the Wizards, Alexthepuffin, Evanmopp, Oshwah, ClueBot, EoGuy, Niceguyedc, Cancilla, Wkboonec, XLinkBot, DoctorHver, Addbot, Plimfix, Yobot, Backslash Forwardslash, AnomieBOT, Mann jess, RevelationDirect, LilHelpa, Xqbot, Betty Logan, GenevieveDH, Bodinagamin, Thustrae, FrescoBot, MilaJECD, Part Time Security, RedBot, Foomandoonian, RjwilmsiBot, -- -- --, Racerx11, Wikipelli, ZéroBot, Jess, ClueBot NG, Catlemur, Snotbot, Widr, HMSSolent, BG19bot, Hawk2987, Headlearn55, LonelyBoy2012, Paspie, MattGrifter, FiredanceThroughTheNight, Thenewaeon, DavidLeighEllis, Getdate, Sam Sailor, MagicatthemovieS, Bssays, Eurodyne, Taivorist, Cjphilo, Gravity2016, Bender the Bot, KolbertBot and Anonymous: 72

- **Maslow's hierarchy of needs** *Source:* https://en.wikipedia.org/wiki/Maslow'{}s_hierarchy_of_needs?oldid=797694451 *Contributors:* AxelBoldt, The Anome, Tarquin, Ap, Ed Poor, Youssefsan, Leandrod, Mrwojo, Patrick, Tillwe, Michael Hardy, Vaughan, Mahjongg, Jahsonic, Tabrez, Skysmith, SebastianHelm, Ahoerstemeier, Snozzwanger, Ronz, Александър, Jonathan Chang, Scott, Andres, Jeandré du Toit, Conti, Nikola Smolenski, Mydogategodshat, Charles Matthews, Dysprosia, E23~enwiki, Nv8200pa, Pietro, Dishayloo, Slawojarek, SD6-Agent, Fredrik, Daelin, Lbs6380, Goethean, Altenmann, Naddy, Sam Spade, Tobycat, Emyth, Pseudonym, Mushroom, Alan Liefting, Sbetten, Ancheta Wis, Dbenbenn, Mat-C, Inter, Lee J Haywood, Tom harrison, Alison, Varlaam, XtinaS, SevenMass, Tagishsimon, Mporch, Andycjp, Xmnemonic, Seba~enwiki, Proberts2003, Antandrus, Ex ottoyuhr, Xinit, Robert Brockway, Jossi, Nzseries1, UltraSheen, Neutrality, GreenReaper, Adashiel, Bluemask, Isaiah, Baf~enwiki, Freakofnurture, DanielCD, Felix Wiemann, Discospinster, Rhobite, Smyth, Autiger, Triskaideka, Arthur Holland, KaiSeun, Bender235, JoeSmack, Bennylin, El C, Lycurgus, Huntster, Vanished user kjij32ro9j4tkse, Sietse Snel, RoyBoy, Aaronbrick, Jpgordon, Bobo192, Longhair, Johnkarp, Nectarflowed, John Vandenberg, Enric Naval, TheProject, Ivansanchez, Jacius, MPerel, Stephen Bain, Eruantalon, Petdance, Alansohn, Gary, Anthony Appleyard, Sheehan, Keenan Pepper, Lacrimulae, Curious1i, WhiteC, Cdc, Mrestko, Hu, AzadMashari, Snowolf, Benna, Fivetrees, Wtmitchell, Melaen, Tedp, Mikeo, Bookandcoffee, Rzelnik, Phi beta, Vanished user dfvkjmet9jweflkmdkcn234, Scarykitty, Kelly Martin, OwenX, Linas, Mindmatrix, TigerShark, Camw, Robert K S, EnSamulili, Psychologesetz, Kmg90, Bkwillwm, Optichan, Jacj, Christopher Thomas, Mandarax, Graham87, CheshireKatz, Josh Parris, Jkatzen, Rjwilmsi, Koavf, Vary, Kazrak, NeonMerlin, ScottJ, Durin, Bhadani, Ttwaring, Keimzelle, SchuminWeb, Psemmusa, Old Moonraker, Dauerad, Gurch, JegaPRIME, Pevernagie, Tedder, SpectrumDT, Chobot, WriterHound, Gwernol, Algebraist, YurikBot, Wavelength, RobotE, Sceptre, Pterantula, Phantomsteve, TheMoot, Petiatil, V Brian Zurita, Briaboru, Splash, SpuriousQ, Nesbit, Vincej, Pseudomonas, Varnav, Wimt, NawlinWiki, Wiki alf, Mipadi, Nirvana2013, Grafen, Jaxl, Fabulous Creature, Nadirsofar, ONEder Boy, Sylvain1972, Ali Karbassi, Anetode, DAJF, Abb3w, Raven4x4x, Mlouns, Froth, Nate1481, Aaron Schulz, Dissolve, Mysid, DeadEyeArrow, Peter Knutsen, Zarboki, Limetom, Robotics1, Fatzebra, Kermit2, Zzuuzz, TheMadBaron, Nikkimaria, Closedmouth, Termspin, Bananafish, Chriswaterguy, Vkmaxwell, Janice Rowe, Rhwentworth, Kramden, Bsod2, Teryx, Rdbrady, Saikiri, CIreland, Robertd, SmackBot, Xkoalax, Moeron, InverseHypercube, Hydrogen Iodide, McGeddon, Jtneill, Apanait, Delldot, Eskimbot, Canthusus, LuisVilla, Nscheffey, Vegasrebel29, Xaosflux, Medicscout, Mifren, Gilliam, Portillo, Urmomma, Chris the speller, Stevenwagner, Liamdaly620, George Rodney Maruri Game, Fizban~enwiki, Stevage, JohnFord64, Uthbrian, Ikiroid, NeoNerd, Patriarch, Baa, Kostmo, Darth Panda, Rlevse, Gracenotes, Laurentsavaete, Emurphy42, Rama's Arrow, Suicidalhamster, Can't sleep, clown will eat me, AntiVan, Kendaniel, Frap, Jaro.p~enwiki, Max David, Hastings Hart, Rrburke, Whpq, RedHillian, Huon, Kukinha, Cybercobra, Downtown dan seattle, TedE, PointyOintment, John D. Croft, Nick125, Harvestman, DMacks, KeithB, Mitchumch, Spanyard, BlackTerror, Ugur Basak Bot~enwiki, DrStein, SashatoBot, Nathanael Bar-Aur L., Mchavez, Rory096, JzG, Zeraeph, Kingfish, J. Finkelstein, Dialectic~enwiki, SilkTork, Kirby1024, DaLaw, Robinwerner, Ghw777, Jjz3d83, Xenodochy, Illythr, Slakr, Avs5221, Mets501, Treyp, O process, Halaqah, GeebsRilie, GarethLewin, Jcbutler, ShakingSpirit, BranStark, BananaFiend, Dreftymac, Beve, Silent reverie86, DafadGoch, Courcelles, Tawkerbot2, Jvol, George100, Lahiru k, Amniarix, JForget, CmdrObot, JohnCD, Kris Schnee, Tschel, Jokes Free4Me, Jesse Viviano, Lentower, Moreschi, Penbat, Johnlogic, WeisheitSuchen, MaxEnt, Nilfanion, Gmcomp, Cydebot, Karimarie, CompKidMan, Kalanithe, Odie5533, Roberta F., Gablejb, Toge1988, Hinate01, Daven200520, Froodiantherapy, Voldemortuet, Rocket000, Mattisse, Thijs!bot, Epbr123, Billysomerville, Mawfive, HappyInGeneral, N5iln, Luigifan, Marek69, Tellyaddict, Sinn, Miller17CU94, NigelR, CharlotteWebb, Heroeswithmetaphors, Openlander, Mentifisto, Hmrox, AntiVandalBot, Luna Santin, Prolog, Danger, Bpbfde4, 3Easy, Dan D. Ric, Roman à clef, Theguy0000, MERC, April's Fool, Tstrobaugh, Rowdymouse, SiobhanHansa, FaerieInGrey, Nomorecyber, Freedomlinux, Da baum, Theunicyclegirl, Bongwarrior, VoABot II, Kuyabribri, Shmuelakam, Mbarbier, Rivertorch, Malcolmxthegreat, Cglenn3932, Froid, Panser Born, Originalname37, Nposs, 28421u2232nfenfcenc, Allstarecho, Glen, DerHexer, JaGa, PoohBear88, Waninge, Lelkesa, Johnbrownsbody, TheRanger, Patstuart, Tojo940, MartinBot, Xumm1du, Drewmutt, Anaxial, Dorvaq, Kyyn, Mschel, Doodlebug106, RockMFR, J.delanoy, Pharaoh of the Wizards, Trusilver, Eliz81, StuIsCool, Mike.lifeguard, Sicanjal, Lunared2, Whitebox, Jerry, Space-Age Mael, Darth Mike, Bslish, Gzkn, Prince of Dharma, Katalaveno, Lis5433wiki, Wadehudson, Cooldawg123, Ryan Postlethwaite, Phetish, DarwinPeacock, Osndok, AntiSpamBot, Lbeaumont, Raining girl, Belovedfreak, NewEnglandYankee, Superdanny303, Rosenknospe, Fheyligh, Zeppalin614, Burzmali, Remember the dot, Jamesofur, DH85868993, Yannaz, ForrestLane42, JavierMC, Useight, Dknix80, CardinalDan, Gcbound, Xnuala, Signalhead, Lights, King Lopez, ABF, Jeff G., Indubitably, TheMindsEye, Jay2332, PureJadeKid, Jessgrogan, DanGLee5000, Philip Trueman, Arulious, TXiKiBoT, Oshwah, Eturk001, Mathiassmith, Qxz, Losethechains, LeaveSleaves, Raymondwinn, Cremepuff222, Bearian, ARUNKUMAR P.R, ACEOREVIVED, MearsMan, Billinghurst, Willdask8ter, Teh roflmaoer, Lova Falk, Falcon8765, Burntsauce, Niknak661, Wikineer, Insanity Incarnate, Thesteffis, Miketear27, Alcmaeonid, Ceranthor, LittleBenW, Logan, Kenjura, StevenJohnston, He1rarchy, Ponyo, Chimin 07, Ttony21, K. Annoyomous, Psbsub, Parhamr, Dawn Bard, Matthew Yeager, Rugburns499, RJaguar3, Sephiroth storm, Yintan, Zlewis09, Calabraxthis, GregoryCJohnson, Formerly the IP-Address 24.22.227.53, Jonson22, Ahniwa, Happysailor, Radon210, Shane hudson, Solace Sylum, Oxymoron83, Jessica DeVoto, Jasenwebb, Steven Crossin, Tombomp, Poindexter Propellerhead, Iain99, The Riddle of Epicurus, DancingPhilosopher, Stfg, StaticGull, Anchor Link Bot, Anarchic Fox, Efe, Asher196, ClueBot, Avenged Eightfold, Beetstra public, Fyyer, The Thing That Should Not Be, Ilya78, Chlingl, GreenLady UK, Parkjunwung, Ndenison, Drmies, HiltonLange, Mild Bill Hiccup, Bgt, DanielDeibler, Boing! said Zebedee, CounterVandalismBot, Lampak, Roflcoptor, Cardinal 1962, Masterpiece2000, Excirial, Alexbot, Jusdafax, M4gnum0n, Hello Control, Kanguole, Detrimental dude, MusicTree3, Simon D M, Sun Creator, Mindstalk, Woowoobeanie, Nmoo, Peter.C, Promethean, Orangees, Sldghmmr, Eustress, Razorflame, Prokopenya Viktor, Cass251, Chigrll, Thingg, Aitias, Horselover Frost, Subash.chandran007, Versus22, The Baroness of Morden, SoxBot III, Vanished user uih38riiw4hjlsd, Spick And Span, Jyothsna01, Alchemist Jack, Jengirl1988, XLinkBot, Stickee, Nepenthes,

33.8. TEXT AND IMAGE SOURCES, CONTRIBUTORS, AND LICENSES

BigEvil15, Adamwalsh16, WikHead, Alexius08, Timlim88, HexaChord, Pinkeyedfreak, Addbot, Djanane, Ennui93, DOI bot, Betterusername, Binary TSO, Annielogue, Migs W, Older and ... well older, Fieldday-sunday, Vishnava, Nmcc89, MrOllie, Download, BFairntrue, Prankster-Turtle, Dyaa, Bassbonerocks, Gpgrazioli, Chzz, Westrim, Tide rolls, Kongyi, QuadrivialMind, Teles, Patent.drafter, Ben Ben, Luckas-bot, Yobot, Ptbotgourou, TaBOT-zerem, Legobot II, Whatifif, Hohhoh, Dandeman2007, Jtw4455, Marshall Williams2, Eric-Wester, Backslash Forwardslash, AnomieBOT, AdjustShift, Yachtsman1, Ulric1313, Flewis, Bluerasberry, Materialscientist, Citation bot, Eskandarany, Ktlane42, SabineK1, Frankenpuppy, Neurolysis, Xqbot, GPJohnson, KiViki, Addihockey10, Capricorn42, Dethlock99, Acebulf, Flying sheep, Tomwsulcer, Sheencarl, Petropoxy (Lithoderm Proxy), Self-esteem wizard, James Wenham, Malerba28, Stroker07, Bnu gaoyong, Speaker1994, Brotherofelijah, Shadowjams, Dou Gweler, Thehelpfulbot, GripTheHusk, Saturn-78, Enriwiki, VI, CanuckK, Liltrigga777, Aaamandaaa, Luttyvella, Luttyvella1, Wireless Keyboard, HamburgerRadio, Citation bot 1, Rsnj, Deadheadned, Siliconmpi, Serols, Meaghan, Manchiu, Jandalhandler, Saayiit, Reconsider the static, Everarddejong, Pm.mohith, Lotje, LilyKitty, Magarcias, Benihana69, Sgravn, Diannaa, Reach Out to the Truth, Ermchicken78, DARTH SIDIOUS 2, Richard Austin McBain, Whisky drinker, RjwilmsiBot, Jojjewestin, DASHBot, HelenGold, CanadianPenguin, John of Reading, Orphan Wiki, Acather96, Immunize, Manekineko3, Philippe.petrinko, Porphyry87, Wikipelli, AvicBot, John Cline, Bongoramsey, Edanehy, Sdp3, TurilCronburg, MaraWhitman, Tom Jacobs, Adaezotron, Nakibhrid, Wikiloop, Porelbiencomun, Tonyaros, James-Panda, Tablethree, Sharonmil, Domitri, Dferruolo, Will Beback Auto, ClueBot NG, Mechanical digger, Peter James, Jack Greenmaven, MIKHEIL, มอใหม, A520, Pillaw, Helpful Pixie Bot, WikiTryHardDieHard, Smcg8374, AvocatoBot, Royalfalcon12, FutureTrillionaire, Ugncreative Usergname, Llepper4, Pulmonological, Zedshort, RudolfRed, Khazar2, USCgregkoz, Oagbatutu, ProfessortIptangmtsu, JYBot, Grzegorznadolski, Everything Is Numbers, KassandraRoyer, Me, Myself, and I are Here, Jodosma, Mason0190, Josh Joaquin, I.am.a.qwerty, FireflySixtySeven, Afifa Afrin, Bookworm729, Olidog, Hayman30, DuckWaffle12, LordGlados, Nøkkenbuer, MusikBot, Permstrump, Dtwedt, DatGuy, InternetArchiveBot, JaneSwifty, GreenC bot, Bender the Bot, Everydaylife.in, Rmk51187, Ofundayo, Raman Kahlon, Alayne4776, Thattil, Ibogomolova2005, SunBot361 and Anonymous: 1500

- **Positive psychology** *Source:* https://en.wikipedia.org/wiki/Positive_psychology?oldid=794237403 *Contributors:* Taw, SimonP, Ewen, Kchishol1970, Vaughan, JakeVortex, Shoehorn~enwiki, Shellreef, Sannse, Jeremymiles, Ronz, Tpbradbury, Fredrik, Brenton, LutzPrechelt, Bfinn, Gamaliel, Andycjp, CryptoDerk, Beland, Rdsmith4, Hugh Mason, Mike Rosoft, DanielCD, Discospinster, Vsmith, Xezbeth, Mal~enwiki, Bender235, Bcjordan, Neko-chan, CanisRufus, Mwanner, Rajah, NickSchweitzer, Tgr, SHIMONSHA, Mac Davis, Bsadowski1, Joelthelion, Kay Dekker, Killing Vector, Saxifrage, Tabletop, Wikiklrsc, Rjwilmsi, Koavf, MZMcBride, FlaBot, Richdiesal, Nihiltres, Metropolitan90, AllyD, Bgwhite, Skoosh, Shell Kinney, Vincej, Afelton, Irrevenant, NawlinWiki, Grafen, Brian Crawford, Elkman, Jmartisk~enwiki, 2over0, Arthur Rubin, Allens, Sardanaphalus, Crystallina, SmackBot, McGeddon, Jtneill, Man with two legs, Gilliam, Ohnoitsjamie, Polaron, Chris the speller, RDBrown, Oli Filth, Uthbrian, DoctorW, Nbarth, Pasado, RT Wolf, Baxter42, Rrburke, Huon, Шизомби, EPM, Kim99, Deepblackwater, Arodb, Cibu, Bo99, Tim bates, Robofish, Ckatz, Bmistler, Mr. Vernon, Beetstra, Lifeartist, Doczilla, AdultSwim, RichardF, Hu12, Iridescent, Iepeulas, RekishiEJ, Dhammapal, CmdrObot, Zagano, Rtv233, Penbat, Memills, Joshnpowell, Vectro, Doctormatt, Vanished user 2340rujowierfj08234irjwfw4, Wayne Vucenic, Gogo Dodo, Anthonyhcole, YechezkelZilber, Dr.enh, Christian75, Scarpy, Iss246, Brobbins, Johann Nepomuk, David2346, Al Lemos, Headbomb, Peace01234, Nick Number, Mph99, Rees11, KarlEd, Colin MacLaurin, Ingolfson, Emmyzen, Narssarssuaq, HypnoSynthesis, MER-C, Epeefleche, The Transhumanist, Michig, JerryKrueger, YK Times, Shumdw, Magioladitis, Swikid, John C. Lewis, Malcolmxthegreat, Deus911, Presearch, Wylfryn, JaGa, S3000, Robert Daoust, Psicol~enwiki, Dataweaver, Snackycakes, Silin2005, Salmon1, Reedy Bot, Titusmars, Jxfawley, 1000Faces, Lbeaumont, Belovedfreak, Szzuk, DadaNeem, Generalist, Psykhosis, Trw193, LeoRomero, VFHwebdev, Idioma-bot, Goodtherapyorg, TXiKiBoT, Davsch65, Tbutlerbowdon, Edit650, BotKung, Bearian, Positivegal, Lova Falk, Sue Rangell, Danbousho, Dreamcatalyst, Blissblog, Flyer22 Reborn, MaynardClark, Nuttycoconut, Sanya3, Dravecky, Wahrmund, XDanielx, J. Ash Bowie, Standardname, ClueBot, Morningstar1814, The Thing That Should Not Be, Drmies, Johnmoff82, Auntof6, Shutterbug433, Alexbot, Mjecclestone, SunnyDisp, Arjayay, Infoculture~enwiki, Sarahmahdaly, SlamMeMore, XLinkBot, Addbot, Alancott, DOI bot, Yoenit, Annielogue, Jncraton, SpillingBot, WikiUserPedia, MrOllie, Download, Tide rolls, Lightbot, HerculeBot, Ben Ben, Luckas-bot, Yobot, TaBOT-zerem, Rsquire3, DanniDK, Dncars1, N1RK4UDSK714, AnomieBOT, Jim1138, Tucoxn, JackieBot, Ornamentalone, Materialscientist, Citation bot, LilHelpa, Xqbot, Aquila89, Plasmon1248, Tomwsulcer, Bilbe, Makeswell, Omnipaedista, Mhotep, Bobauthor, Smw9135, Benjamin Davison, DocPsych, Zhaoyukun, Touchatou, Wpoeop, A.amitkumar, FrescoBot, Waffler2009, Sisyphustkd, AlexanderKaras, D'ohBot, T@Di, Citation bot 1, Drfloody, Tom.Reding, Wally2121, MondalorBot, Wfsf, Sundaysw, Trappist the monk, Annajonag, Lam Kin Keung, Sternenmeer, Pamdfitz, Celenie, Dinamik-bot, Ahougen, Naturalpsychology, DeanBrettle, RjwilmsiBot, Born2bgratis, Toobin, Titantalent, Tesseract2, P2prules, Plantforestsoil, John of Reading, Oliverlyc, MindtheWiki, Fgtfound, Jheggers, GoingBatty, Subvisser5, Dcirovic, Hrld11, Swhitbo, ZéroBot, PS., Jimmydemesa, Mewmew125, Indomitableal, Wvufanaz, Bika f, Laney2060, Anita Hodder, Bex256, Marinadb, AManWithNoPlan, Amyblankson, AlexJohnTorres12, Milinyte, Calgarypsychology, Okidok, Anita5192, Miradre, Aliciarogowski123, William Loudermilk, ClueBot NG, EcoSleek, Sjbalsama, Intheflow, Abrambk, Joefromrandb, Scooby199, Aaronwayneodonahue, Snotbot, Plusorminuszero, RachelAB, Healthexpert, CaroleHenson, Sharanbngr, BlackPlatinumChowChow, MerlIwBot, Helpful Pixie Bot, MI755604, Fkaratum, Positive270, Flow11, Dylhill, Jeraphine Gryphon, Bookish899, BG19bot, Danny0miller, Smcg8374, Frze, Mark Arsten, Monberl, CitationCleanerBot, Joshua Jonathan, Mooshka215, Funkylasse, KateWoodhouse, Winston Trechane, FeralOink, Sydactive, FordPrefect1979, EricEnfermero, BattyBot, Parrottreats, Charlene-attard, Db4wp, Luigi Del Piero, Bluenile1, Rhombus11, Louey37, Arcandam, Khazar2, Mand1958, JCJC777, Iamozy, Omaro23, Computerbg, Dexbot, Silvertrunk, Jayant2164, BuddhaSoup, RPgzLp, Liaper1, Quick trick, Catheryn.yqz, SFK2, Alsala, Me, Myself, and I are Here, Carolyn16, Hillbillyholiday, VickiMae, Ellyssap, Vargovic1, PositivePsych-Professor, MrLukeDevlin, I am One of Many, Doctor Vulcano, Pdecalculus, Biogeographirst, Brittany Jackson, Jenniwey, Israelsands, Sallybibb, New worl, MaieshaR, Ugog Nizdast, Slimgamslim, Esalwin, FireflySixtySeven, Josehamiltonbr, Twhanks66, PhilosophyBrayton, Z!zek, Positivepsychologyphd, Monkbot, Equilibrium103, Pospsych, Psy270culture, Adamreinman, Inexxa, NQ-Alt, Gavmarklund, Archaeologist03, Narky Blert, Ebisabeti, RunIowa, Lingveno, Kapil.yadav231, CyanoTex, I enjoy sandwiches, Matthebl, Dagannt, Gracea235, Hokinsc, Meganmcculloch, Communist-party-van, Big fat asthmatic cat, Ira Leviton, Vvase, Ronwiki5, Moonriver54, Vcustudentinquiringminds, Bluebird1, Viybel, Tardispower, Evelyn Mak, Freshawake, Bender the Bot, Pleasepsychme, EpicMan, The Happy Librarian, Mandymax825, Mareema, Chocter, Hyperbolick, Likuilaws, Maddieaalund, Tiffanyizer, Spicypr45, Vbereznuk, Bestdude, Wildbenjaminmason, MrPositivity and Anonymous: 298

- **Posthumanism** *Source:* https://en.wikipedia.org/wiki/Posthumanism?oldid=790918391 *Contributors:* Lee Daniel Crocker, Vicki Rosenzweig, The Anome, Patrick, Gdvorsky, Ahoerstemeier, Evercat, Sethmahoney, Wik, Tpbradbury, Pakaran, Hadal, Lussmu~enwiki, Deego, Loremaster, Trevor MacInnis, Noisy, GrantHenninger, Dbachmann, Mjk2357, Tmh, Jason One, Hoary, Schaefer, Firsfron, Derktar, Chrontius, Toussaint, Rjwilmsi, Viznut, Bgwhite, YurikBot, RobotE, Deku-shrub, Juha Raipola~enwiki, KnightRider~enwiki, SmackBot, Classicfilms, Hmains, Lotusduck, OrphanBot, Savidan, Dacoutts, Metamagician3000, TenPoundHammer, AdamFJohnson, Andymiah, Megawattbulbman, Melinda19, Ken Gallager, Gregbard, Da66yy, Letranova, Thijs!bot, Guy Macon, Skomorokh, Giler, Some thing, Magioladitis, -Fire-, Anarchia, Parneix, Drewucsc, Fences and windows, Natg 19, Lightmouse, Sanya3, A Laughton, CharlesGillingham, Bowei Huang 2, JJRhetorical, Martarius, Drmies, Editor2020, DumZiBoT, Multisell, Merodack, Good Olfactory, Svea Kollavainen, Verbal, Yobot, Legobot II, GateKeeper, AnomieBOT, Citation bot, J04n, GrouchoBot, Omnipaedista, FrescoBot, Paine Ellsworth, Cannolis, I dream of horses, Postsuperfly, Curious1949, Supervidin, RMGunton, Rangoon11, Bemland, Helpful Pixie Bot, Jeraphine Gryphon, BG19bot, Ewigekrieg, DPL bot, Khazar2, AK456, Wmental, Dexbot, Wilder Wein 21, FiredanceThroughTheNight, Dooypages, YiFeiBot, Paul2520, Fixture, Ceosad, Mohanbhan, Junbibi, KasparBot, CAPTAIN RAJU, ObscuraScientia, Neon Meadows, Laurhur, UltR, Jgm nw, Mbarrington and Anonymous: 89

33.8.2 Images

- File:'No_Prayer_Breakfast'_event_run_by_the_British_Humanist_Association.jpg *Source:* https://upload.wikimedia.org/wikipedia/commons/6/60/%27No_Prayer_Breakfast%27_event_run_by_the_British_Humanist_Association.jpg *License:* CC BY-SA 3.0 *Contributors:* Own work *Original artist:* Andrew West, of the British Humanist Association

- File:25_Wendy_PB_Picture.jpg *Source:* https://upload.wikimedia.org/wikipedia/commons/c/cb/25_Wendy_PB_Picture.jpg *License:* Public domain *Contributors:* Own work *Original artist:* Sgerbic

- File:5_steps_to_a_happier_life.jpg *Source:* https://upload.wikimedia.org/wikipedia/commons/9/9c/5_steps_to_a_happier_life.jpg *License:* CC BY-SA 3.0 *Contributors:* Own work *Original artist:* LeoRomero

- File:AHLC_FINAL_LOGO.jpg *Source:* https://upload.wikimedia.org/wikipedia/commons/5/52/AHLC_FINAL_LOGO.jpg *License:* CC BY-SA 3.0 *Contributors:* Own work *Original artist:* Kvonnegut

- File:A_coloured_voting_box.svg *Source:* https://upload.wikimedia.org/wikipedia/en/0/01/A_coloured_voting_box.svg *License:* Cc-by-sa-3.0 *Contributors:* ? *Original artist:* ?

- File:Agricola,_Rudolf,_Nordisk_familjebok.png *Source:* https://upload.wikimedia.org/wikipedia/commons/5/5c/Agricola%2C_Rudolf%2C_Nordisk_familjebok.png *License:* Public domain *Contributors:* ? *Original artist:* ?

- File:Albrecht_Dürer_104.jpg *Source:* https://upload.wikimedia.org/wikipedia/commons/4/4d/Albrecht_D%C3%BCrer_-_Selbstbildnis_im_Pelzrock_-_Alte_Pinakothek.jpg *License:* Public domain *Contributors:* The Yorck Project: *10.000 Meisterwerke der Malerei.* DVD-ROM, 2002. ISBN 3936122202. Distributed by DIRECTMEDIA Publishing GmbH. *Original artist:* Albrecht Dürer

- File:Ambox_important.svg *Source:* https://upload.wikimedia.org/wikipedia/commons/b/b4/Ambox_important.svg *License:* Public domain *Contributors:* Own work based on: Ambox scales.svg *Original artist:* Dsmurat, penubag

- File:Ambox_question.svg *Source:* https://upload.wikimedia.org/wikipedia/commons/1/1b/Ambox_question.svg *License:* Public domain *Contributors:* Based on Image:Ambox important.svg *Original artist:* Mysid, Dsmurat, penubag

- File:American_Humanist_Association_President_David_Niose.jpg *Source:* https://upload.wikimedia.org/wikipedia/commons/3/34/American_Humanist_Association_President_David_Niose.jpg *License:* GFDL *Contributors:* American Humanist Association *Original artist:* Leslie A. Zukor

- File:Andrew_Copson_Voltaire_Lecture.jpg *Source:* https://upload.wikimedia.org/wikipedia/commons/2/29/Andrew_Copson_Voltaire_Lecture.jpg *License:* CC BY-SA 3.0 *Contributors:* Andrew Copson *Original artist:* British Humanist Association

- File:Ariane_Sherine_and_Richard_Dawkins_at_the_Atheist_Bus_Campaign_launch.jpg *Source:* https://upload.wikimedia.org/wikipedia/commons/d/d3/Ariane_Sherine_and_Richard_Dawkins_at_the_Atheist_Bus_Campaign_launch.jpg *License:* CC BY 2.0 *Contributors:* Atheist Bus Campaign Launch *Original artist:* Zoe Margolis

- File:Aristotle_Altemps_Inv8575.jpg *Source:* https://upload.wikimedia.org/wikipedia/commons/a/ae/Aristotle_Altemps_Inv8575.jpg *License:* Public domain *Contributors:* Jastrow (2006) *Original artist:* After Lysippos

- File:Atheism_template.svg *Source:* https://upload.wikimedia.org/wikipedia/commons/d/d0/Atheism_template.svg *License:* CC BY-SA 3.0 *Contributors:*

- Atom_of_Atheism-Zanaq.svg *Original artist:* Atom_of_Atheism-Zanaq.svg: User:Zanaq

- File:Back_pews_&_stained_glass_NYSEC_Concert_hall_jeh.jpg *Source:* https://upload.wikimedia.org/wikipedia/commons/b/b5/Back_pews_%26_stained_glass_NYSEC_Concert_hall_jeh.jpg *License:* CC0 *Contributors:* Own work *Original artist:* Jim.henderson

- File:CFI_Lecture_Hall_2.jpg *Source:* https://upload.wikimedia.org/wikipedia/commons/b/b9/CFI_Lecture_Hall_2.jpg *License:* CC BY-SA 3.0 *Contributors:* Own work *Original artist:* Sgerbic

- File:CFI_Library.jpg *Source:* https://upload.wikimedia.org/wikipedia/commons/7/74/CFI_Library.jpg *License:* CC BY-SA 3.0 *Contributors:* Own work *Original artist:* Sgerbic

- File:CFI_logo.svg *Source:* https://upload.wikimedia.org/wikipedia/en/a/af/CFI_logo.svg *License:* Fair use *Contributors:*
Center for Inquiry
Original artist: ?

- File:Cambridge_Humanists,_July_2010.JPG *Source:* https://upload.wikimedia.org/wikipedia/commons/8/8d/Cambridge_Humanists%2C_July_2010.JPG *License:* CC BY-SA 3.0 *Contributors:* Own work *Original artist:* Ardfern

- File:Capuchin_monkeys_sharing.jpg *Source:* https://upload.wikimedia.org/wikipedia/commons/f/f8/Capuchin_monkeys_sharing.jpg *License:* CC BY 2.5 *Contributors:* Powell K: Economy of the Mind. PLoS Biol 1/3/2003: e77. http://dx.doi.org/10.1371/journal.pbio.0000077 *Original artist:* (Photo courtesy of Frans de Waal.)

33.8. TEXT AND IMAGE SOURCES, CONTRIBUTORS, AND LICENSES

- **File:Center_for_Inquiry_Front.jpg** *Source:* https://upload.wikimedia.org/wikipedia/commons/f/fa/Center_for_Inquiry_Front.jpg *License:* CC BY-SA 3.0 *Contributors:* Own work *Original artist:* Sgerbic
- **File:Commons-logo.svg** *Source:* https://upload.wikimedia.org/wikipedia/en/4/4a/Commons-logo.svg *License:* PD *Contributors:* ? *Original artist:* ?
- **File:Conscience_and_law.jpg** *Source:* https://upload.wikimedia.org/wikipedia/commons/d/d6/Conscience_and_law.jpg *License:* Public domain *Contributors:* Image:Justitia auf Gericht 2006-02-05 (2).JPG *Original artist:* Johannes Otto Först (cropped by Marcel Douwe Dekker)
- **File:Council_for_Secular_Humanism_logo.png** *Source:* https://upload.wikimedia.org/wikipedia/en/6/62/Council_for_Secular_Humanism_logo.png *License:* Fair use *Contributors:*
  The logo is from the CSH website website. CSH website *Original artist:* ?
- **File:Da_Vinci_Vitruve_Luc_Viatour.jpg** *Source:* https://upload.wikimedia.org/wikipedia/commons/2/22/Da_Vinci_Vitruve_Luc_Viatour.jpg *License:* Public domain *Contributors:*
  Leonardo Da Vinci - Photo from www.lucnix.be. 2007-09-08 (photograph). Photograpy:
  Original artist: Leonardo da Vinci
- **File:Epictetus.jpg** *Source:* https://upload.wikimedia.org/wikipedia/commons/9/90/Epictetus.jpg *License:* Public domain *Contributors:* ? *Original artist:* ?
- **File:Ethical_Culture_Society_&_school_jeh.jpg** *Source:* https://upload.wikimedia.org/wikipedia/commons/d/d2/Ethical_Culture_Society_%26_school_jeh.jpg *License:* Public domain *Contributors:* Own work *Original artist:* Jim.henderson.
- **File:Fabian_society_1886.jpg** *Source:* https://upload.wikimedia.org/wikipedia/en/d/d4/Fabian_society_1886.jpg *License:* PD-US *Contributors:*
  Original publication: Poster
  Immediate source: http://encyclopedian.blogspot.co.uk/2012/05/fabian-society.html *Original artist:*
  Fabian Society
  (Life time: na)
- **File:Felix-Adler-Hine.jpeg** *Source:* https://upload.wikimedia.org/wikipedia/commons/9/91/Felix-Adler-Hine.jpeg *License:* Public domain *Contributors:* Library of Congress, Prints & Photographs Division, LC-DIG-nclc-04844 (color digital file from b&w original print), archival TIFF version (54 MB), cropped and converted to JPEG with the GIMP 2.4.5, image quality 85. *Original artist:* Lewis Wickes Hine (1874–1940)
- **File:Flickr_-_The_U.S._Army_-_Comprehensive_Soldiers_Fitness_(1)cropped.jpg** *Source:* https://upload.wikimedia.org/wikipedia/commons/4/44/Flickr_-_The_U.S._Army_-_Comprehensive_Soldiers_Fitness_%281%29cropped.jpg *License:* Public domain *Contributors:* This file was derived from Flickr - The U.S. Army - Comprehensive Soldiers Fitness (1).jpg: <a href='//commons.wikimedia.org/wiki/File:Flickr_-_The_U.S._Army_-_Comprehensive_Soldiers_Fitness_(1).jpg' class='image'><img alt='Flickr - The U.S. Army - Comprehensive Soldiers Fitness (1).jpg' src='https://upload.wikimedia.org/wikipedia/commons/thumb/5/50/Flickr_-_The_U.S._Army_-_Comprehensive_Soldiers_Fitness_%281%29.jpg/50px-Flickr_-_The_U.S._Army_-_Comprehensive_Soldiers_Fitness_%281%29.jpg' width='50' height='32' srcset='https://upload.wikimedia.org/wikipedia/commons/thumb/5/50/Flickr_-_The_U.S._Army_-_Comprehensive_Soldiers_Fitness_%281%29.jpg/75px-Flickr_-_The_U.S._Army_-_Comprehensive_Soldiers_Fitness_%281%29.jpg 1.5x, https://upload.wikimedia.org/wikipedia/commons/thumb/5/50/Flickr_-_The_U.S._Army_-_Comprehensive_Soldiers_Fitness_%281%29.jpg/100px-Flickr_-_The_U.S._Army_-_Comprehensive_Soldiers_Fitness_%281%29.jpg 2x' data-file-width='3980' data-file-height='2564' /></a>
  Original artist: Flickr_-_The_U.S._Army_-_Comprehensive_Soldiers_Fitness_(1).jpg: The U.S. Army
- **File:Flock_of_Seagulls_(eschipul).jpg** *Source:* https://upload.wikimedia.org/wikipedia/commons/1/18/Flock_of_Seagulls_%28eschipul%29.jpg *License:* CC BY-SA 2.0 *Contributors:* running with the seagulls *Original artist:* Ed Schipul from Houston, TX, US
- **File:Folder_Hexagonal_Icon.svg** *Source:* https://upload.wikimedia.org/wikipedia/en/4/48/Folder_Hexagonal_Icon.svg *License:* Cc-by-sa-3.0 *Contributors:* ? *Original artist:* ?
- **File:Fr._Luigi_Taparelli.jpg** *Source:* https://upload.wikimedia.org/wikipedia/en/b/bd/Fr._Luigi_Taparelli.jpg *License:* PD-US *Contributors:*
  Original publication: Unknown
  Immediate source: http://goodjesuitbadjesuit.blogspot.com.au/2010/11/social-justice-before-liberation.html *Original artist:*
  Unknown
  (Life time: 1862)
- **File:Francois_Rabelais_-_Portrait.jpg** *Source:* https://upload.wikimedia.org/wikipedia/commons/8/8d/Francois_Rabelais_-_Portrait.jpg *License:* Public domain *Contributors:* http://www.banqueimages.crcv.fr/fullscreenimage.aspx?rank=1&numero=MV4046 *Original artist:* Anonymous
- **File:Frans_Hals_-_Portret_van_René_Descartes.jpg** *Source:* https://upload.wikimedia.org/wikipedia/commons/7/73/Frans_Hals_-_Portret_van_Ren%C3%A9_Descartes.jpg *License:* Public domain *Contributors:* André Hatala [e.a.] (1997) *De eeuw van Rembrandt*, Bruxelles: Crédit communal de Belgique, ISBN 2-908388-32-4. *Original artist:* After Frans Hals
- **File:Freedom_of_Thought_Report_2016_cover_image.png** *Source:* https://upload.wikimedia.org/wikipedia/commons/5/53/Freedom_of_Thought_Report_2016_cover_image.png *License:* CC BY-SA 4.0 *Contributors:* Own work *Original artist:* Bob.Churchill
- **File:Goya_Caprichos3.jpg** *Source:* https://upload.wikimedia.org/wikipedia/commons/a/a2/Goya_Caprichos3.jpg *License:* Public domain *Contributors:* Web Gallery of Art: <a href='http://www.wga.hu/art/g/goya/9g/91/43capric.jpg' data-x-rel='nofollow'><img alt='Inkscape.svg' src='https://upload.wikimedia.org/wikipedia/commons/thumb/6/6f/Inkscape.svg/20px-Inkscape.svg.png' width='20' height='20' srcset='https://upload.wikimedia.org/wikipedia/commons/thumb/6/6f/Inkscape.svg/30px-Inkscape.svg.png 1.5x, https://upload.wikimedia.org/wikipedia/commons/thumb/6/6f/Inkscape.svg/40px-Inkscape.svg.png 2x' data-file-width='60' data-file-height='60'

/></a> Image <a href='http://www.wga.hu/html/g/goya/9g/91/43capric.html' data-x-rel='nofollow'><img alt='Information icon.svg' src='https://upload.wikimedia.org/wikipedia/commons/thumb/3/35/Information_icon.svg/20px-Information_icon.svg.png' width='20' height='20' srcset='https://upload.wikimedia.org/wikipedia/commons/thumb/3/35/Information_icon.svg/30px-Information_icon.svg.png 1.5x, https://upload.wikimedia.org/wikipedia/commons/thumb/3/35/Information_icon.svg/40px-Information_icon.svg.png 2x' data-file-width='620' data-file-height='620' /></a> Info about artwork *Original artist:* Francisco de Goya

- **File:HVD_Berlin_Wallstraße_2014.jpg** *Source:* https://upload.wikimedia.org/wikipedia/commons/5/5a/HVD_Berlin_Wallstra%C3%9Fe_2014.jpg *License:* CC BY-SA 3.0 de *Contributors:* Own work *Original artist:* de:Benutzer: HeywoodFloyd

- **File:Happy_Human_black.svg** *Source:* https://upload.wikimedia.org/wikipedia/commons/4/44/Happy_Human_black.svg *License:* CC-BY-SA-3.0 *Contributors:* Own work *Original artist:* User:Kontos

- **File:Head_of_Socrates_in_Palazzo_Massimo_alle_Terme_(Rome).JPG** *Source:* https://upload.wikimedia.org/wikipedia/commons/0/01/Head_of_Socrates_in_Palazzo_Massimo_alle_Terme_%28Rome%29.JPG *License:* CC BY-SA 4.0 *Contributors:* Own work *Original artist:* Livioandronico2013

- **File:Holyoake2.JPG** *Source:* https://upload.wikimedia.org/wikipedia/commons/e/ea/Holyoake2.JPG *License:* Public domain *Contributors:* http://uts.cc.utexas.edu/~{}laurel/cooproots/holyoake.html
  Original artist: ?

- **File:Humanism.jpg** *Source:* https://upload.wikimedia.org/wikipedia/commons/3/3d/Humanism.jpg *License:* CC BY-SA 3.0 *Contributors:* Own work *Original artist:* Kvonnegut

- **File:HumanismSymbol.svg** *Source:* https://upload.wikimedia.org/wikipedia/commons/f/f5/HumanismSymbol.svg *License:* GFDL *Contributors:* Own work utilizando ésta imagen *Original artist:* Andres Rojas

- **File:Humanist_Society_Scotland.png** *Source:* https://upload.wikimedia.org/wikipedia/commons/a/ab/Humanist_Society_Scotland.png *License:* CC BY-SA 4.0 *Contributors:* Own work *Original artist:* Franko2nd

- **File:IHEYO_logo.png** *Source:* https://upload.wikimedia.org/wikipedia/commons/d/dc/IHEYO_logo.png *License:* Public domain *Contributors:* Facebook *Original artist:* IHEYO

- **File:Immanuel_Kant_(painted_portrait).jpg** *Source:* https://upload.wikimedia.org/wikipedia/commons/4/43/Immanuel_Kant_%28painted_portrait%29.jpg *License:* Public domain *Contributors:* /History/Carnegie/kant/portrait.html *Original artist:* Unidentified painter

- **File:Jeremy_Bentham_by_Henry_William_Pickersgill_detail.jpg** *Source:* https://upload.wikimedia.org/wikipedia/commons/c/c8/Jeremy_Bentham_by_Henry_William_Pickersgill_detail.jpg *License:* Public domain *Contributors:*
  National Portrait Gallery: NPG 413
  Original artist: Henry William Pickersgill (died 1875)

- **File:JohnStuartMill.jpg** *Source:* https://upload.wikimedia.org/wikipedia/commons/7/75/JohnStuartMill.jpg *License:* Public domain *Contributors:* ? *Original artist:* ?

- **File:John_Stuart_Mill_by_London_Stereoscopic_Company,_c1870.jpg** *Source:* https://upload.wikimedia.org/wikipedia/commons/9/99/John_Stuart_Mill_by_London_Stereoscopic_Company%2C_c1870.jpg *License:* Public domain *Contributors:* Hulton Archive *Original artist:* London Stereoscopic Company

- **File:KurtzGardner.jpg** *Source:* https://upload.wikimedia.org/wikipedia/commons/d/d9/KurtzGardner.jpg *License:* CC BY-SA 3.0 *Contributors:* Own work *Original artist:* RobertoTenore

- **File:Levi_Fragell,_print.jpg** *Source:* https://upload.wikimedia.org/wikipedia/commons/c/c4/Levi_Fragell%2C_print.jpg *License:* CC BY-SA 2.0 *Contributors:* Flickr: Levi Fragell, print *Original artist:* Arnfinn Pettersen

- **File:Lock-green.svg** *Source:* https://upload.wikimedia.org/wikipedia/commons/6/65/Lock-green.svg *License:* CC0 *Contributors:* en:File:Free-to-read_lock_75.svg *Original artist:* User:Trappist the monk

- **File:Logic_portal.svg** *Source:* https://upload.wikimedia.org/wikipedia/commons/7/7c/Logic_portal.svg *License:* CC BY-SA 3.0 *Contributors:* Own work *Original artist:* <a href='//commons.wikimedia.org/wiki/File:Watchduck.svg' class='image'><img alt='Watchduck.svg' src='https://upload.wikimedia.org/wikipedia/commons/thumb/d/d8/Watchduck.svg/40px-Watchduck.svg.png' width='40' height='46' srcset='https://upload.wikimedia.org/wikipedia/commons/thumb/d/d8/Watchduck.svg/60px-Watchduck.svg.png 1.5x, https://upload.wikimedia.org/wikipedia/commons/thumb/d/d8/Watchduck.svg/80px-Watchduck.svg.png 2x' data-file-width='703' data-file-height='806' /></a> Watchduck (a.k.a. Tilman Piesk)

- **File:Logo_of_Humanist_Disaster_Recovery.jpg** *Source:* https://upload.wikimedia.org/wikipedia/commons/6/68/Logo_of_Humanist_Disaster_Recovery.jpg *License:* CC BY-SA 4.0 *Contributors:* Foundation Beyond Belief/American Humanist Association *Original artist:* Foundation Beyond Belief/American Humanist Association

- **File:Logo_sociology.svg** *Source:* https://upload.wikimedia.org/wikipedia/commons/a/a6/Logo_sociology.svg *License:* Public domain *Contributors:* Own work *Original artist:* Tomeq183

- **File:Map_of_discrimination_against_the_non-religious_in_the_IHEU_Freedom_of_Thought_Report.png** *Source:* https://upload.wikimedia.org/wikipedia/commons/c/cf/Map_of_discrimination_against_the_non-religious_in_the_IHEU_Freedom_of_Thought_Report.png *License:* CC BY-SA 4.0 *Contributors:* Own work *Original artist:* Bob.Churchill

- **File:Maria_Ossowska.jpg** *Source:* https://upload.wikimedia.org/wikipedia/commons/5/58/Maria_Ossowska.jpg *License:* Public domain *Contributors:* ? *Original artist:* User Wujaszek on pl.wikipedia

- **File:MaslowsHierarchyOfNeeds.svg** *Source:* https://upload.wikimedia.org/wikipedia/commons/3/33/MaslowsHierarchyOfNeeds.svg *License:* CC BY-SA 4.0 *Contributors:* Own work using Inkscape, based on Maslow's paper, A Theory of Human Motivation. *Original artist:* FireflySixtySeven

33.8. TEXT AND IMAGE SOURCES, CONTRIBUTORS, AND LICENSES 239

- **File:Modern_Utiitarianism_by_Birks.png** *Source:* https://upload.wikimedia.org/wikipedia/en/d/d3/Modern_Utiitarianism_by_Birks.png *License:* Public domain *Contributors:* ? *Original artist:* ?

- **File:Nickell_in_office.jpg** *Source:* https://upload.wikimedia.org/wikipedia/commons/2/2c/Nickell_in_office.jpg *License:* CC BY-SA 3.0 *Contributors:* Own work *Original artist:* Sgerbic

- **File:Nicolas_P._Rougier'{}s_rendering_of_the_human_brain.png** *Source:* https://upload.wikimedia.org/wikipedia/commons/7/73/Nicolas_P._Rougier%27s_rendering_of_the_human_brain.png *License:* GPL *Contributors:* http://www.loria.fr/~{}rougier *Original artist:* Nicolas Rougier

- **File:Nietzsche187a.jpg** *Source:* https://upload.wikimedia.org/wikipedia/commons/1/1b/Nietzsche187a.jpg *License:* Public domain *Contributors:* https://s-media-cache-ak0.pinimg.com/originals/04/10/0b/04100baec90c105729b47f33c371476b.jpg *Original artist:* Photography by Friedrich Hartmann (1822-1902) in Basel

- **File:Open_book_nae_02.svg** *Source:* https://upload.wikimedia.org/wikipedia/commons/9/92/Open_book_nae_02.svg *License:* CC0 *Contributors:* OpenClipart *Original artist:* nae

- **File:People_icon.svg** *Source:* https://upload.wikimedia.org/wikipedia/commons/3/37/People_icon.svg *License:* CC0 *Contributors:* OpenClipart *Original artist:* OpenClipart

- **File:Peter_Singer_MIT_Veritas.jpg** *Source:* https://upload.wikimedia.org/wikipedia/commons/5/52/Peter_Singer_MIT_Veritas.jpg *License:* CC BY 3.0 *Contributors:* Own work *Original artist:* Joel Travis Sage

- **File:Pintoricchio_012.jpg** *Source:* https://upload.wikimedia.org/wikipedia/commons/9/9e/Pintoricchio_012.jpg *License:* Public domain *Contributors:* The Yorck Project: *10.000 Meisterwerke der Malerei.* DVD-ROM, 2002. ISBN 3936122202. Distributed by DIRECTMEDIA Publishing GmbH. *Original artist:* Pinturicchio

- **File:Portal-puzzle.svg** *Source:* https://upload.wikimedia.org/wikipedia/en/f/fd/Portal-puzzle.svg *License:* Public domain *Contributors:* ? *Original artist:* ?

- **File:Psi2.svg** *Source:* https://upload.wikimedia.org/wikipedia/commons/6/6c/Psi2.svg *License:* Public domain *Contributors:* No machine-readable source provided. Own work assumed (based on copyright claims). *Original artist:* No machine-readable author provided. Gdh~commonswiki assumed (based on copyright claims).

- **File:Question_book-new.svg** *Source:* https://upload.wikimedia.org/wikipedia/en/9/99/Question_book-new.svg *License:* Cc-by-sa-3.0 *Contributors:*
Created from scratch in Adobe Illustrator. Based on Image:Question book.png created by User:Equazcion *Original artist:*
Tkgd2007

- **File:Rare_Book_Room_3.jpg** *Source:* https://upload.wikimedia.org/wikipedia/commons/2/27/Rare_Book_Room_3.jpg *License:* CC BY-SA 3.0 *Contributors:* Own work *Original artist:* Sgerbic

- **File:Richard_Dawkins_speaking_at_the_British_Humanist_Association_Annual_Conference.jpg** *Source:* https://upload.wikimedia.org/wikipedia/commons/7/70/Richard_Dawkins_speaking_at_the_British_Humanist_Association_Annual_Conference.jpg *License:* CC BY-SA 3.0 *Contributors:* Own work *Original artist:* Andrew West, of the British Humanist Association

- **File:Sanzio_01.jpg** *Source:* https://upload.wikimedia.org/wikipedia/commons/9/94/Sanzio_01.jpg *License:* Public domain *Contributors:* Stitched together from vatican.va *Original artist:* Raphael

- **File:Sanzio_01_Plato_Aristotle.jpg** *Source:* https://upload.wikimedia.org/wikipedia/commons/9/98/Sanzio_01_Plato_Aristotle.jpg *License:* Public domain *Contributors:* Web Gallery of Art: <a href='http://www.wga.hu/art/r/raphael/4stanze/1segnatu/1/athens1.jpg' data-x-rel='nofollow'><img alt='Inkscape.svg' src='https://upload.wikimedia.org/wikipedia/commons/thumb/6/6f/Inkscape.svg/20px-Inkscape.svg.png' width='20' height='20' srcset='https://upload.wikimedia.org/wikipedia/commons/thumb/6/6f/Inkscape.svg/30px-Inkscape.svg.png 1.5x, https://upload.wikimedia.org/wikipedia/commons/thumb/6/6f/Inkscape.svg/40px-Inkscape.svg.png 2x' data-file-width='60' data-file-height='60' /></a> Image <a href='http://www.wga.hu/html/r/raphael/4stanze/1segnatu/1/athens1.html' data-x-rel='nofollow'><img alt='Information icon.svg' src='https://upload.wikimedia.org/wikipedia/commons/thumb/3/35/Information_icon.svg/20px-Information_icon.svg.png' width='20' height='20' srcset='https://upload.wikimedia.org/wikipedia/commons/thumb/3/35/Information_icon.svg/30px-Information_icon.svg.png 1.5x, https://upload.wikimedia.org/wikipedia/commons/thumb/3/35/Information_icon.svg/40px-Information_icon.svg.png 2x' data-file-width='620' data-file-height='620' /></a> Info about artwork *Original artist:* Raphael

- **File:Shishapangma.jpg** *Source:* https://upload.wikimedia.org/wikipedia/commons/e/e9/Shishapangma.jpg *License:* Public domain *Contributors:* Swinelin *Original artist:* Swinelin at English Wikipedia

- **File:Skeptic'{}s_Toolbox_2012.JPG** *Source:* https://upload.wikimedia.org/wikipedia/commons/9/99/Skeptic%27s_Toolbox_2012.JPG *License:* CC BY-SA 3.0 *Contributors:* Own work *Original artist:* Sgerbic

- **File:Social_Justice_Week_Cardboard_Village.JPG** *Source:* https://upload.wikimedia.org/wikipedia/commons/6/6e/Social_Justice_Week_Cardboard_Village.JPG *License:* Public domain *Contributors:* Own work *Original artist:* Jorfer

- **File:Socrates.png** *Source:* https://upload.wikimedia.org/wikipedia/commons/c/cd/Socrates.png *License:* Public domain *Contributors:* Transferred from en.wikipedia to Commons. *Original artist:* The original uploader was Magnus Manske at English Wikipedia Later versions were uploaded by Optimager at en.wikipedia.

- **File:Socrates_BM_GR1973.03-27.16.jpg** *Source:* https://upload.wikimedia.org/wikipedia/commons/1/10/Socrates_BM_GR1973.03-27.16.jpg *License:* CC BY 2.5 *Contributors:* Marie-Lan Nguyen (2011) *Original artist:* ?

- **File:Sonja_Eggerickx.jpg** *Source:* https://upload.wikimedia.org/wikipedia/commons/e/ef/Sonja_Eggerickx.jpg *License:* CC BY-SA 3.0 *Contributors:* Own work *Original artist:* HeywoodFloyd

- **File:Stanton_Coit_001.jpg** *Source:* https://upload.wikimedia.org/wikipedia/commons/5/53/Stanton_Coit_001.jpg *License:* Public domain *Contributors:* Photograph courtesy of Bishopsgate Library *Original artist:* Unknown<a href='https://www.wikidata.org/wiki/Q4233718' title='wikidata:Q4233718'><img alt='wikidata:Q4233718' src='https://upload.wikimedia.org/wikipedia/commons/thumb/f/ff/Wikidata-logo.svg/20px-Wikidata-logo.svg.png' width='20' height='11' srcset='https://upload.wikimedia.org/wikipedia/commons/thumb/f/ff/Wikidata-logo.svg/30px-Wikidata-logo.svg.png 1.5x, https://upload.wikimedia.org/wikipedia/commons/thumb/f/ff/Wikidata-logo.svg/40px-Wikidata-logo.svg.png 2x' data-file-width='1050' data-file-height='590' /></a>
- **File:Symbol_book_class2.svg** *Source:* https://upload.wikimedia.org/wikipedia/commons/8/89/Symbol_book_class2.svg *License:* CC BY-SA 2.5 *Contributors:* Mad by Lokal_Profil by combining: *Original artist:* Lokal_Profil
- **File:Tabash_toast_at_CFI_Student_Leadership_Conference_2013.JPG** *Source:* https://upload.wikimedia.org/wikipedia/commons/c/c8/Tabash_toast_at_CFI_Student_Leadership_Conference_2013.JPG *License:* CC BY-SA 3.0 *Contributors:* Own work *Original artist:* Sgerbic
- **File:Templo_Positivista_em_Porto_Alegre_-_Posse_de_Erlon_Jacques_de_Oliveira_02.JPG** *Source:* https://upload.wikimedia.org/wikipedia/commons/9/90/Templo_Positivista_em_Porto_Alegre_-_Posse_de_Erlon_Jacques_de_Oliveira_02.JPG *License:* CC BY-SA 3.0 *Contributors:* Own work *Original artist:* Lechatjaune
- **File:Templo_positivista.jpg** *Source:* https://upload.wikimedia.org/wikipedia/commons/1/11/Templo_positivista.jpg *License:* CC BY-SA 3.0 *Contributors:* User:Tetraktys *Original artist:* User:Tetraktys
- **File:Text_document_with_red_question_mark.svg** *Source:* https://upload.wikimedia.org/wikipedia/commons/a/a4/Text_document_with_red_question_mark.svg *License:* Public domain *Contributors:* Created by bdesham with Inkscape; based upon Text-x-generic.svg from the Tango project. *Original artist:* Benjamin D. Esham (bdesham)
- **File:The_punishment_of_Bessus_by_Andre_Castaigne_(1898-1899).jpg** *Source:* https://upload.wikimedia.org/wikipedia/commons/5/58/The_punishment_of_Bessus_by_Andre_Castaigne_%281898-1899%29.jpg *License:* Public domain *Contributors:* http://www.alexanderstomb.com/main/imageslibrary/alexander/castbessuscrucif.jpg *Original artist:* Andre Castaigne
- **File:Tom_Flynn_at_lectern,_banner-5.jpg** *Source:* https://upload.wikimedia.org/wikipedia/commons/8/8b/Tom_Flynn_at_lectern%2C_banner-5.jpg *License:* CC BY-SA 4.0 *Contributors:* Own work *Original artist:* Monica Harmsen
- **File:Tree_of_life.svg** *Source:* https://upload.wikimedia.org/wikipedia/commons/0/09/Tree_of_life.svg *License:* CC-BY-SA-3.0 *Contributors:* No machine-readable source provided. Own work assumed (based on copyright claims). *Original artist:* No machine-readable author provided. Vanished user fijtji34toksdcknqrjn54yoimascj assumed (based on copyright claims).
- **File:Trithemiusmoredetail.jpg** *Source:* https://upload.wikimedia.org/wikipedia/commons/6/69/Trithemiusmoredetail.jpg *License:* Public domain *Contributors:* Own work *Original artist:* CSvBibra
- **File:Vista_Login_Manager_Cropped.svg** *Source:* https://upload.wikimedia.org/wikipedia/commons/1/15/Vista_Login_Manager_Cropped.svg *License:* GPL *Contributors:* Image:Vista-Login Manager2.png, from [1] *Original artist:* Sa-Ki at DeviantArt, traced by User:Stannered
- **File:Wiki_letter_w_cropped.svg** *Source:* https://upload.wikimedia.org/wikipedia/commons/1/1c/Wiki_letter_w_cropped.svg *License:* CC-BY-SA-3.0 *Contributors:* This file was derived from Wiki letter w.svg: <a href='//commons.wikimedia.org/wiki/File:Wiki_letter_w.svg' class='image'><img alt='Wiki letter w.svg' src='https://upload.wikimedia.org/wikipedia/commons/thumb/6/6c/Wiki_letter_w.svg/50px-Wiki_letter_w.svg.png' width='50' height='50' srcset='https://upload.wikimedia.org/wikipedia/commons/thumb/6/6c/Wiki_letter_w.svg/75px-Wiki_letter_w.svg.png 1.5x, https://upload.wikimedia.org/wikipedia/commons/thumb/6/6c/Wiki_letter_w.svg/100px-Wiki_letter_w.svg.png 2x' data-file-width='44' data-file-height='44' /></a>
Original artist: Derivative work by Thumperward
- **File:Wikibooks-logo.svg** *Source:* https://upload.wikimedia.org/wikipedia/commons/f/fa/Wikibooks-logo.svg *License:* CC BY-SA 3.0 *Contributors:* Own work *Original artist:* User:Bastique, User:Ramac et al.
- **File:Wikidata-logo.svg** *Source:* https://upload.wikimedia.org/wikipedia/commons/f/ff/Wikidata-logo.svg *License:* Public domain *Contributors:* Own work *Original artist:* User:Planemad
- **File:Wikiquote-logo.svg** *Source:* https://upload.wikimedia.org/wikipedia/commons/f/fa/Wikiquote-logo.svg *License:* Public domain *Contributors:* Own work *Original artist:* Rei-artur
- **File:Wikisource-logo.svg** *Source:* https://upload.wikimedia.org/wikipedia/commons/4/4c/Wikisource-logo.svg *License:* CC BY-SA 3.0 *Contributors:* Rei-artur *Original artist:* Nicholas Moreau
- **File:Wikiversity-logo-Snorky.svg** *Source:* https://upload.wikimedia.org/wikipedia/commons/1/1b/Wikiversity-logo-en.svg *License:* CC BY-SA 3.0 *Contributors:* Own work *Original artist:* Snorky
- **File:Wikiversity-logo.svg** *Source:* https://upload.wikimedia.org/wikipedia/commons/9/91/Wikiversity-logo.svg *License:* CC BY-SA 3.0 *Contributors:* Snorky (optimized and cleaned up by verdy_p) *Original artist:* Snorky (optimized and cleaned up by verdy_p)
- **File:Wiktionary-logo-en-v2.svg** *Source:* https://upload.wikimedia.org/wikipedia/commons/9/99/Wiktionary-logo-en-v2.svg *License:* CC-BY-SA-3.0 *Contributors:* ? *Original artist:* ?
- **File:Wiktionary-logo-v2.svg** *Source:* https://upload.wikimedia.org/wikipedia/en/0/06/Wiktionary-logo-v2.svg *License:* CC-BY-SA-3.0 *Contributors:* ? *Original artist:* ?
- **File:William_H_Childs_house.jpg** *Source:* https://upload.wikimedia.org/wikipedia/commons/6/68/William_H_Childs_house.jpg *License:* CC BY 2.5 *Contributors:* Transferred from en.wikipedia to Commons. *Original artist:* Fordmadoxfraud at English Wikipedia
- **File:Young_Humanists.png** *Source:* https://upload.wikimedia.org/wikipedia/commons/5/56/Young_Humanists.png *License:* Public domain *Contributors:* Facebook *Original artist:* Humanists UK
- **File:இசுரேலிய_குடியரசுத்_தலைவர்.JPG** *Source:* https://upload.wikimedia.org/wikipedia/commons/d/de/%E0%AE%87%E0%AE%B2%E0%AE%A3%E0%AF%8D%E0%AE%9F%E0%AE%A9%E0%AF%8D_%E0%AE%A4%E0%AE%BF%E0%AE%B0%E0%AF%81%E0%AE%B5%E0%AE%B3%E0%AF%8D%E0%AE%B3%E0%AF%81%E0%AE%B5%E0%AE%B0%E0%AF%8D.JPG *License:* CC BY-SA 4.0 *Contributors:* Own work *Original artist:* சஒெலின் george

33.8.3 Content license

- Creative Commons Attribution-Share Alike 3.0

www.ingramcontent.com/pod-product-compliance
Lightning Source LLC
Chambersburg PA
CBHW082323220526
45470CB00008B/2386